About Pearson

Pearson is the world's learning company, with presence across 70 countries worldwide. Our unique insights and world-class expertise comes from a long history of working closely with renowned teachers, authors and thought leaders, as a result of which, we have emerged as the preferred choice for millions of teachers and learners across the world.

We believe learning opens up opportunities, creates fulfilling careers and hence better lives. We hence collaborate with the best of minds to deliver you class-leading products, spread across the Higher Education and Test Preparation spectrum.

Superior learning experience and improved outcomes are at the heart of everything we do. This product is the result of one such effort.

Your feedback plays a critical role in the evolution of our products and you can contact us – reachus@pearson.com. *We look forward to it.*

Biophysics and Molecular Biology

Tools and Techniques

Fifth edition

Biophysics and Molecular Biology

Tools and Techniques

Fifth edition

Pranav Kumar
Former faculty,
Department of Biotechnology,
Jamia Millia Islamia,
New Delhi, India

Pathfinder Publication

Pranav Kumar
Former faculty,
Department of Biotechnology,
Jamia Millia Islamia,
New Delhi, India

Biophysics and Molecular Biology

Tools and Techniques

ISBN 978-81-198-9662-2

First Impression

Published by Pearson India Education Services Pvt. Ltd, CIN: U72200TN2005PTC057128.

Head Office: 1st Floor, Berger Tower, Plot No. C-001A/2, Sector 16B, Noida - 201 301, Uttar Pradesh, India.
Registered Office: Featherlite, 'The Address' 5th Floor, Survey No 203/10B, 200 Ft MMRD Road,
Zamin Pallavaram, Chennai – 600 044.
Website: in.pearson.com
Email: companysecretary.india@pearson.com

Printer: Saurabh Printers Pvt. Ltd.

Preface

The field of biophysics and molecular biology continues to be one of the most exciting and dynamic areas of science. Over the past few decades, spectacular progress in this field has occurred due to the conceptual synthesis of ideas from biology, physics, chemistry, mathematics, statistics, and computer science.

The primary goal of this textbook is to teach students about the theoretical principles and applications of key biophysical and molecular methods used in biochemistry and molecular biology. I aim to present the subject from a conceptual perspective, covering a substantial theoretical basis to facilitate the understanding of key experimental techniques. This equips students to make appropriate choices and efficiently use these techniques. While there have been numerous major advances in molecular biology in the past few years, I have focused on selected topics that provide the basic principles for understanding the structure and functional relationships of molecular biology.

The most significant feature of this book is its clear, up-to-date, and accurate explanations of mechanisms, rather than a mere description of facts and events. The question of what to include and what to omit is crucially important for today's authors. With such a broad array of potential topics and techniques available, selecting those that students should experience and master becomes challenging. However, there are techniques and concepts that most of us would agree form a 'core' in biophysics and molecular biology. I have resisted the temptation to describe more and more techniques, adding detail without increasing the understanding of basic concepts. I hope that this textbook will prove useful to both teachers and students. Finally, I have provided a concise list of selected references (research papers, reviews, and books) so that curious readers can trace fundamentals and ideas to their roots. These references are arranged in alphabetical order.

Although the chapters of this book can be read independently of one another, they are arranged in a logical sequence. Each page is carefully laid out to position related text, figures, and tables near one another, minimizing the need for page turning while reading a topic. I have given equal importance to both text and illustrations as well.

Acknowledgements

There is an old proverb that says you never really learn a subject until you teach it. We now know that you learn a subject even better when you write about it. Preparing this text has provided me with a wonderful opportunity to share my knowledge with students. Thanks go first and foremost to our students. In preparing this book, I have relied heavily on and benefited greatly from the advice and constructive criticism of numerous colleagues. I am particularly grateful to Ajay Kumar for his enthusiastic editing of the complete manuscript. I would also like to thank Prakash Vardhan and Harleen Kaur for their invaluable contributions. This book is a team effort, and producing it would be impossible without the outstanding people at Pathfinder Publication. It was a pleasure to work with many other dedicated and creative individuals at Pathfinder Publication during the production of this book, especially Pradeep Verma.

Pranav

Contents

Biophysics

Molecular Biology

Abbreviations

μm	micrometer
Å	angstrom
A_{260}	absorbance at 260 nm
Ab	antibody
AD	activation domain
ADA	adenosine deaminase
ADP	adenosine 5′-diphosphate
AFLP	amplified fragment length polymorphism
AFM	atomic force microscopy
Ag	antigen
AMP	adenosine 5′-monophosphate
ATP	adenosine 5′-triphosphate
bis	bisacrylamide N,N′-methylenebisacrylamide
bp	base pair
BrdU	5-bromo-2-deoxyuridine
cccDNA	covalently closed circular DNA
CD	circular dichroism
cDNA	complementary DNA
CHEF	contour-clamped homogeneous electric field
CM	carboxymethyl
CNBr	cyanogen bromide
cpm	counts per minute
CRISPRs	clustered regularly interspaced short palindromic repeats
Da	dalton
dATP	deoxyadenosine triphosphate
DAPI	4′,6-diamidino-2-phenylindole dihydrochloride
DBD	DNA binding domain
DBM	diazobenzyloxymethyl
ddNTP	dideoxynucleoside triphosphate
DEAE	diethylaminoethyl
DMS	dimethyl sulfate
dNTP	deoxynucleoside triphosphate
ELISA	enzyme-linked immunosorbent assay
EMSA	electrophoretic mobility shift assay

ER	electromagnetic radiation
EtBr	ethidium bromide
FACS	fluorescence activated cell sorter
FISH	fluorescence in situ hybridization
FITC	fluorescein isothiocyanate
FLIP	fluorescence loss in photobleaching
FRAP	fluorescence recovery after photobleaching
FRET	fluorescence (Förster) resonance energy transfer
FSC	forward scatter
GC	gas chromatography
GFP	green fluorescent protein
GLC	gas-liquid chromatography
GSC	gas-solid chromatography
HAT	hypoxanthine-aminopterin-thymidine
HGPRT	hypoxanthine-guanine phosphoribosyl transferase
HPLC	high performance liquid chromatography
IEF	isoelectric focusing
IFE	immunofixation electrophoresis
Ig	immunoglobulin
IR	infrared
kb	kilobase
kcal	kilocalorie
K_d	partition or distribution coefficient
kDa	kilodalton
LC	liquid chromatography
mAb	monoclonal antibody
MALDI	matrix-assisted laser desorption/ionization
Mb	megabase pair
MRI	magnetic resonance imaging
MS	mass spectrometry
NA	numerical aperture
nm	nanometer
NMR	nuclear magnetic resonance
ORD	optical rotatory dispersion
PAGE	polyacrylamide gel electrophoresis
PCR	polymerase chain reaction
PE	phycoerythrin
PFGE	pulsed-field gel electrophoresis

PI	propidium iodide
PMT	photomultiplier tube
RACE	rapid amplification of cDNA ends
RAPD	random amplification of polymorphic DNA
RCF	relative centrifugal field
R_f	relative front
RFLP	restriction fragment length polymorphism
RIA	radioimmunoassay
RPM	revolution per minute
RT-PCR	reverse transcription polymerase chain reaction
SDS	sodium dodecylsulfate
SEM	scanning electron microscope
SP	sulfopropyl
SPR	surface plasmon resonance
SSC	side scatter
STM	scanning tunneling electron microscopy
Taq	*Thermus aquaticus*
TdT	terminal deoxynucleotidyl transferase
TEM	transmission electron microscope
TEMED	N, N, N′, N′-tetramethylethylenediamine
TLC	thin layer chromatography
T_m	melting temperature
TMS	tetramethylsilane
TOF	time-of-flight
Tris	tris(hydroxymethyl)aminomethane
UV	ultraviolet
Vis	visible

1

Chromatography

Chromatography is a physical method for the separation of compounds present in a sample. Tswett, a Russian botanist (referred to as the father of *chromatography*), is credited for developing chromatography. He employed the technique to separate various plant pigments such as chlorophylls and xanthophylls, bypassing solutions of these compounds through a glass column packed with finely divided calcium carbonate. The separated species appeared as colored bands on the column, which accounts for the name he chose for the method (Greek *chroma* meaning 'color' and *graphein* meaning 'writing').

Chromatography is a separation process in which the components of a sample are distributed between two phases, one of which is stationary (termed as **stationary phase**) while the other moves (termed as **mobile phase**). The sample to be examined (called the *solute* or *analyte*) is allowed to interact with the mobile phase and stationary phase. These two phases could be a solid and a liquid, or a gas and a liquid or a liquid and another liquid. The stationary phase, which may be a solid or a liquid supported on a solid, does not move. The mobile phase moves the sample through or along the stationary phase in a definite direction. The mobile phase may be a *liquid* (**liquid chromatography**) or a *gas* (**gas chromatography**). In gas chromatography, the term *carrier gas* may be used for the mobile phase. All chromatographic methods involve passing a mobile phase through a stationary phase. Substances separated by a chromatographic system must have different relative affinities for these two phases. Thus, a substance with a relatively higher affinity for the stationary phase moves slowly through the chromatographic system than does a substance with a lower affinity. This difference in mobility ultimately leads to the physical separation of the components present in a sample. The nature of the specific mobile and stationary phases determines which substances travel more quickly or slowly and is how they are separated.

> The international union of pure and applied chemistry (IUPAC) has defined chromatography as: A method, used primarily for the separation of the components of a sample, in which the components are distributed between two phases, one of which is stationary while the other moves. The stationary phase may be a solid, or a liquid supported on a solid, or a gel. The stationary phase may be packed in a column, spread as a layer, or distributed as a film, etc.; in these definitions, chromatographic bed is used as a general term to denote any of the different forms in which the stationary phase may be used. The mobile phase may be gas or liquid.

Analyte
An analyte (or solute) is the chemical entity being analyzed.

Eluent
An eluent is a solvent used to carry the components of a mixture through a stationary phase. It is an alternative term used for the mobile phase.

Eluate
The mobile phase that exits the column is termed the eluate.

Elution
It is a process in which solutes are washed through a stationary phase by the movement of a mobile phase.

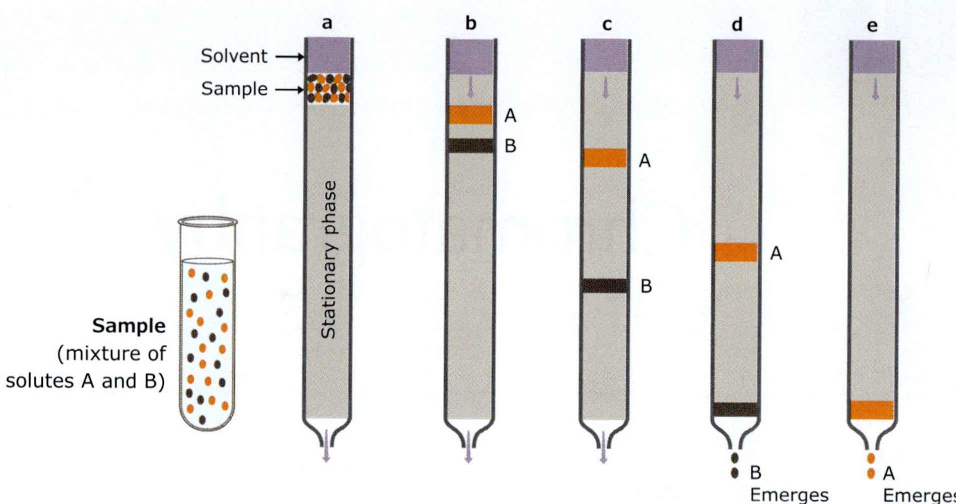

Figure 1.1 The figure shows the separation of solutes A and B present in a mixture by chromatography. A continuous flow of solvent carries a mixture of solutes A and B. The solvent carries the two solutes down the column (b). Solutes A and B have different relative affinities for stationary and mobile phases. Thus, a solute with a relatively higher affinity for the stationary phase moves slowly through the chromatographic system than does a solute with a lower affinity. This difference in mobility ultimately leads to the physical separation of the different solutes present in a sample. Here, solute B has a lower affinity with the stationary phase. Hence, after some time, solute B is moving at a much faster rate than A (c). Finally, solute B emerges first (d), while solute A finally emerges in (e).

Classification of chromatographic methods

Chromatographic methods can be classified in three fundamental ways:

- Based on the shape of the chromatographic bed.
- Based on the physical state of the mobile and stationary phases.
- Based on the mechanism of separation.

Based on the shape of the chromatographic bed

Based on the shape of the chromatographic bed, there are two types of chromatography — *planar* chromatography and *column* chromatography.

In **planar chromatography**, the stationary phase is spread on a flat, planar surface. The plane can be a paper impregnated by a substance acting as a stationary phase (**paper chromatography, PC**) or a thin layer of a substance acting as a stationary phase spread on a glass, metal or plastic plate (**thin layer chromatography, TLC**). Planar chromatography is also termed **open-bed chromatography**. In thin layer chromatography, the stationary phase is a thin layer of silica gel or alumina on a glass, metal or plastic plate. Most commonly, silica gel is used as a stationary phase. In silica gel, the silicon atoms are joined via oxygen atoms in a giant covalent structure. The other commonly used stationary phase is alumina (aluminium oxide).

Note: Only liquid chromatography can be performed either in columns or on planar surfaces; gas chromatography, on the other hand, can be performed *only* in columns.

In **column chromatography**, the stationary bed is within a tube. The particles of the solid stationary phase or support coated with a liquid stationary phase may fill the whole inside volume of the tube (**packed column**) or be concentrated on or along the inside tube wall leaving an open, unrestricted path for the mobile phase in the middle part of the tube (**open-tubular column**). In gas chromatography, the stationary phase is always packed in a column in order to contain the mobile phase, a gas.

Based on the physical state of the mobile and stationary phases

Based on the physical nature of *mobile phase*, there are two types of chromatography – **Gas chromatography** and **liquid chromatography**. Further based on the physical nature of stationary phase, there are two subgroups of gas chromatography by naming the mobile phase followed by the stationary phase – *gas–solid* chromatography and *gas–liquid* chromatography. In similar fashion, liquid chromatography can be sub-grouped into *liquid-solid* chromatography and *liquid-liquid* chromatography.

Based on the mechanism of separation

The chromatographic techniques use several types of mechanisms to separate analytes. Based on the mechanism of separation, chromatographic techniques can be partition, adsorption, size exclusion, affinity and ion exchange chromatography.

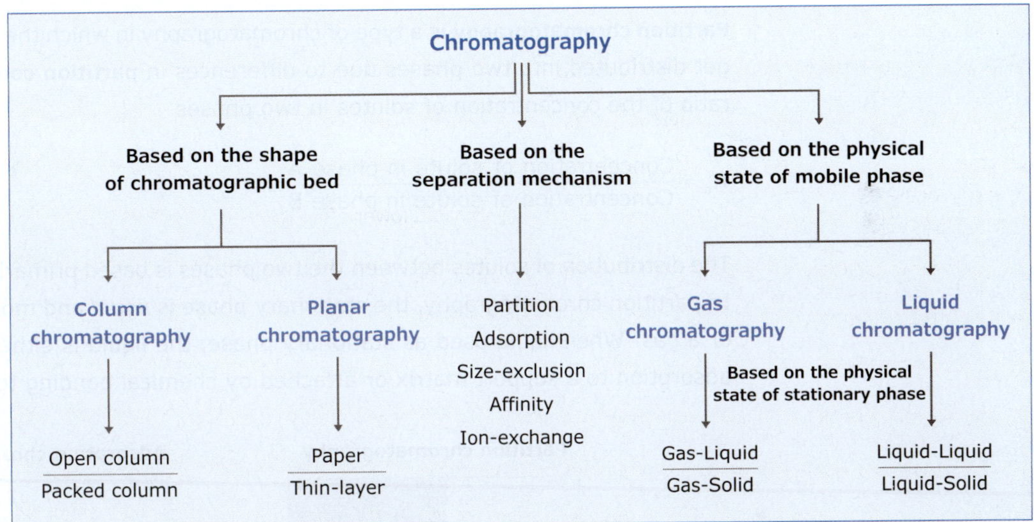

1.1 Adsorption and partition chromatography

Chromatographic techniques can be classified into two main categories: *partition* and *adsorption* chromatography - depending on how the solutes interact with the stationary phase. The mode of interaction between the sample components and the two phases can be classified into two types, although many separation processes are combinations of both. If the sample is attracted to the surfaces of the phases, commonly to the surface of a solid stationary phase, the process is called **adsorption**. Alternatively, if the sample diffuses into the interior of the stationary phase, the process called **absorption** (term **partition** is recommended by the IUPAC).

Adsorption chromatography is a type of chromatography in which the separation of components present in a mixture is based on the relative differences in adsorption of components to the stationary phase present in the chromatography column. The components of the mixture travel with different rates due to differences in their non-covalent interactions with stationary phase. The adsorption involves weak non-covalent interactions such as ionic interactions, van der Waals interactions, hydrogen bondings and hydrophobic forces between the components of the mixture and the stationary phase. For example, polar compounds adsorb strongly to the polar stationary phase while non-polar compounds adsorb strongly to the non-polar stationary phase. The more polar a molecule, the more strongly it will be adsorbed by a polar stationary phase. Similarly, the more non-polar a molecule, the more strongly it will be adsorbed by non-polar stationary phase. Due to differences in degree of adsorption, less tightly bound compounds

Chromatography represents a separation technique; whereas a chromatograph is a system for performing chromatography. The chart displaying the time dependent change in signal intensity as a result of the separation is called a chromatogram.

Martin and Synge in 1944 developed the methodology of partition chromatography and were honoured with Nobel Prize.

elute out by the mobile phase earlier than the tightly bonded ones. Hence during separation of components of the mixture, when we use a polar stationary phase, polar components elute out late due to greater adsorption and non-polar components get out of the column or elute out first. This is exactly reverse on using a non-polar stationary phase.

Adsorption chromatography consists of a solid stationary phase (known as **adsorbent**). The mobile phase is either a liquid (*liquid-solid chromatography*) or a gas (*gas-solid chromatography*). The adsorbents can be polar or non-polar molecules. Most commonly used adsorbents are polar, either *acidic* (e.g. silica) or *basic* (e.g. alumina) molecules. The adsorptive effects of the polar adsorbents are often due to the presence of hydroxyl groups and the formation of hydrogen bonds with the solute molecules. The strength of these bonds and hence the degree of adsorption increases as the polarity of the solute molecule increases. Charcoal is a non-polar adsorbent that binds large and non-polar molecules.

Partition chromatography is a type of chromatography in which the components of the mixture get distributed into two phases due to differences in **partition coefficients** (K_d), which is the ratio of the concentration of solutes in two phases.

$$K_d = \frac{\text{Concentration of solute in phase A}}{\text{Concentration of solute in phase B}}$$

The distribution of solutes between the two phases is based primarily on solubility differences. In partition chromatography, the stationary phase is *liquid* and mobile phase is either a *liquid* or a *gas*. When liquid used as stationary phase, the liquid is either held in place by physical adsorption to a support matrix or attached by chemical bonding to a support matrix.

Adsorption chromatography
Separation is based mainly on differences between the adsorption affinities of the sample components for the surface of an active solid stationary phase.

Partition chromatography

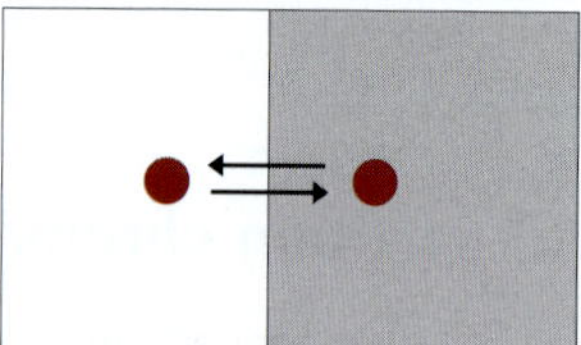

Separation is based on solute partitioning between two liquid phases. (Relative solubility)

Adsorption chromatography

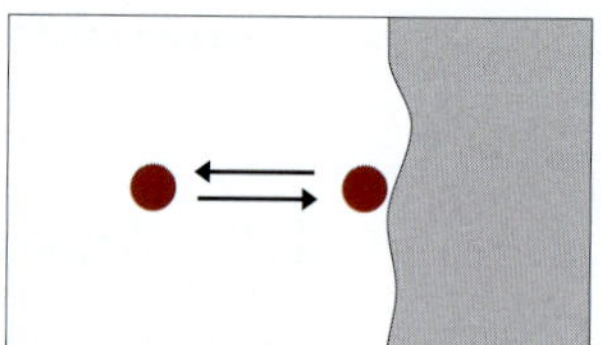

The stationary phase is a solid. Separation is due to a series of adsorption/desorption steps.

Partition chromatography
Separation is based mainly on differences between the solubilities of the sample components in the stationary phase (gas-liquid chromatography), or on differences between the solubilities of the components in the mobile and stationary phases (liquid-liquid chromatography).

Liquid–liquid partition chromatography

In liquid-liquid partition chromatography (or simply as *partition chromatography*), separation is based on solute partitioning between stationary liquid phase and the mobile liquid phase. Substances which are more soluble in the mobile phase will pass rapidly through the system while those which favour the stationary phase will be retarded. It is of two types:

Normal-phase partition chromatography: An elution procedure in which the stationary phase is more polar than the mobile phase. During elution least polar analyte (solute) is eluted first and the most polar last.

Reversed-phase partition chromatography: An elution procedure in which the mobile phase is significantly more polar than the stationary phase. In this case, most polar solutes elute first and least polar elute last.

Paper chromatography is a type of partition chromatography. In paper chromatography, the end of the paper is dipped into a solvent mixture consisting of aqueous and organic components. The solvent soaks into the paper by capillary action because of the fibrous nature of the paper. The aqueous component of the solvent binds to the cellulose of the paper and thereby forms a stationary phase with it. The organic component of the solvent continues migrating, thus forming the mobile phase. The rates of migration of the various substances being separated are governed by their relative solubilities in the polar stationary phase and the non-polar mobile phase. During the separation process, a given solute is distributed between the mobile and stationary phases according to its **partition coefficient**. The nonpolar molecules move faster than polar ones. The migration rate of a substance during paper chromatography is usually expressed as the dimensionless term R_f (Relative front), which is the ratio of the distance traveled by substance and solvent front.

$$R_f = \frac{\text{Distance traveled by substance}}{\text{Distance traveled by solvent front}}$$

Naturally the R_f can be calculated only in those instances when the solvent is not allowed to leave the end of the paper sheet.

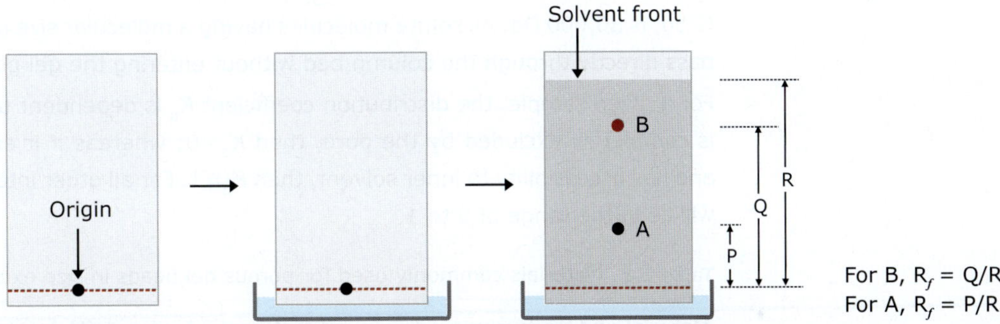

Paper chromatograms can be developed either by *ascending* or *descending* solvent flow. There is little difference in the quality of the chromatograms and the choice is usually a matter of personal preference. Descending chromatography has two advantages: 1. it is faster because gravity aids the flow and 2. for quantitative separations of materials with very small R_f values, which therefore require long runs, the solvent can run off the paper.

1.2 Size exclusion chromatography

Size exclusion chromatography or molecular sieve chromatography separates molecules on the basis of size and shape. A column matrix filled with porous gel beads, made up of an insoluble and hydrated polymer such as polyacrylamide (Sephacryl or BioGel P) or dextran (Sephadex) or agarose (Sepharose) acts as a stationary phase. Size exclusion chromatography includes: **gel permeation chromatography** and **gel filtration chromatography**. Gel permeation chromatography uses organic mobile solvent while gel filtration chromatography uses aqueous mobile solvent to separate and characterize molecules.

The basis of size exclusion chromatography is very simple. If a solution containing molecules of various sizes is passed through the column, molecules smaller than the pores can enter the pores in the beads whereas larger molecules cannot. So larger molecules move faster and elute first. Since smaller molecules can enter the pores present in the beads, they have longer path and longer retention time than larger molecules that cannot enter the pores.

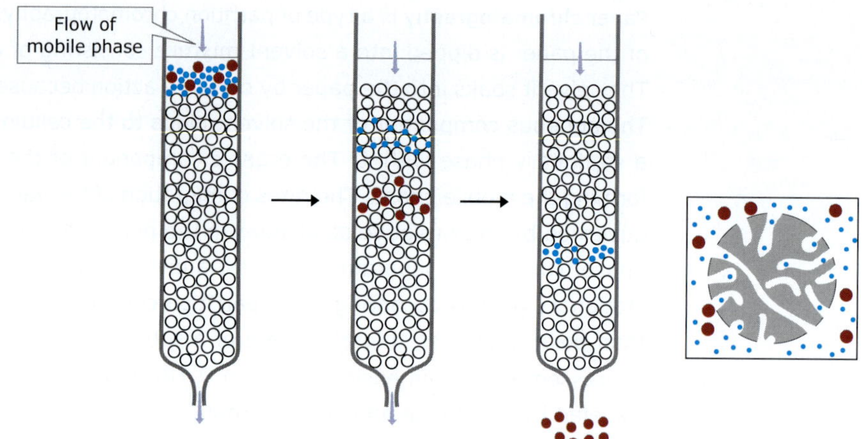

Figure 1.2 When the sample passes through the porous gel beads, small sample molecules can enter the pores, causing them to flow slower through the column. Large molecules which cannot enter the pores, pass through the column at a faster rate than the smaller ones. Correct pore sizes and solvents are crucial for a good separation.

Thus, if a mixture of proteins is applied to a column and then washed with an appropriate buffer, the first proteins to emerge from the column are those that are too large to enter the pores of the gel beads. Other proteins are eluted in decreasing order of their molecular size.

The molecular mass of the smallest molecule unable to penetrate the pores of a given gel is said to be the gel's **exclusion limit.** For example, the exclusion limit of a typical gel, Sephadex G-50, is 30,000 Da. All solute molecules having a molecular size greater than this value would pass directly through the column bed without entering the gel pores.

For a given sample, the distribution coefficient K_d is dependent upon its size. If the molecule is completely excluded by the pore, then $K_d = 0$; whereas if it enters into the porous beads and has accessibility to inner solvent, then $K_d = 1$. For all other intermediate sizes, the K_d value will lie in the range of 0 to 1.

Table 1.1 Materials commonly used for porous gel beads in size exclusion chromatography

Material and trade name	Fractionation range* (Molecular mass in Da)
Dextran	
Sephadex G-10	0 – 700
Sephadex G-25	1000 – 5000
Sephadex G-50	1500 – 30,000
Sephadex G-75	3000 – 70,000
Sephadex G-100	4000 – 150,000
Sephadex G-150	5000 – 300,000
Polyacrylamide	
Bio-gel P-2	100 – 1800
Bio-gel P-6	1000 – 6000
Bio-gel P-60	3000 – 60,000
Bio-gel P-150	15,000 – 150,000
Bio-gel P-300	60,000 – 400,000
Agarose	
Sepharose 2B	$2 \times 10^6 - 25 \times 10^6$
Sepharose 4B	$3 \times 10^5 - 3 \times 10^6$
Sepharose 6B	$10^4 - 20 \times 10^6$

**The molecular mass listed are for globular proteins.*

Size measurements by size exclusion chromatography

In order to obtain size information about a solute from a size exclusion chromatography experiment, the column must first be characterized in terms of the volumes accessible to analytes. The total volume (V_T) of a size exclusion chromatography column is divided into three parts:

1. The volume external to the packing material i.e. **void volume**, V_0;
2. The volume contained within the porous beads i.e. **internal volume**, V_i;
3. The volume occupied by the packing material itself i.e. **bed volume**, V_g.

Therefore, $\mathbf{V_T = V_0 + V_i + V_g}$

The values of the V_0 and V_i are determined experimentally by measuring the elution volumes of, respectively, a large solute that is totally excluded from the interior of the porous bead and a small solute that has access to all pores of the gel bead. The elution volume of a given solute, V_e, is the volume of solvent required to elute the solute from the column after it has first contacted the gel. The elution volume, V_e, of a solute that is partially included in the pores of the gel bead can be related to the void and internal volumes of a column by the following equation:

$$\mathbf{V_e = V_0 + \sigma V_i}$$

where, σ is the *partition coefficient* of the solute.

It is the partition coefficient, σ, which describes how much of the internal volume is available for the solute ($0 < \sigma < 1$). When σ is compared with the values measured for solutes of known size, it provides information about the molecular size of an unknown solute. If a series of solutes of known size is subjected to size exclusion chromatography, a linear relationship between partition coefficient and size is observed.

1.3 Ion exchange chromatography

Ion exchange chromatography is applicable for the separation of charged molecules. In this chromatographic technique, ionic solutes display reversible electrostatic interactions with a charged stationary phase. The stationary solid phase commonly consists of an insoluble matrix with covalently attached anions or cations (called **ion exchangers**). Solutes entering the column may be negatively charged, positively charged, or neutral under the experimental conditions. Solute ions of the opposite charge in the mobile liquid phase bind reversibly to the ion exchanger by electrostatic interactions. The strength of interactions depends on the size of the charge and the charge density (amount of charge per unit volume) of the solute. The greater the charge or the charge density, the stronger the interaction. Neutral solutes show little or no affinity for the stationary phase and move with the eluting buffer. The bound solutes can be released by eluting the column with a buffer of increased ionic strength or pH. An increase in buffer ionic strength releases bound solutes by displacement. Increasing the buffer pH decreases the strength of the interaction by reducing the charge on the solute or on the resin.

Ion exchanger

Ion exchangers are made up of two parts – an insoluble matrix and chemically bonded charged groups within and on the surface of the matrix. An ion exchanger is classified as *cationic* or *anionic* depending on whether it exchanges cations or anions.

Cation exchanger (also called *acidic ion exchanger*): It is used for cation separation.

Anion exchanger (also called *basic ion exchanger*): It is used for anion separation.

Each type of exchanger is also classified as *strong* or *weak* according to the ionizing strength of the functional group. An exchanger with a quaternary amino group is, therefore, a strongly basic anion exchanger, whereas primary or secondary aromatic or aliphatic amino groups would lead to a weakly basic anion exchanger. A strongly acidic cation exchanger contains the sulfonic acid group.

Table 1.2 Commonly used ion exchangers

Name	Type	Functional group
Anion exchanger		
DEAE-cellulose	Weakly basic	Diethylaminoethyl (DEAE)
QAE-Sephadex	Strongly basic	Quaternary aminoethyl (QAE)
Q-Sepharose	Strongly basic	Quaternary ammonium (Q)
Cation exchanger		
CM-cellulose	Weakly acidic	Carboxymethyl (CM)
SP-Sepharose	Strongly acidic	Sulfopropyl (SP)
SOURCE S	Strongly acidic	Methylsulphate (S)

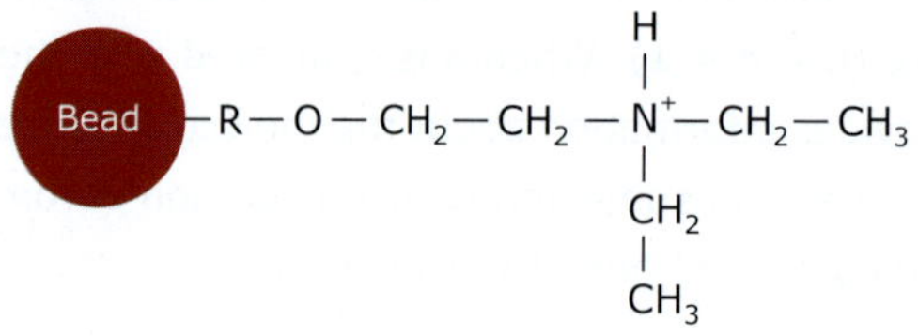

Bead $-R-O-CH_2-CH_2-N^+(H)(CH_2-CH_3)-CH_2-CH_3$

DEAE: pK_a = 9.5
Anion exchanger → binds to negatively charged molecules (anions)

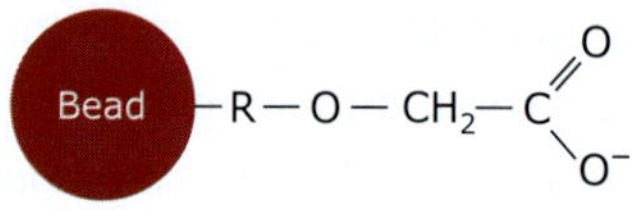

Bead $-R-O-CH_2-C(=O)O^-$

CM: pK_a = 4.0
Cation exchanger → binds to positively charged molecules (cations)

Figure 1.3 Ion exchangers – diethylaminoethyl (DEAE) and carboxymethyl (CM). The positive charge of DEAE attracts negatively charged molecules. CM is suitable for binding with positively charged molecules.

A solute in a given sample, which has the opposite charge to that of the charged group of the ion exchanger binds to the column. Separation of charged solutes occurs because different solutes have different degree of interaction with the ion-exchanger due to differences in their charges, charge density and distribution of charge on their surfaces. If a solute has a net positive charge at pH 7, it will usually bind to ion exchanger (cation exchanger) containing carboxylate groups, whereas a negatively charged protein will not. The bound molecules can be eluted by altering the pH of the eluting buffer or by increasing the salt concentration of the eluting buffer. A positively charged protein bound to cation exchanger can be eluted by increasing the salt concentration in the eluting buffer because cations present in the buffer compete with positively charged groups on the protein for binding to the ion exchanger. Proteins that have a low density of net positive charge will tend to emerge first, followed by those having a higher charge density.

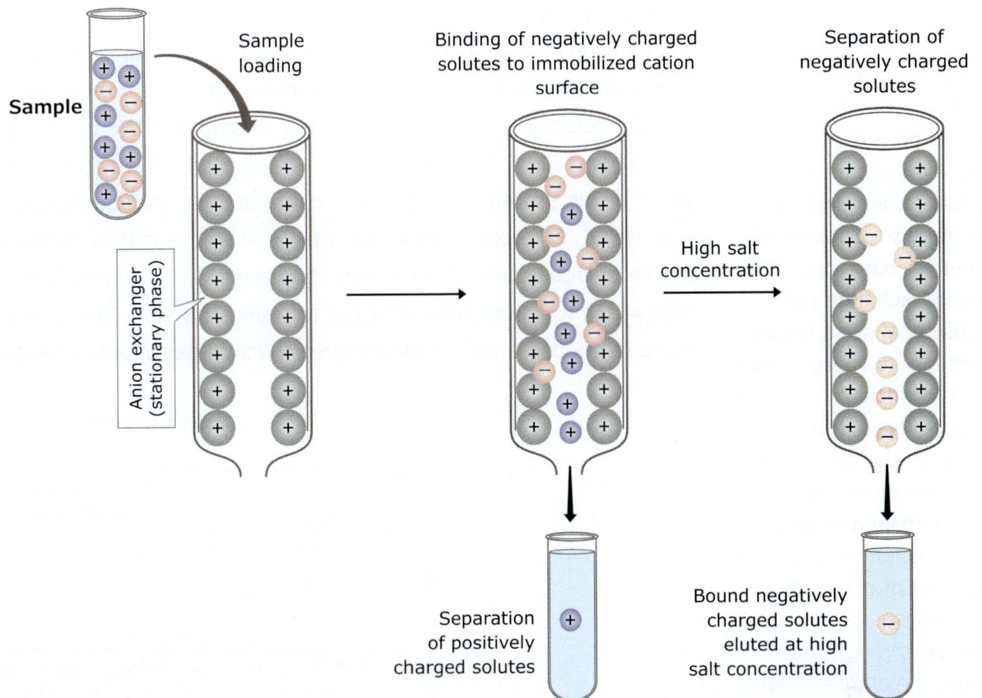

Figure 1.4 Column materials used for ion exchange chromatography contain charged groups covalently linked to the surface of an insoluble matrix. The charged groups of the matrix can be positively or negatively charged. Anion exchangers have positively charged groups that will attract negatively charged anions. When a mixture of solutes is loaded into the anion exchanger, negatively charged solutes bind to the exchanger.

Selection of the ion exchanger

Before a proper choice of ion exchanger can be made, the nature of the solutes to be separated must be considered. The choice of ion exchanger (whether to use a cationic or anionic exchanger) for the purification of a biomolecule largely depends on the isoelectric point, pI, of the biomolecule.

If the solute molecule has only one type of charged group, the choice is simple. A solute that has a positive charge will bind to a cationic exchanger and vice versa. However, many solutes have more than one type of ionizing group and may have both negatively and positively charged groups. The net charge on such molecules depends on pH. At the isoelectric point, the solute has no net charge and would not bind to any type of ion exchanger. At a pH value above the pI of a solute, it will have a net negative charge and adsorb to an anion exchanger. Below the pI, the solute has a net positive charge and will adsorb to a cation exchanger.

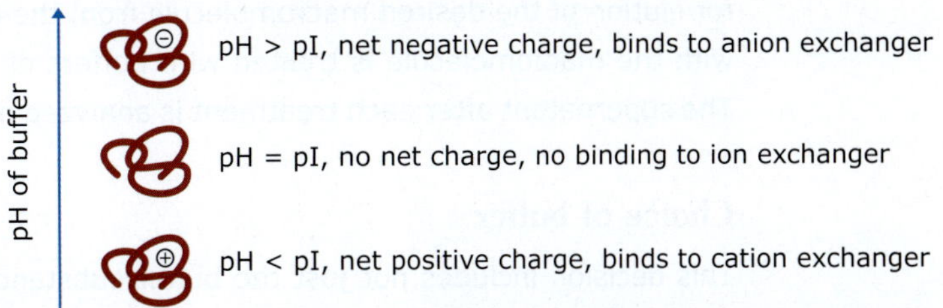

Figure 1.5 The net charge on a protein is influenced by the pH of its solvent. At pH=pI, the protein has zero net charge and, therefore, will not bind to a cation exchanger or an anion exchanger. Adjusting the pH above or below the pI of the protein will lead to a net charge and protein binding to either an anion exchanger (pH > pI) or a cation exchanger (pH < pI).

Ion exchange chromatography can also be used to separate DNA from a cell extract. It is based on the interaction between anion exchanger (DEAE) and negatively charged phosphates of the DNA backbone. The anion-exchange resin consists of silica beads with a high charge density. When the cell extract passes through the column, all the negatively charged molecules bind to the resin and retained in the column. If the salt solution of gradually increasing concentration is passed through the column, the different types of molecule will elute in the sequence protein, RNA and finally DNA.

In principle, solutes with both negatively and positively charged groups should bind to both anionic and cationic exchangers. However, when one is dealing with large biomolecules, the pH *range of stability* must also be evaluated. The range of stability refers to the pH range in which the biomolecule is not denatured. For example, if the pI of a protein is 4 then, in most cases, it is advisable to choose an ion exchanger which binds to the protein at a pH > 4. Since at pH > 4 this protein is negatively charged, the ion exchanger has to be an anion exchanger, e.g. DEAE. One could also use a pH < 4 and a cation exchanger, but many proteins are not stable or aggregate under these conditions. If, in contrast, the protein we want to purify has a pI = 10, it is positively charged at a pH around 7. Thus, in general for this protein type we have to choose a cation exchanger, which is negatively charged at neutral pH.

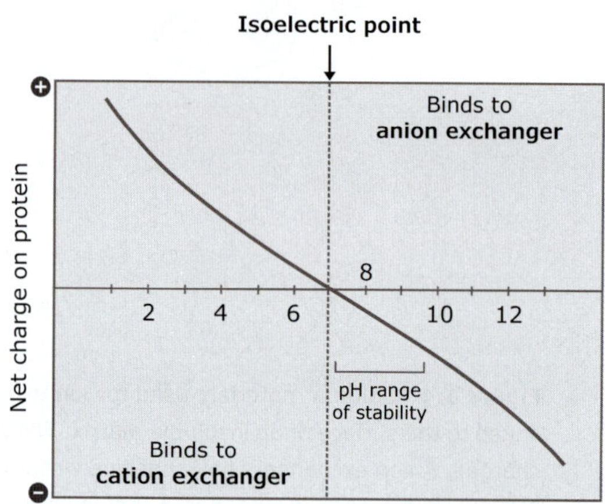

Figure 1.6 Diagram shows how the net charge of a hypothetical protein changes as a function of pH. Below the isoelectric point, the molecule has a net positive charge and would be bound to a cation exchanger. Above the isoelectric point, the net charge is negative, and the protein would bind to an anion exchanger. Superimposed on this graph is the pH range of stability for the hypothetical protein. The range of stability refers to the pH range in which the biomolecule is not denatured. Because it is stable in the range of pH 7.0-9.0, the ion exchanger of choice is an anionic exchanger. In most cases, the isoelectric point of the protein is not known. The type of ion exchanger must be chosen by trial and error.

In most cases, the isoelectric point of the protein is not known. The type of ion exchanger must be chosen by trial and error as follows. Small samples of the solute mixture in buffer are equilibrated for 10 to 15 minutes in separate test tubes, one with each type of ion exchanger. The tubes are then centrifuged or let stand to sediment the ion exchanger. Check each supernatant for the presence of solute. If a supernatant has a relatively low level of added solute, that ion exchanger would be suitable for use. This simple test can also be extended to find conditions for elution of the desired macromolecule from the ion exchanger. The ion exchanger charged with the macromolecule is treated with buffers of increasing ionic strength or changing pH. The supernatant after each treatment is analyzed as before for release of the macromolecule.

Choice of buffer

This decision includes not just the buffer substance but also the pH and the ionic strength. Buffer ions will, of course, interact with ion-exchange resins. Buffer ions with a charge opposite to that on the ion exchanger compete with solute for binding sites and greatly reduce the capacity of the column. Cationic buffers should be used with anionic exchangers; anionic buffers should be used with cationic exchangers.

The pH chosen for the buffer depends first of all on the range of stability of the macromolecule to be separated. Second, the buffer pH should be chosen so that the desired macromolecule will bind to the ion exchanger. In addition, the ionic strength should be relatively low to avoid damping of the interaction between solute and ion exchanger. Buffer concentrations in the range 0.05 to 0.1 M are recommended.

Problem

Suppose, you have determined that your partially purified protein is stable and active between pH 5 and pH 7.5. On either side of that pH range, the protein is no longer active. Now, you want to do a quick experiment to determine proper pH to do the ion-exchange chromatography. You mixed a bit of the crude preparation with a small amount of the ion-exchange resin DEAE-Sepharose in a series of buffer solutions that have a pH between 5 and 7.5. Next, you pelleted the resin and assayed the supernatant for the presence of your protein. Finally, you chosen this information to pick the appropriate pH to do the ion-exchange chromatography. You have completed the first two steps and have obtained the results shown in figure 1.7.

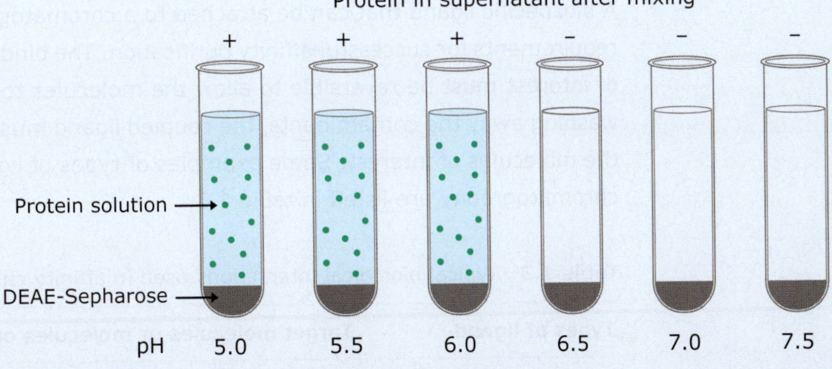

Figure 1.7 Samples of the protein were mixed with DEAE-Sepharose beads in buffers at a range of pH values and then the mixtures were centrifuged to pellet the beads. The presence of the protein in the supernatant is indicated by (+) and its absence is indicated by (–).

a. For the chromatography, should you pick a pH at which the protein binds to the beads (pH 6.5 to 7.5) or a pH where it does not bind (pH 5 to 6)?

b. Should you pick a pH close to the boundary (i.e. pH 6 or 6.5) or far away from the boundary (i.e. pH 5 or pH 7.5)?

Solution

a. You want to pick a pH at which your protein binds to the DEAE-Sepharose beads. If it does not bind to the beads, the protein will pass through the column with all the other proteins that do not bind. Ion-exchange chromatography is carried out under conditions in which the protein interacts with the beads.

b. In general, you want to pick a pH at which the protein of interest binds to the column, but does not bind too strongly. Thus, pH 6.5 would be the best choice for your initial studies. If you pick too high a pH, the protein may bind to the column too strongly, requiring harsh conditions to remove it.

1.4 Affinity chromatography

Affinity chromatography is a technique enabling purification of a biomolecule with respect to biological function or individual chemical structure. The substance to be purified is specifically and reversibly adsorbed to a *ligand* (binding substance), immobilized by a covalent bond to a chromatographic bed material (i.e. matrix). Samples are applied under favourable conditions for their specific binding to the ligand. Substances of interest are consequently bound to the ligand while unbound substances are washed away. Recovery of substances of interest can be achieved by changing experimental conditions to favour desorption. The operation of affinity chromatography involves the following steps:

- Choice of an appropriate ligand.
- Immobilization of the ligand onto a support matrix.
- Binding of the molecules of interest with the ligand.
- Removal of non-specifically bound molecules.
- Elution of the molecules of interest in a purified form.

A biospecific ligand that can be attached to a chromatography matrix covalently is one of the requirements for successful affinity purification. The binding between the ligand and molecules of interest must be reversible to allow the molecules to be removed in an active form. After washing away the contaminants, the coupled ligand must retain its specific binding affinity for the molecules of interest. Some examples of types of ligands that are usually used in affinity chromatography are listed in table 1.3.

Table 1.3 Typical biological interactions used in affinity chromatography

Types of ligand	Target molecules or molecules of interest
Enzyme	Substrate analogue, inhibitor, cofactor
Antibody	Antigen
Lectin	Polysaccharide, glycoprotein, cell surface receptor, cell
Nucleic acid	Complementary base sequence, nucleic acid binding protein
Avidin	Biotin
Calmodulin	Calmodulin-binding molecule
Poly(A)	RNA containing poly(U) sequences
Glutathione	Glutathione-S-transferase or GST fusion proteins
Proteins A and G	Immunoglobulins

For example, the eukaryotic mRNA with poly (A) tail can be separated from other types of RNA molecules by oligo (dT)-cellulose affinity chromatography. Poly (A) tails form stable interaction with short chains of oligo (dT) that are attached to the support matrices. High salt is added to the chromatography buffer to stabilize the nucleic acid duplexes as only a few dT-A base pairs are formed. A low-salt buffer is used after non-polyadenylated RNAs have been washed from the matrix. This buffer helps to destabilize the double-stranded structures and elute the poly (A) RNAs from the resin.

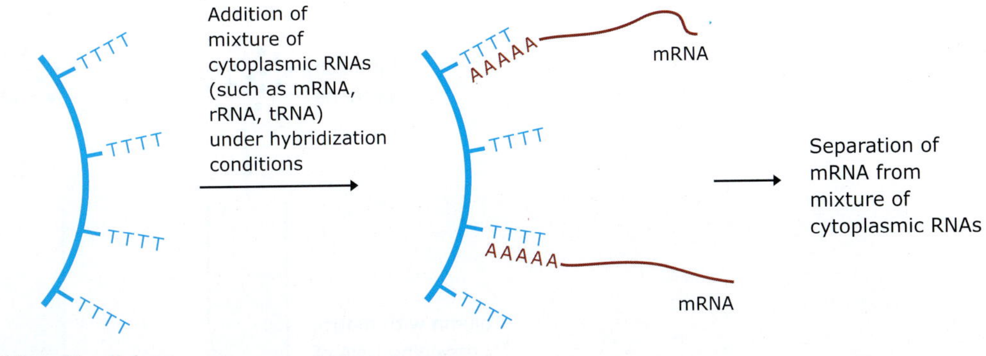

Figure 1.8 Isolation of mRNA by affinity chromatography. mRNA is isolated from the cytoplasmic mixture of RNAs using the oligo (dT)-cellulose in a column. Cytoplasmic RNAs such as tRNA and rRNA which are not bound to the matrix beads can wash away and then mRNA can be eluted from the column using a low-salt buffer.

Choice of ligand

Choice of a suitable ligand is the most important feature of affinity chromatography. A number of factors to be considered when selecting a ligand. These factors are:

Specificity: The ligand should recognize only the molecule of interest to be purified.

Reversibility: The ligand should form a reversible complex with the molecule of interest.

Stability: The ligand should be stable to the condition to be used for immobilization as well as the conditions of use.

Size: The ligand should be large enough such that it contains several groups able to interact with the molecules of interest resulting in sufficient affinity.

Affinity: Binding affinity is the strength of the binding interaction between a molecule of interest to its ligand. Binding affinity is typically measured in terms of *equilibrium dissociation constant* (K_d). The smaller the K_d value, the greater the binding affinity of the ligand for its target. The equilibrium dissociation constant is the inverse of the *equilibrium association constant* (K_a). The interaction of a molecule of interest (M) and a ligand (L) can be described by the equation:

$$M + L \underset{k_{off}}{\overset{k_{on}}{\rightleftharpoons}} M-L \quad [k_{on} = \text{association rate constant and } k_{off} = \text{dissociation rate constant}]$$

For this reaction, the K_d (unit in concentration) or K_a (unit in 1/concentration) is defined by:

$$K_a = \frac{[M-L]}{[M]\,[L]} = \frac{k_{on}}{k_{off}} \quad \text{and} \quad K_d = \frac{1}{K_a} = \frac{[M]\,[L]}{[M-L]} = \frac{k_{off}}{k_{on}}$$

A small value for K_d means that the equilibrium favors the complex – there is a high affinity of the molecule of interest for the ligand.

DNA affinity chromatography

DNA affinity chromatography facilitates the purification of sequence-specific DNA-binding proteins. In this method a double-stranded oligonucleotide of the correct sequence is chemically synthesized and linked to an insoluble matrix such as agarose. The matrix with the oligonucleotide attached is then used to construct a column that selectively binds proteins that recognize the particular DNA sequence. The whole process is explained in the following figure 1.9.

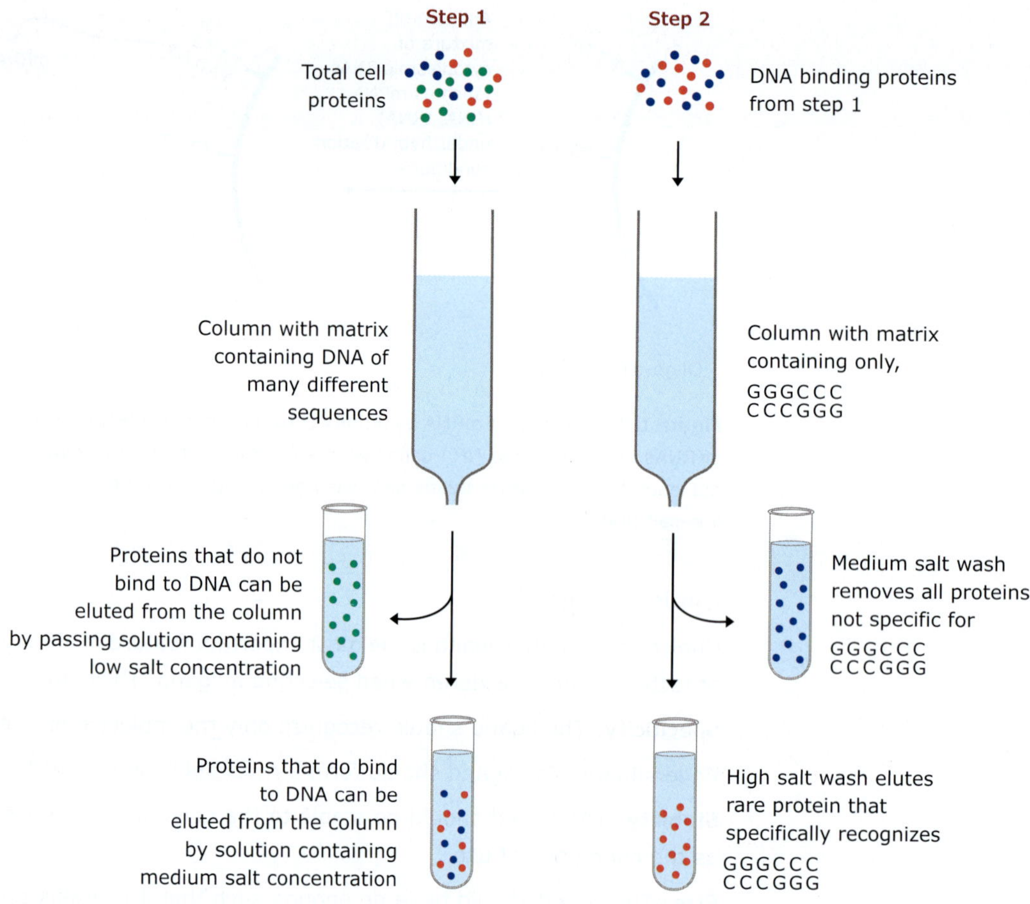

Figure 1.9 DNA affinity chromatography. In the first step, total cell proteins are passed through the column containing a huge number of different DNA sequences. In this process, the proteins that can bind DNA are separated from the remainder of the cellular proteins. Most sequence-specific DNA-binding proteins have a weak (nonspecific) affinity for bulk DNA and are, therefore, retained on the column. This affinity is largely due to ionic attractions, and the proteins can be washed off the DNA by a solution that contains a moderate concentration of salt. In the second step, the mixture of DNA-binding proteins is passed through a column that contains only DNA of a particular sequence. Typically, all the DNA-binding proteins will stick to the column, mostly by nonspecific interactions. These are again eluted by solutions of moderate salt concentration, leaving in the column only those proteins that bind specifically and, therefore, very tightly to the particular DNA sequence. These remaining proteins can be eluted from the column by solutions containing a very high concentration of salt.

1.5 High performance liquid chromatography

High performance liquid chromatography (HPLC) is a column chromatography. It is liquid chromatographic technique i.e. mobile phase is liquid. Instead of a solvent (mobile phase) being allowed to drip through a column under gravity, it is forced through under high pressure. It yields high performance and high speed compared with traditional column chromatography because the mobile phase is pumped with high pressure. The stationary phase may be solid or liquid. The HPLC can function in several chromatographic modes: adsorption, partition, ion exchange and size exclusion. Actually, some of these modes appear to cause separation in combination. If the stationary phase is solid, the technique is called adsorption chromatography. If the stationary phase is a liquid, the sample is partitioned between the stationary and mobile liquid phase.

This is called *partition chromatography*. There are two variants of partition chromatography depending on the relative polarity of the solvent and the stationary phase. In *normal phase*, the stationary phase is polar, and the mobile phase is non-polar. In *reversed phase*, the stationary phase is non-polar, and the mobile phase is polar. Reversed phase HPLC is the most commonly used forms of HPLC.

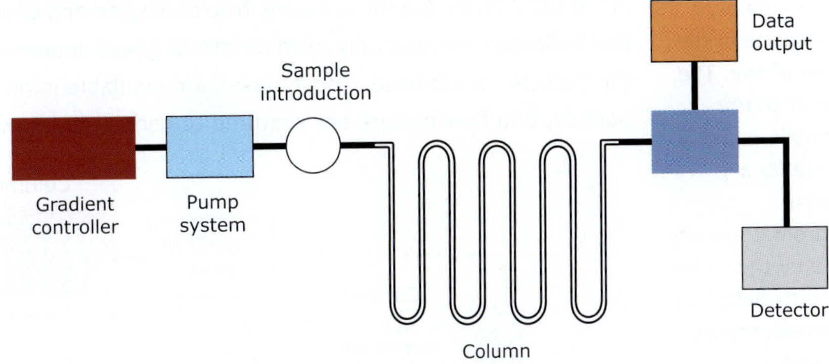

Figure 1.10 A schematic diagram of a typical high-pressure liquid chromatograph. The basic components are a gradient controller, high-pressure pump, packed column, detector and recorder. A computer is used to control the process and to collect and analyze data. The purpose of the pump is to provide a constant, reproducible flow of solvent through the column. Two types of pumps are available – constant pressure and constant volume. HPLC columns are prepared from stainless steel or glass-Teflon tubing. Typical column inside diameters are 2.1, 3.2 or 4.5 mm for *analytical* separations and up to 30 mm for *preparative* applications. The length of the column can range from 5 to 100 cm, but 10 to 20 cm columns are common.

HPLC is basically a highly improved form of liquid column chromatography. Instead of a solvent being allowed to drip through a column under gravity, it is forced through under high pressures of up to 400 atmospheres. That makes it much faster. It also allows us to use very small size particles as column packing material which gives a much greater surface area for interactions between the stationary phase and the molecules flowing past it. This allows a much better separation of the components of the mixture.
The advantages of HPLC are the result of two major advances: 1. the development of stationary supports with very small particle sizes and large surface areas and 2. the improvement of elution rates by applying high pressure to the solvent flow.

The parameters used to describe a HPLC column refer to the nature, type and size of its packing material and the dimensions of the column used. In all forms of chromatography, a measure of column efficiency is *resolution.* Resolution indicates how well solutes are separated. In HPLC, the increased resolution, as compared to classical column chromatography, is primarily the result of adsorbents of very small particle sizes and large surface areas. Since the smaller the particle size, the lower the flow rate; therefore, it is not feasible to use very small gel beads in liquid column chromatography as low flow rates lead to increased analysis time. In HPLC, increased flow rates are obtained by applying a pressure difference across the column. A combination of high pressure and adsorbents of small particle size leads to the high resolving power and short analysis time characteristic of HPLC.

Advantages of HPLC

High speed	Analysis times measured in minutes or seconds
High resolution	Columns tightly packed with small, uniform particles
High sensitivity	Parts-per-million (ppm) to sub-parts-per-billion (ppb) detection limits
High accuracy	High precision sampling devices and good standards yield accurate analysis

1.6 Gas chromatography

In gas chromatography (GC), the mobile phase is a carrier gas. The stationary phase is either a *solid* adsorbent termed **gas-solid chromatography** (GSC), or a *liquid* on an inert support termed **gas-liquid chromatography** (GLC). The *gas-liquid chromatography* is usually shortened to *gas chromatography*. GC is used to analyze volatile substances in the gas phase. Its use is, therefore, confined to analytes that are volatile, but thermally stable. It is the only form of

Gas-liquid chromatography is a direct extension of liquid-liquid partition chromatography, since the vapour in the carrier gas is exactly analogous to the solute in the mobile liquid phase. The distribution of solute between mobile phase liquid and stationary phase liquid in partition chromatography is described by the same law as the distribution of vapour between mobile phase gas and stationary liquid in gas-liquid chromatography, so that gas-liquid chromatography has often been called 'gas-liquid partition chromatography,' though international convention no longer favours this name.

chromatography that does not utilize the mobile phase for interacting with the analyte. GSC is based on a solid stationary phase in which retention of analytes occurs because of physical adsorption. In GLC, the analyte is partitioned between a gaseous mobile phase and a liquid phase immobilized on the surface of an inert solid packing or on the walls of a capillary tube. The mobile-phase gas in GC is called the **carrier gas**. Different types of carrier gases are used in the GC. Carrier gas must be dry, free of oxygen and chemically inert. Helium, nitrogen, argon and hydrogen are generally used as carrier gases depending upon the desired performance and the detector being used. These gases are available in pressurized tanks. Pressure regulators, gauges, and flow meters are required to control the flow rate of the gas.

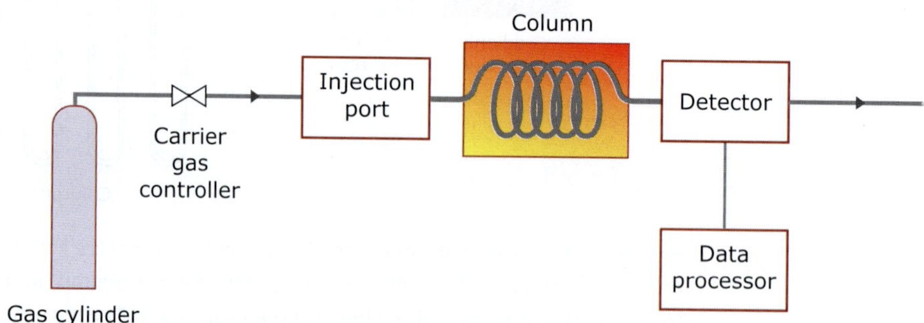

Figure 1.11 Basic components of a gas chromatograph. The carrier gas passes through the sample injection port, column and detector before finally being purged to the outside. The sample injection port, column, and detector are each held within independent isothermal ovens and various temperatures are maintained for each part depending on the sample being analyzed. Sample is introduced into the sample injection port. When liquids or solids sample are used, the chamber is heated to vaporize the sample. A detector is then used to detect the various eluting components.

Comparison of gas chromatography and liquid chromatography

The two most important parameters in *gas chromatography* (GC) are the nature of the stationary phase and temperature. In *liquid chromatography* (LC) they are the nature of the stationary phase, the nature of the mobile phase and to some extent temperature. While temperature is important in both techniques, it is more critical in GC. Also, in LC, both the mobile phase and stationary phase interact with the analytes, whereas for GC, only the stationary phase does so.

References

Berg JM, Tymoczko JL and Stryer L (2007), *Biochemistry*, 6th ed. W.H. Freeman and Company.

Chang R (2005), *Physical Chemistry for the Biosciences*, University Science Books.

Dorsey JG, Foley JP, Cooper WT, Barford RA and Barth HG (1990), Liquid chromatography: theory and methodology. *Analytical Chemistry* 62:324R–356R.

Freifelder D (1982), *Physical Biochemistry*, 2nd ed. W.H. Freeman and Company.

Holler FJ and Crouch SR (2014), *Fundamentals of analytical chemistry*, 9th ed. Cengage Learning.

Lodish H, Berk A, Kaiser CA et al (2008), *Molecular Cell Biology*, 6th ed. W.H. Freeman and Company

Miller JM (2005), *Chromatography: Concepts and Contrasts*, 2nd ed. John Wiley and Sons Inc.

Skoog DA, Holler FJ and Crouch SR (2007), *Principles of instrumental analysis*, 6th ed. Thomson Publishing, USA.

2

Electrophoresis

Electrophoresis (*Electro* refers to the energy of electricity and *Phoresis*, from the Greek verb *phoros*, means to carry across) is a technique for separating or resolving charged molecules (such as amino acids, peptides, proteins, nucleotides, and nucleic acids) in a mixture under the influence of an applied electric field. Charged molecules in an electric field move or migrate, at a speed determined by their *charge to mass* ratio. According to the laws of electrostatics, an ion with charge 'Q' in an electric field of strength 'E' will experience an electric force, $F_{electrical}$

$$F_{electrical} = Q.E$$

The resulting migration of the charged molecule through the solution is opposed by a frictional force, $F_{frictional}$

$$F_{frictional} = V.f$$

where, V is the rate of migration of charged molecule and f is its *frictional coefficient*

The **frictional coefficient** measures the resistance encountered by the molecule in moving through the solvent. It depends on the size and shape of the migrating molecule and the viscosity of the medium. For a spherical molecule, it is given by the Stokes' law;

$$f = 6\pi\eta r$$

where, η is the viscosity of the solvent and r is the radius of the molecule

In the constant electric field, the force on charged molecule balances each other;

$$Q.E = V.f$$

so that each charged molecule moves with a constant characteristic velocity.

The migration of the charged molecule in the electric field is generally expressed in terms of **electrophoretic mobility** (μ), which is the ratio of the migration rate of a charged molecule to the applied electric field:

$$\mu = \frac{V}{E} = \frac{Q}{f}$$

The SI unit of electrophoretic mobility is $m^2\ s^{-1}\ V^{-1}$.

So according to the above equation, electrophoretic mobility is directly proportional to the charge and inversely proportional to the viscosity of the medium, size and shape of the molecule.

Electrophoresis is of two types – *moving boundary* (or free boundary) *electrophoresis* and *zone electrophoresis.*

Moving boundary electrophoresis is the electrophoresis in a free solution, without a supporting media. It was developed by Tiselius in 1937. To separate the different charged molecules present in a mixture, the sample (dissolved in a buffer solution that serves as an electrolyte and maintains the desired pH) is placed in a glass tube connected to electrodes. When an electrical potential is applied across the tube, the charged molecules migrate toward one or the other electrode. Because different charged molecules migrate at different rates, a number of *interfaces* or *boundaries* are formed between the leading edge of each charged molecules and the remaining mixture.

In **zone electrophoresis**, a sample is constrained to move in some kind of inert matrix such as filter paper moistened with buffer (**paper electrophoresis**) or a gel (**gel electrophoresis**). A matrix is required because the electric current passing through the electrophoresis solution generates heat, which causes diffusion and convective mixing of the bands in the absence of a supporting matrix. The kind of supporting matrix used depends on the type of molecules to be separated and on the desired basis for separation.

2.1 Gel electrophoresis

Gel electrophoresis is a simple, rapid and sensitive analytical tool for separating charged molecules. Under the influence of an electrical field charged molecules migrate in the direction of the electrode bearing the opposite charge. Most biomolecules carry a net charge at any pH other than at their isoelectric point. These charged biomolecules have different electrophoretic mobilities due to their differences in charge, shape and size. In gel electrophoresis, gel serves as porous matrix and behave like a molecular sieve. Typically, a gel matrix provides resistance to the migration of the charged molecules that are being subjected to the electrical force. There are two basic types of materials used to make gels: *agarose* and *polyacrylamide*.

Agarose is a natural colloid extracted from seaweed. It is a linear polysaccharide made up of the basic repeating unit **agarobiose**, which comprises alternating units of *galactose* and *3,6-anhydrogalactose*. Agarose gels have a relatively large pore size and are used primarily to separate larger molecules with a molecular mass greater than 200 kDa. Agarose is usually used at concentrations between 1% and 3%. Agarose gels are used for the electrophoresis of both proteins and nucleic acids. Because the pores of an agarose gel are large, agarose is used to separate macromolecules such as nucleic acids, large proteins and protein complexes.

In gel electrophoresis separation of charged molecules depends on:

- Size, shape and charge of the molecules
- Properties of the gel
- Strength of the electric field
- Features of the buffer
- Temperature

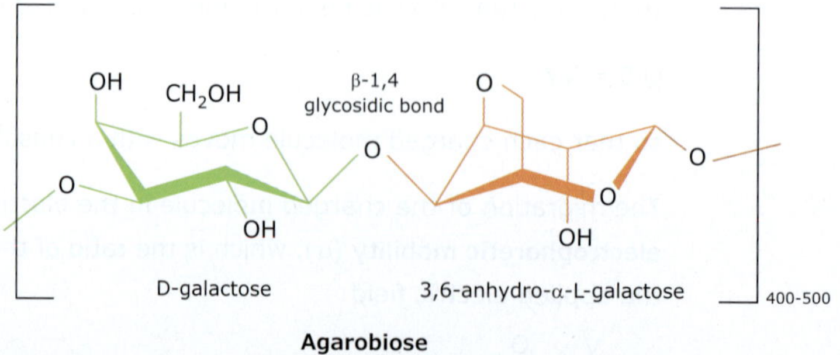

Figure 2.1 Agarose is a linear polysaccharide made up of the basic repeat unit *agarobiose*, which comprises alternating units of galactose and 3,6-anhydro-α-L-galactose.

A **polyacrylamide gel** consists of chains of acrylamide monomers (CH_2=CH–CO–NH_2) crosslinked with *N, N′*-methylenebisacrylamide units (CH_2=CH–CO–NH–CH_2–NH–CO–CH=CH_2), the latter commonly called **bisacrylamide**. Bisacrylamide ('bis' for short) is two units of acrylamide connected by a methylene bridge. The polymerization of acrylamide is a free radical catalyzed reaction. The reaction is initiated by the quaternary amine **TEMED** (N, N, N′, N′-tetramethylethylenediamine), which induces free radical formation from *ammonium persulphate* (APS). In the case of photochemical polymerization, *riboflavin* is used as a source of free radicals. Photochemical polymerization is used when the ionic strength in the gel must be very low because only a minute amount of riboflavin is required. It is also used if the protein studied is sensitive to ammonium persulphate or the by-products of peroxide-initiated polymerization. The free radicals transfer electrons to the acrylamide/bisacrylamide monomers, radicalizing them and causing them to react with each other to form the polyacrylamide chain. In the absence of bisacrylamide, the acrylamide would polymerize into long strands, not a porous gel. But as the diagram shows (figure 2.2), bisacrylamide cross-links the acrylamide chains and this is what gives rise to the formation of the porous gel matrix. The amount of crosslinking, and therefore the pore size and consequent separation properties of the gel can be controlled by varying the ratio of acrylamide to bisacrylamide. It should be noted that acrylamide and bisacrylamide are a *neurotoxin* in their monomer form.

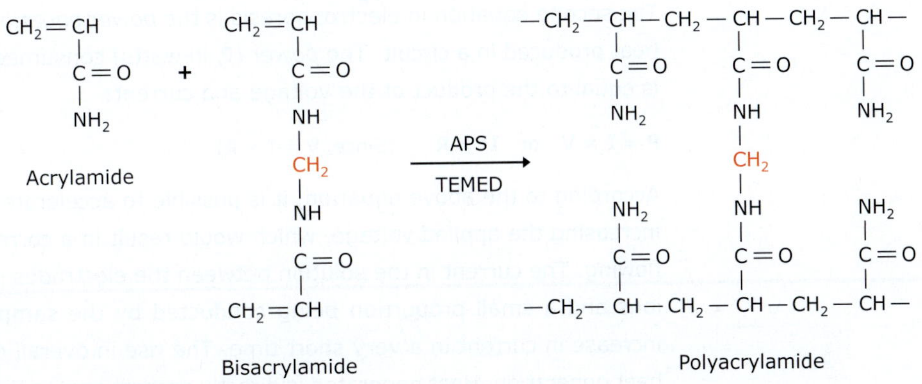

Acrylamide + Bisacrylamide —(APS, TEMED)→ Polyacrylamide

Figure 2.2 Formation of a polyacrylamide gel. In this reaction, acrylamide is used as the monomer and *N, N′*-methylenebisacrylamide (bisacrylamide) is used as a cross-linking agent. The reaction of these two agents is begun by adding ammonium persulfate (APS), where persulfate forms sulfate radicals that cause the acrylamide and bisacrylamide to combine. TEMED is added to this mixture which induces free radical formation from *ammonium persulphate*. The size of the pores that are formed in the polyacrylamide gel is related to how much bisacrylamide is used with respect to acrylamide.

The resolving power and pore size of a gel depends on the concentrations of acrylamide and bisacrylamide. The size of the pores in a polyacrylamide gel is determined by two parameters: the total percentage concentration of the monomers (acrylamide plus bisacrylamide) in grams per 100 mL (%T) and the weight percentage of crosslinker bisacrylamide relative to the total amount of monomers (%C). By varying these two parameters, the pore size of the gel can be changed. %T and %C can be expressed as:

$$\%T\ (w/v) = \frac{\text{Mass of acrylamide in grams} + \text{Mass of bisacrylamide in grams}}{\text{Volume in milliliters}} \times 100\%$$

$$\%C\ (w/w) = \frac{\text{Mass of bisacrylamide in grams}}{\text{Mass of acrylamide in grams} + \text{Mass of bisacrylamide in grams}} \times 100\%$$

When C remains constant and T increases, the pore size decreases. When T remains constant and C increases, the pore size follows a parabolic function: at high and low values of C the pores are large.

The acrylamide concentration of the gel normally varies from 5% to 25%. Lower percentage (concentration) gels are better for resolving very high molecular weight molecules, while much higher percentage is needed to resolve smaller. Polyacrylamide gel is used to separate most proteins, ranging in molecular weight from <5000 to $>200,000$ and polynucleotides from <5 to ~3000 base pairs in size.

Electrical parameters: voltage, current and power

To separate samples by electrophoresis, an electrical field is applied so that charged samples migrate toward the electrode. The driving force behind the electrophoresis is the *voltage* (V, in volts) applied across the electrodes. This leads to *current flow* (I, in amperes) through the gel, which has an intrinsic *resistance* (R, in ohms). These electrical parameters influence the sample migration and resolution of its constituent. Ohm's law describes the relationship between these three parameters:

$$V = I \times R$$

The second equation in electrophoresis is the *power equation*, which describes the amount of heat produced in a circuit. The *power* (P, in watts) consumed by an electrical current element is equal to the product of the voltage and current:

$$P = I \times V \quad \text{or} \quad I^2 \times R \qquad (\text{Since, } V = I \times R)$$

According to the above equation, it is possible to accelerate an electrophoretic separation by increasing the applied voltage, which would result in a corresponding increase in the current flowing. The current in the solution between the electrodes is conducted mainly by the buffer ions and a small proportion being conducted by the sample ions. High voltage causes an increase in current in a very short time. The rise in overall current flow increases power and heat generation. Heat generated is directly proportional to the power consumed by the system and is dependent upon buffer conductivity, applied voltage, and resistance. The high amount of the heat and current built up in the process lead to the melting of the gel. To overcome any heating problem, the electrophoresis should be performed at very low power (low current). But this can lead to poor separations as a result of the increased amount of diffusion resulting from long separation time. Compromise conditions, therefore, have to be found with reasonable power settings, to give acceptable separation times, and an appropriate cooling system, to remove liberated heat. A general recommendation is to set electrical running parameters high enough to separate samples efficiently without generating excessive heat.

Understanding the relationships between power, voltage, current, resistance and heat is central to understanding the factors that influence the efficiency and efficacy of electrophoresis.

2.2 Gel electrophoresis of proteins

Protein electrophoresis involves the movement of protein molecules in an electric field. During electrophoresis, protein molecules present in a sample can be separated on the basis of differences in the size (mass), shape and net charge of the protein molecules.

Gel

When electrophoresis is performed in acrylamide or agarose gels, the gel serves as a size-selective sieve during separation. As proteins move through a gel in response to an electric field, the gel's pore size allows smaller proteins to travel more rapidly than larger proteins. Acrylamide is the material of choice for preparing electrophoretic gels to separate proteins. Acrylamide mixed with bisacrylamide forms a crosslinked polymer network when the polymerizing agent, ammonium persulfate is added. TEMED catalyzes the polymerization reaction by promoting the production of free radicals by ammonium persulfate. **Low-percentage gels** are used to resolve large proteins, and **high-percentage gels** are used to resolve small proteins. Gels that have a single acrylamide percentage (concentration) throughout the gel are referred to as **uniform acrylamide concentration gels** (also called *single-percentage gels*). Gels having different acrylamide percentage through the gel, a low percentage of acrylamide at the top and a high percentage at the bottom are referred to as **gradient acrylamide gels**. The advantage of using a gradient gel is that it allows the separation of a broader range of proteins than a uniform acrylamide concentration gels.

Buffer systems

The separation and migration patterns of proteins in gel electrophoresis are determined by the chemical composition and pH of the buffer system. During electrophoresis, the buffer ions are carried through the gel just like the sample ions: negatively charged ions towards the anode and positively charged ones towards the cathode. Three basic types of buffers are required: **gel casting buffer** (to cast the gel), **sample buffer** (to prepare the sample) and **running buffer** (to fill the electrode reservoirs).

The buffer in which a sample is suspended prior to loading onto a gel is called **sample buffer**. SDS-PAGE sample buffer typically contains denaturing agents (including reducing agents and SDS), tracking dye and glycerol. The most common tracking dye for sample loading buffers is **bromophenol blue**. It is a negatively charged, low-molecular-weight dye included in the sample buffer. It migrates at the buffer-front, enabling one to monitor the progress of electrophoresis.

The **running buffers** (or *electrode buffers*) contain ions that conduct current through the gel. This buffer provides the ions for the electrical current in an electrophoresis run.

There are two types of buffer systems used in protein gel electrophoresis: *continuous* and *discontinuous* buffer systems. A **continuous buffer system** utilizes only one buffer type for the gel and running i.e. the same buffer in the gel, sample and electrode reservoirs. This provides a uniform separation matrix. It is most often used for nucleic acid electrophoresis and rarely used for protein gel electrophoresis. Proteins separated using a continuous buffer system tend to be diffuse and poorly resolved. A **discontinuous buffer system** utilizes a different gel buffer and running buffer. This system may also use two gel layers of different pore sizes and different buffer composition. A large-pore gel called a **stacking gel** is layered on top of a **separating** (or resolving) **gel**. The two gel layers are each made with a different buffer. Electrophoresis using a discontinuous buffer system results in higher resolution. Most common PAGE applications utilize discontinuous buffer systems.

Denaturing and nondenaturing electrophoresis

Electrophoresis can be performed under the denaturing or nondenaturing conditions. In **denaturing electrophoresis**, electrophoresis of proteins is performed under denaturing conditions using an anionic detergent such as sodium dodecylsulfate (SDS). SDS denatures (loss of three-dimensional structure) the protein. It also provides a uniform negative charge to protein, so protein can migrate towards positive electrode during electrophoresis. **Nondenaturing (native) electrophoresis** is performed under nondenaturing (native) conditions using buffer systems that maintain the native protein conformation, subunit interaction and biological activity. During native electrophoresis, proteins are separated based on their charge to mass ratios. Proteins move towards positive electrode due to its own charge. In this case, the electrophoresis buffer does not contain SDS. Also, the sample loading buffer does not have SDS and reducing agents and samples are not boiled.

Discontinuous PAGE

In *continuous gel system*, gel consists of a single acrylamide solution with a uniform pH throughout. The uniform separation matrix yields diffuse and poorly resolved protein bands.

In *discontinuous gel system*, polyacrylamide gel is divided into two regions – **stacking gel** (or *spacer gel*) and **separating gel** (or *running gel* or *resolving gel*). These gels are formed from solutions of different acrylamide concentration. In this system, a small, low-percentage large pore stacking gel is layered on top of a separating gel. The *separating gel* differs from the *stacking gel* due to greater concentrations of acrylamide; this results in smaller pore sizes and provides the sieving effect. The two gel layers are made with different buffers. Proteins first migrate quickly through the large-pore stacking gel and then are slowed as they enter the small-pore resolving gel. As a result, proteins concentrate into a small zone in the stacking gel before entering the separating gel.

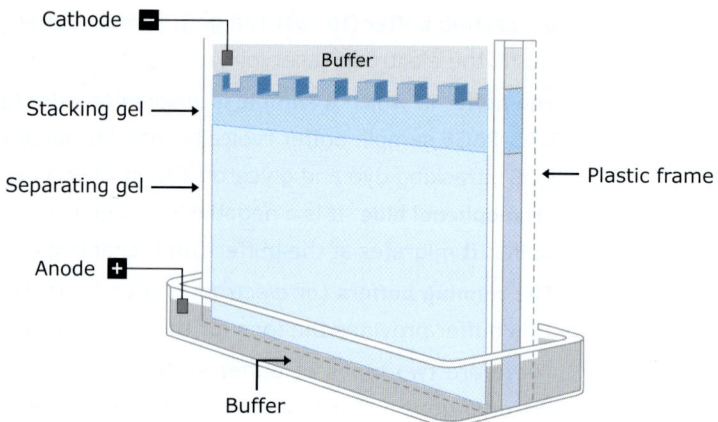

Figure 2.3 Discontinuous PAGE system.

Commonly used electrophoresis buffers for *resolution of proteins*

Stacking gel buffer

0.5 M Tris-HCl

0.4% SDS

pH 6.8

Separating gel buffer

1.5 M Tris-HCl

0.4% SDS

pH 8.8

Tris-glycine running buffer

25 mM Tris-base

192 mM glycine

0.1% SDS

pH 8.3

Discontinuous (multiphasic) systems employ different buffers for tank (running) and gel, and often two different buffers-stacking and separating- within the gel. Discontinuous systems concentrate (or stack) the samples into a very narrow zone prior to separation, which results in improved band sharpness and resolution. The gel is divided into an upper *stacking gel* of low percentage of acrylamide and low pH (6.8) and a *separating gel* with a pH of 8.8 and much smaller pores (higher percentage of acrylamide). Both, the stacking and the separating gels, contain only chloride as the mobile anion, while the upper and lower tank filled with the running buffer contains glycine as its anion, at a pH of 8.3.

This technique uses three different electrophoresis buffers

1. Running buffer of pH 8.3
2. Stacking gel buffer of pH 6.8
3. Resolving or separating gel buffer of pH 8.8

How does a discontinuous gel system work? What is the mechanism of stacking protein sample in the discontinuous gel system?

As we know, glycine can exist in three different charged states - positive, neutral or negative - depending on the pH. In the **electrode** or **running buffer** at pH 8.3, glycine exists primarily in two forms, a zwitterion and an anion. When the voltage is applied, the negatively-charged glycine ions in the **electrode buffer** are forced to enter the stacking gel. The pH of the stacking gel is 6.8 and at this pH, glycine switches predominantly to the zwitterionic state. Glycine has a pI 6.7, it has almost no net charge at pH 6.8. Thus, glycine has low mobility in the stacking gel and move very slowly in the electric field. The chloride ions present in the stacking gel buffer, on the other hand, move much more quickly in the electric field and they form an ion front that migrates ahead of the glycine. The mobility of chloride ion is more than the protein in the sample. The glycine moves slower than the protein sample and as a result protein samples get sandwiched between fast-moving chloride ions (**leading ions**) and slow-moving glycinate ions (**trailing ions**). Therefore, the relative mobility of ions in the stacking gel is chloride ions > protein sample > glycinate ions. The stacking gel also has a very large pore size, which allows the proteins to move freely and concentrate, or stack, under the effect of the electric field.

Now, when the ionic front reaches the separating gel with pH 8.8, the anionic glycine concentration increases. Second, the higher gel concentration in the separating gel begins to impede the protein molecules. Consequently, glycine overpasses the proteins, and a glycine chloride boundary moves ahead. The protein sample encounters both an increase in pH and a decrease in pore size. The increase in pH would, of course, tend to increase electrophoretic mobility, but the smaller pores decrease mobility. Protein molecules separate as per their size (or molecular weight). The relative rate of movement of anions in the separating gel is chloride ions > glycinate ions > protein sample.

The separation of sample components in the separating gel occurs as described in an earlier section on gel electrophoresis. Discontinuous electrophoresis solves two problems of protein electrophoresis. It prevents aggregation and precipitation of protein during the entry from liquid sample to gel matrix and promotes well-defined bands.

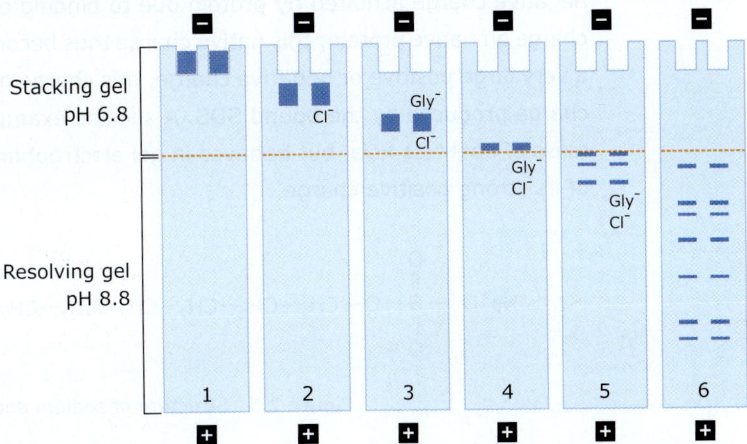

Figure 2.4 Migration of proteins and buffer ions in a denaturing discontinuous PAGE. 1. Denatured sample proteins are loaded into the wells; 2. Voltage is applied and the samples move into the gel. The chloride ions in the stacking gel buffer run faster than the SDS-bound proteins and form an ion front (*leading ion*). 3. Protein samples get sandwiched between fast-moving chloride ion (*leading* ion) and slow-moving glycinate (*trailing* ions). 4. The proteins are stacked between the chloride and glycinate ions fronts. 5. Movement of the proteins into the resolving gel is met with increased resistance; the smaller pore size resolving gel begins to separate the proteins. 6. The individual proteins are separated into band patterns ordered according to their molecular weights.

In discontinuous PAGE, discontinuity is based on four parameters:

- the gel structure
- the pH value of the buffer
- the ionic strength of the buffer
- the nature of the ions in the gel and in the electrode buffer

Discontinuous PAGE can be native or denaturing. In discontinuous native PAGE, proteins are prepared in nonreducing, nondenaturing sample buffer, and electrophoresis is also performed in the absence of denaturing and reducing agents. In discontinuous denaturing PAGE, electrophoresis is performed in the presence of denaturing detergent sodium dodecyl sulfate (**SDS**), called **SDS-PAGE**.

2.2.1 SDS–PAGE

Electrophoretic separation of proteins is most commonly performed in polyacrylamide gels. The relative movement of proteins through a polyacrylamide gel depends on the *charge density* (charge per unit of mass), *mass* (or size) and *shape* of the molecules. If two proteins have the same mass and shape, the one with the greater charge density will move faster through the gel. Similarly, if two proteins having the same charge density and shape, the one of smaller mass or size will migrate faster than the large size protein. The shape is also a factor because compact globular proteins move more rapidly than elongated fibrous proteins of comparable mass. In **SDS-PAGE,** proteins are treated with the negatively charged anionic detergent **sodium dodecyl sulfate** (also called *sodium lauryl sulfate*) before and during gel electrophoresis. SDS imparts a large net negative charge on protein. Proteins treated with SDS have similar charge-to-mass ratio i.e. binds to proteins to make them uniformly negatively charged. This is because the amount of SDS bound per unit weight of the protein is constant ~1.4 g of SDS per gram of protein (a stoichiometry of about one SDS molecule per two amino acids). The amount of SDS in the complex depends only on the size of the protein, not on charge or sequence. The negative charge acquired by protein due to binding of SDS is usually much greater than the charge on native protein; this native charge thus becomes insignificant. If the protein itself has a very large positive or negative charge, this charge may not be negligible compared with the charge produced by the bound SDS. A specific example is histone H1 which has a molecular mass of about 21 kDa, but behaves in gel electrophoresis like a molecule of 30 kDa, because of its strong positive charge.

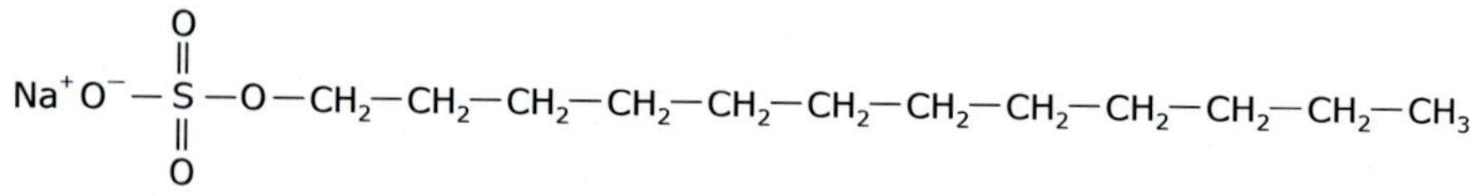

$$Na^+O^- - \overset{\overset{\displaystyle O}{\|}}{\underset{\underset{\displaystyle O}{\|}}{S}} - O - CH_2 - CH_2 - CH_2 - CH_2 - CH_2 - CH_2 - CH_2 - CH_2 - CH_2 - CH_2 - CH_2 - CH_3$$

Figure 2.5 Structure of sodium dodecyl sulfate (SDS).

SDS bound polypeptides are flexible rod-shaped with a uniform negative charge per unit length. In SDS-PAGE, migration is, therefore, determined not by the intrinsic electric charge of polypeptides but by molecular weight because SDS treatment eliminates the effect of differences in charge density and shape. Thus, when a current is applied, all SDS-bound proteins in a sample will migrate through the gel toward the positively charged electrode. Proteins with less mass travel more quickly through the gel than those with greater mass because of the sieving effect of the gel matrix.

Reducing and non-reducing SDS-PAGE: SDS denatures proteins, causing multimeric proteins to dissociate into their subunits, and all polypeptide chains are forced into extended conformations. In **reducing SDS-PAGE**, along with SDS, protein is also treated with **reducing agents** such as β-mercaptoethanol, dithiothreitol to break intrachain or interchain disulfide bonds between cysteine residues. In a multimeric protein, subunits are dissociated when treated with SDS and a reducing agent. Thus, reducing SDS-PAGE gives monomeric mass. A protein shows a band corresponding to 100kDa in native-PAGE but 25kDa in reducing SDS-PAGE suggest the protein is an oligomeric protein with four identical subunits. In **nonreducing SDS-PAGE**, reducing agents are not used.

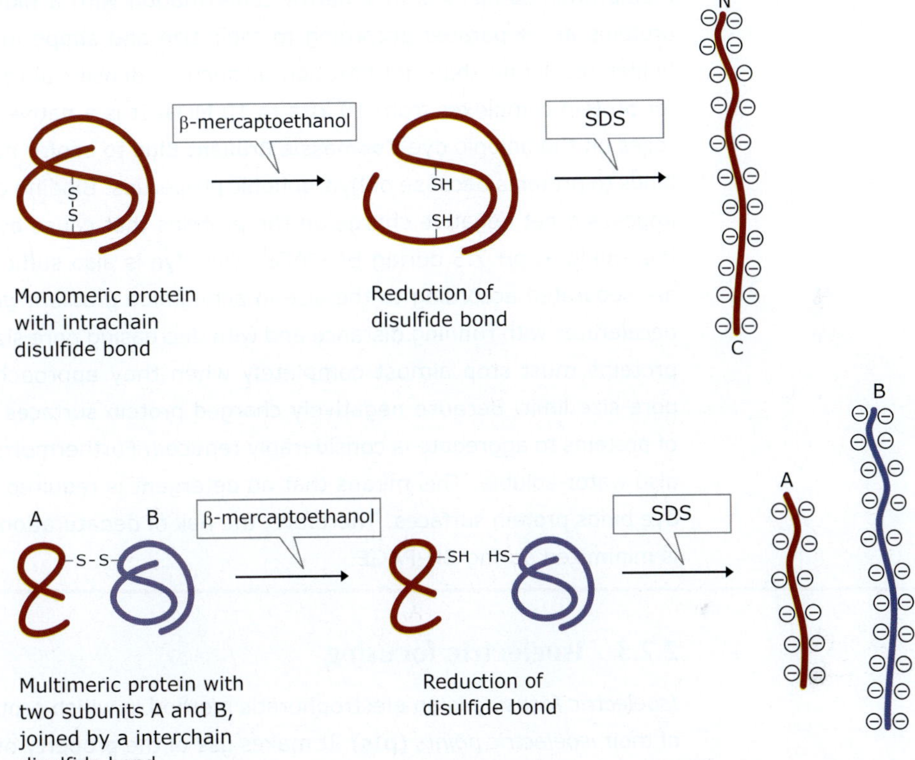

Figure 2.6 Effect of SDS on the conformation and charge of a protein. The protein is denatured and coated with a molecule with a uniform charge per unit length. The intrinsic charge of the protein is neutralized. Addition of reducing agents such as β-mercaptoethanol reduces any disulfide bonds.

2.2.2 Native PAGE

In SDS-PAGE, the gel is cast in a buffer containing SDS, an anionic detergent. The SDS denatures proteins in the sample. Usually, a *reducing agent* such as β-mercaptoethanol, dithiothreitol is also added to break intrachain and interchain disulfide bonds and ensure that no tertiary or quaternary protein structure remains. Consequently, when these samples are electrophoresed, proteins separate according to mass alone.

In native PAGE, proteins are separated according to the net charge, size and shape of their native structure. Most protein molecules carry a net charge at any pH other than their isoelectric point and hence electrophoretic migration occurs because most proteins carry a net negative charge in alkaline running buffers. The higher the negative charge density, the faster a protein will migrate. At the same time, the frictional force of the gel matrix creates a sieving effect, regulating the movement of proteins according to their size and three-dimensional shape. Small proteins face only a small frictional force, while larger proteins face a larger frictional force. Thus, in native PAGE, proteins are prepared in nonreducing, nondenaturing sample buffer, and electrophoresis is performed in the absence of denaturing and reducing agents. The native charge is preserved. It separates proteins based upon their charge, mass and shape in their native state and can resolve proteins of the same molecular weight.

Blue Native PAGE

Blue native PAGE (**BN-PAGE**) is a technique that allows separation of proteins present in multiprotein complexes in a native conformation with a higher resolution. By this method, proteins are separated according to their size and shape in a polyacrylamide gel. It has a higher resolution than gel filtration or sucrose density ultracentrifugation and can be used for protein complexes from 10 kDa to 10 MDa. It is a native protein separation method that relies on the anionic dye **Coomassie Brilliant Blue** to confer negative charge. The anionic dye binds to proteins because of hydrophobic properties. Binding a large number of dye molecules imposes a net negative charge on the proteins that cause even basic proteins to migrate to the anode at pH 7.5 during BN-PAGE. This dye is also sufficiently soluble in water. Proteins are separated according to the size in acrylamide *gradient gels*. Protein migration gradually decelerates with running distance and with decreasing pore size of the gradient gel. Individual proteins must stop almost completely when they approach their size-dependent specific pore-size limit. Because negatively charged protein surfaces repel each other, the tendency of proteins to aggregate is considerably reduced. Furthermore, binding of dye makes proteins also water-soluble. This means that no detergent is required in the BN gels once Coomassie dye binds protein surfaces. Therefore, the risk of denaturation of detergent-sensitive proteins is minimized during BN-PAGE.

2.2.3 Isoelectric focusing

Isoelectric focusing is an electrophoretic method in which proteins are separated on the basis of their *isoelectric points* (**pIs**). It makes use of the property of proteins that their net charges are determined by the pH of their local environments. Proteins carry positive, negative or zero net electrical charge, depending on the pH of their surroundings. The net charge of any particular protein is the sum of all of its positive and negative charges. These are determined by the ionizable acidic and basic side chains of the constituent amino acids and prosthetic groups of the protein. For each protein species, there is a pH at which the molecule has no net charge. This pH is called the pI. Proteins show considerable variation in pIs, but pI values usually fall in the range of pH 3-12 with mostly having pIs between pH 4 to 7. If the number of acidic groups in a protein exceeds the number of basic groups, the pI of that protein will be at a low pH value and the protein is classified as being *acidic*. When the basic groups outnumber the acidic groups in a protein, the pI will be high with the protein classified as *basic*. Proteins are positively charged in solutions at pH values below their pI and negatively charged above their pI. Thus, at pH values below the pI of a particular protein, it will migrate toward the cathode during electrophoresis. Similarly, at pH values above its pI, a protein will move toward the anode. A protein at its pI will not move in an electric field. Thus, in a pH gradient, under the influence of an electric field, a protein will move to the position in the gradient where its net charge is zero. A protein with a positive net charge will migrate toward the cathode, becoming progressively less positively charged as it moves through the pH gradient until it reaches its pI. Similarly, a protein with a net negative charge will migrate toward the anode, becoming less negatively charged until it also reaches zero net charge. Thus, under the influence of an electric field charged proteins migrate through the pH gradient until they reach their pI. In this way, proteins are focused, into sharp bands in the pH gradient at their individual characteristic pI values. This is called **isoelectric focusing** (IEF). This technique can readily resolve proteins that differ in pI by as little as 0.01, which means that proteins differing by one net charge can be separated.

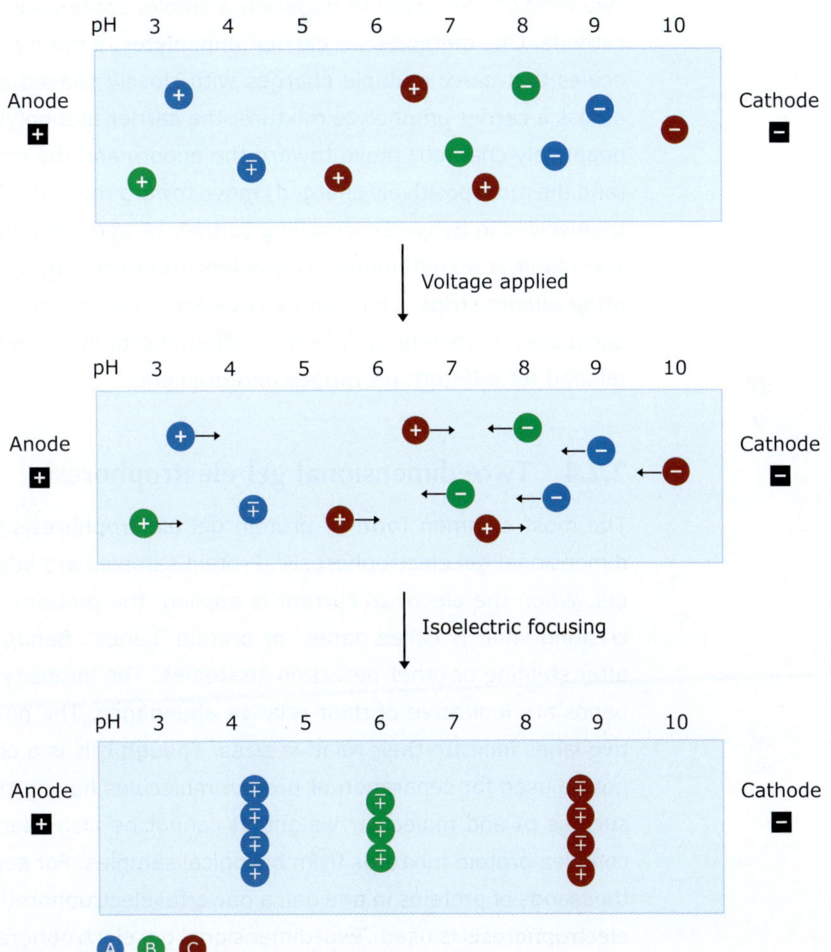

Figure 2.7 Isoelectric focusing (IEF). A solution of *ampholytes* (amphoteric electrolytes) is first electrophoresed through a gel. The migration of these substances in an electric field establishes a pH gradient. The pH gradient formed by ampholyte molecules under the influence of an electric field is indicated. The gradient increases from acidic (pH 3) at the anode to basic (pH 10) at the cathode. A mixture of proteins - A (pI = 4.3), B (pI = 6) and C (pI = 8.8) – is loaded in this pH gradient. A pH gradient is established in a gel before loading the sample. When a mixture of proteins is loaded and subjected to an electric field, they initially move toward the electrode with the opposite charge. The electric field drives the protein toward the cathode when it is positively charged and toward the anode when it is negatively charged, as shown by the arrows. When a protein moves through a pH gradient, its net charge changes in response to the pH it encounters. During migration through the pH gradient, the protein either pick up or lose protons. As it migrates, its net charge changes. Eventually, the protein arrives at the point where the pH gradient is equal to its pI. At this point because the net charge is zero, it stops migrating. In this way, proteins are focused, into sharp bands in the pH gradient at their individual characteristic pI values. Focusing is a steady-state mechanism with regard to pH. Proteins approach their respective pI values at differing rates, but remain relatively fixed at those pH values for extended periods. By contrast, proteins in conventional electrophoresis continue to move through the medium until the electric field is removed.

IEF is carried out in an acrylamide gel and can be run in either a *native* or a *denaturing* mode. **Native IEF** retains protein structure and enzymatic activity. The use of native IEF, however, is often limited by the fact that many proteins are not soluble at low ionic strength or have low solubility close to their isoelectric point. **Denaturing IEF** is performed in the presence of high concentrations of urea, which dissociates proteins into individual subunits and abolishes secondary and tertiary structures. In denaturing IEF, *urea* is the denaturant of choice, as this uncharged compound can solubilize many proteins not otherwise soluble under IEF conditions.

Two methods are used to generate a stable, continuous pH gradient between the anode and cathode. One method uses **carrier ampholytes**, a mixture of polyanionic and polycationic molecules that carry multiple charges with closely spaced pI values. When a voltage is applied across a carrier ampholyte mixture, the carrier ampholytes with the lowest pI (and the most negatively charged) move toward the anode, and the carrier ampholytes with the highest pI (and the most positively charged) move toward the cathode. The other carrier ampholytes align themselves in between according to their pI and will determine the pH of their environment. The result is a continuous pH gradient from pH 3 to 10. Another method uses **immobilized pH gradients strips** – formed by covalently grafting buffering groups to a polyacrylamide gel backbone. A gradient of different buffering groups generates a stable pH gradient that can be tailored for different pH ranges and gradients.

2.2.4 Two-dimensional gel electrophoresis

The most common form of protein gel electrophoresis is the analysis of samples by 'one-dimensional gel electrophoresis'. Protein samples are added to sample wells at the top of the gel. When the electrical current is applied, the proteins move down through the gel matrix, creating what is called 'lanes' of protein 'bands'. Bands are easily compared to each other after staining or other detection strategies. The intensity of staining and thickness of protein bands are indicative of their relative abundance. The positions of bands within their respective lanes indicate their relative sizes. Though this is a common electrophoretic method, can not be used for separation of protein molecules having the same physicochemical properties such as pI and molecular weight. It cannot be also used for separation and fractionation of complex protein mixtures from biological samples. For separation and analysis of hundreds to thousands of proteins in one gel, a powerful electrophoretic method called **two-dimensional gel electrophoresis** is used. Two-dimensional gel electrophoresis is based on separating a mixture of proteins according to two properties, one in each dimension. The *first dimension* separates proteins according to their native isoelectric point (pI) using a form of electrophoresis called **isoelectric focusing** (IEF) and the *second dimension* separates by mass using **SDS-PAGE**. Two-dimensional gel electrophoresis provides the highest resolution for protein analysis and is an important technique in proteomic research, where the resolution of thousands of proteins on a single gel is sometimes necessary.

First-dimension separation: This is performed by denaturing IEF. Using this technique, proteins are separated on the basis of their pI. In IEF, proteins are electrophoresed into a pH gradient. As the proteins move through the gradient, they encounter a point where the pH is equal to their pI and they stop migrating. Because of differences in pI, different proteins stop (focus) at different points in the gradient.

Equilibration: A conditioning step is applied to proteins separated by IEF prior to the second-dimension separation. This process reduces disulfide bonds and alkylates the resultant sulfhydryl groups of the cysteine residues. Concurrently, proteins are coated with SDS for separation on the basis of mass.

Second-dimension separation: This part is performed by SDS-PAGE. Proteins that have been separated on an IEF gel can then be separated in a second dimension based on their size or mass. To accomplish this, the IEF gel is placed lengthwise on a second polyacrylamide gel saturated with SDS. When an electric field is imposed, the proteins migrate from the IEF gel into the SDS gel and then separate according to their mass. The sequential resolution of proteins by their charge and mass can achieve excellent separation of cellular proteins.

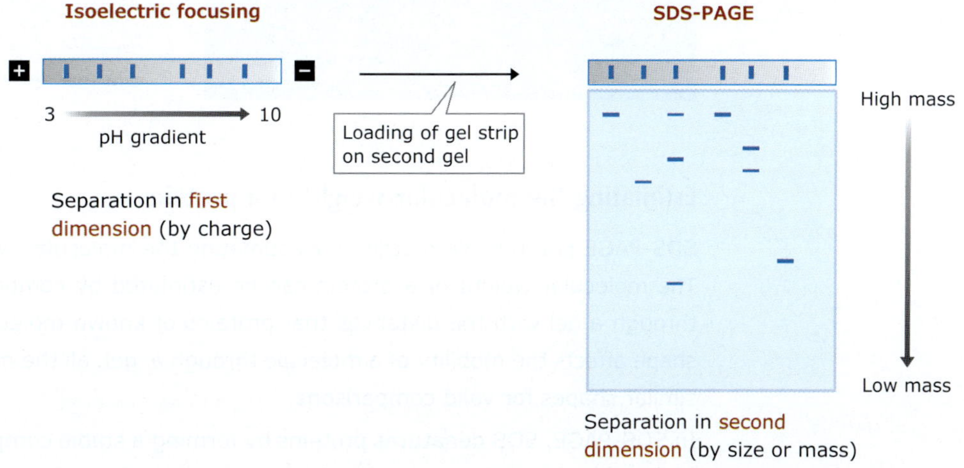

Figure 2.8 Two-dimensional gel electrophoresis. The first dimension in a 2D gel electrophoresis involves the separation of proteins according to their isoelectric point (pI) by isoelectric focusing (IEF). IEF works by applying an electric field to protein within a pH gradient. The isoelectric focusing gel is then laid horizontally on a second, slab-shaped gel, and the proteins are separated by SDS polyacrylamide gel electrophoresis. SDS-PAGE is performed perpendicular to the direction of the isoelectric focusing. Horizontal separation reflects differences in pI; vertical separation reflects differences in mass. Thus, the bands that appear on the gel have been separated first by their isoelectric points and then by size.

2.2.5 Sample visualization in the gel

After the electrophoresis run is complete, the gel must be analyzed qualitatively or quantitatively. Because most proteins are not directly visible, the gel must be processed to determine the location and amount of the separated molecules. Once protein bands have been separated by gel electrophoresis, they can be blotted (transferred) to a membrane for analysis by western blotting. Alternatively, they can be visualized directly in the gel using various staining or detection methods. Proteins are often visualized by staining (**gel staining**).

Coomassie brilliant blue (the lower detectable limit about 0.1–0.5 μg of protein) is the most widely used dye which binds to proteins but not to the gel itself. The dye binds nonspecifically to virtually all proteins. In acidic buffer conditions, Coomassie dye binds to basic and hydrophobic residues of proteins, changing in color from dull reddish-brown to intense blue.

Silver staining is the most sensitive method for permanent visible staining of proteins in polyacrylamide gels. In the case of silver staining, silver ions are reduced to metallic silver on the protein, where the silver is deposited to give a brown-black color. The silver ions interact and bind with carboxylic acid groups (Asp and Glu), imidazole (His), sulfhydryls (Cys) and amines (Lys). The silver staining is about 100 times more sensitive than Coomassie brilliant blue, with a lower detection limit of about 0.5 ng of proteins.

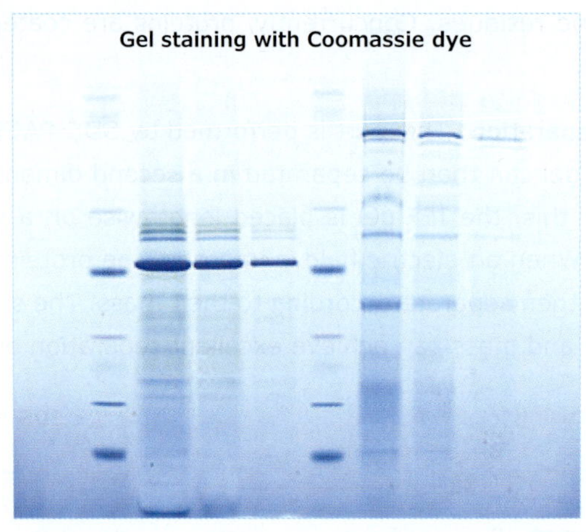

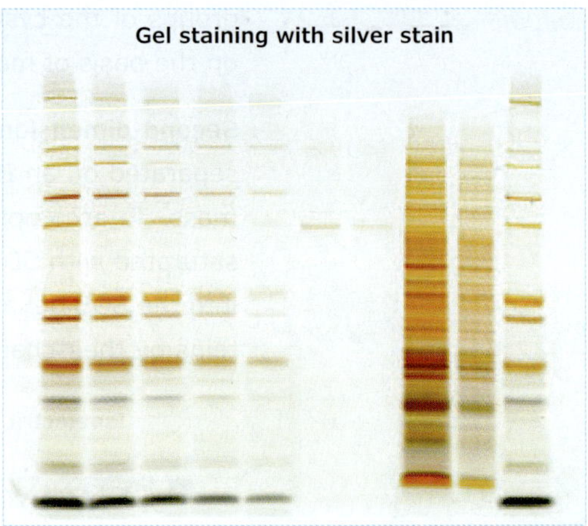

Estimating the molecular weight of a protein

SDS-PAGE is a reliable method for estimating the molecular weight of an unknown protein. The molecular weight of a protein can be estimated by comparing the distance it migrates through a gel with the distances that proteins of known molecular weight migrate. Because shape affects the mobility of a molecule through a gel, all the molecules in one gel must have similar shapes for valid comparisons.

In SDS-PAGE, SDS denatures proteins by forming a stable complex that removes most native folded structure. The electrophoretic mobility of a protein coated with SDS on an SDS-PAGE is inversely proportional to the logarithm of its molecular weight. When a set of standard proteins of known molecular weight (called **molecular weight markers**) are run alongside samples in the same gel, they separate into a series of bands and provide a reference by which the molecular weight of sample proteins can be determined. The molecular weight markers are used to generate a curve correlating molecular weight and migration in the gel, from which the molecular weight of the unknown sample proteins can be determined.

To determine the molecular weight of an unknown sample protein, we should separate the sample on the same gel with a set of molecular weight markers. After separation, we should then determine the **relative migration distance** (R_f) of the molecular weight markers and the unknown sample protein. R_f is defined as the migration distance of a protein band divided by the migration distance of the ion front. Because the ion front can be difficult to locate, mobilities are normalized to the tracking dye that migrates only slightly behind the ion front:

$$R_f = \frac{\text{Migration distance of protein band}}{\text{Migration distance of dye front}}$$

Based on the values obtained for the molecular weight markers, the log of the molecular weight of an SDS-denatured polypeptide and its relative migration distance (R_f) is plotted into a graph. If proteins are fully denatured and the gel percentage is appropriate for the molecular weight range of the sample, the plot will be linear. The standard curve is sigmoid at extreme molecular weight values because, at high molecular weight, the sieving effect of the matrix is so large that molecules are unable to penetrate the gel. At low molecular weight, the sieving effect is negligible and proteins migrate almost freely. To determine the molecular weight of the unknown protein band, we interpolate the value from this graph.

The **molecular weight** or *relative molecular mass* (M_r) is defined as the ratio of a molecule's mass to that of 1/12 the mass of carbon-12 (^{12}C). Since molecular weight is a ratio, it is dimensionless—it has no associated units. **Molecular mass**, not a ratio, is defined as the mass of one molecule and is generally expressed in daltons (Da). One can say that myosin has a molecular mass of 205000 Da or molecular weight of 205000.

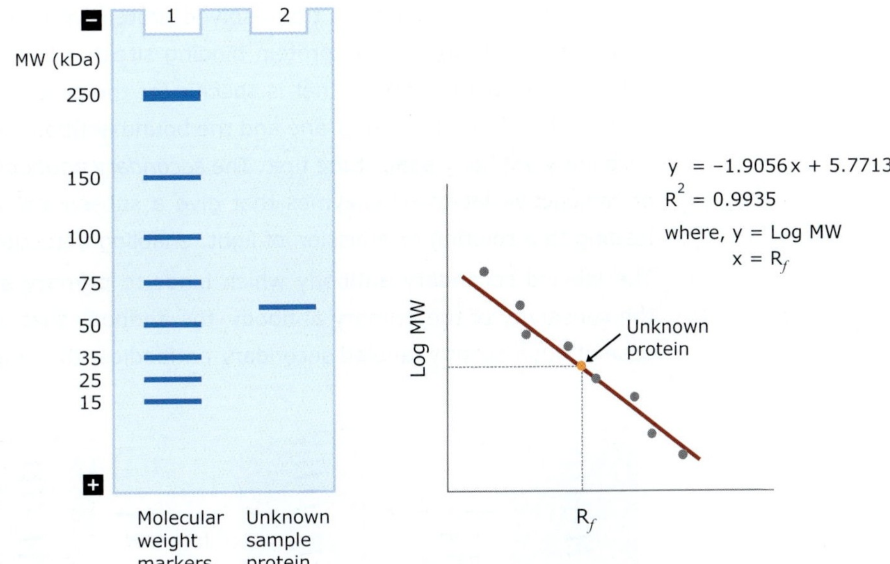

Figure 2.9 Estimating the molecular weight of an unknown protein by electrophoresis. The electrophoretic mobility of many proteins in SDS-PAGE is inversely proportional to the logarithm of their molecular weight (MW). First, molecular weight markers and unknown sample protein are run on a SDS-PAGE. The migration distance is measured from the top of the resolving gel to each molecular weight marker band and to the dye front. For each band of the molecular weight marker, the R_f value is calculated. This step is repeated for the unknown protein band in the sample. The log molecular weight value is plotted as a function of R_f. By using the formula for the slope of the line, $y = mx + c$, the molecular weight of the unknown protein is determined.

Immunoblotting

Separation of a mixture of proteins by electrophoretic techniques usually results in a complex pattern of protein bands or zones. Specific proteins can often be identified using an *immunoblotting* technique (also known as **Western blotting**). This technique requires an antibody against the test protein. After the initial separation by an electrophoretic technique in a gel, the proteins are transferred (or *blotted*) from the gel to a membrane, usually nitrocellulose or polyvinylidene difluoride (PVDF). Proteins are transferred from the gel to a membrane by applying an electric field perpendicular to the gel. During the transfer, the gel is at the negative electrode (cathode) side and the membrane at the positive electrode (anode) side. Proteins that are coated with negatively charged SDS will move from the negative side, *the gel*, to the positive side, *the membrane*.

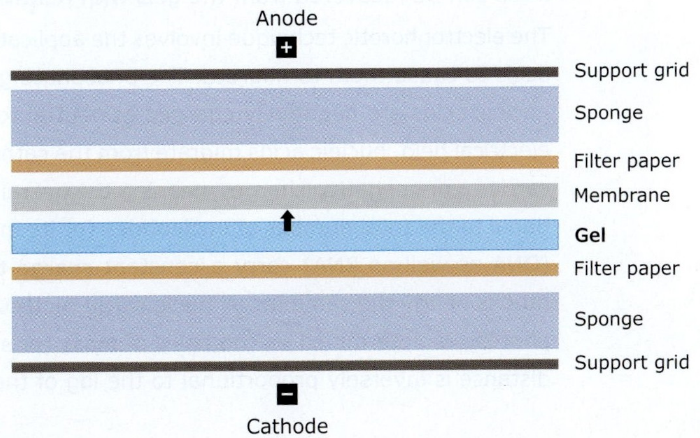

Figure 2.10 Schematic diagram showing the position of the gel, transfer membrane and direction of protein in relation to the electrode position.

The membrane that contains the resolved proteins will be first incubated with an unrelated protein to block nonspecific protein binding sites on the membrane, and then treated with a suitable primary antibody that is specific for the protein of interest. Excess antibodies are then washed from the membrane and the bound antibody, which remains, is detected using a secondary antibody against the first. The *secondary antibodies* are conjugated with fluorescent or radioactive labels or enzymes that give a subsequent reaction with an applied reagent, leading to a coloring or emission of light, enabling detection.

The labeled secondary antibody which binds to primary antibody is often used to enhance the sensitivity of the primary antibody–the antibody that reacts with the target molecule. In case of fluorescently labeled secondary antibodies, the amplification is usually several-fold.

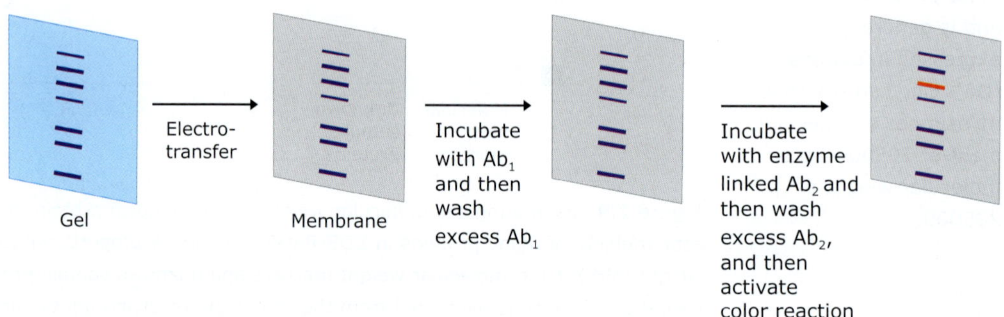

Figure 2.11 Immunoblotting using antibodies to detect specific proteins. Proteins are separated through a polyacrylamide gel, prior to blotting onto a membrane. Non-specific protein binding sites are blocked on the membrane - using solubilized milk powder - before the primary antibody (Ab_1) is added. The primary antibody will specifically bind to the antigen against which it was raised. A labeled secondary antibody (Ab_2) is then added to detect the location of the primary antibody. The secondary antibody is often labeled with an enzyme whose activity, in the presence of appropriate substrates, results in a color change on the membrane.

2.3 Gel electrophoresis of nucleic acids

Gel electrophoresis is a common laboratory technique used to identify, quantify, and separation and analysis of nucleic acids. Typically a gel matrix in gel electrophoresis serves as a *molecular sieve* and provides resistance to migration of the nucleic acid molecules that are being subjected to the electrical force. Using gel electrophoresis, nucleic acid fragments in the range of approximately 0.1–25 kbp can be separated for analysis, and separated nucleic acids can be recovered from the gels with relatively high purity and efficiency.

The electrophoretic technique involves the application of an electrical field to mixtures of nucleic acids to cause them to migrate. The phosphate groups of the sugar-phosphate backbones of nucleic acids are negatively charged at neutral to basic pH. Therefore, when subjected to an electrical field, nucleic acids migrate from the cathode toward the anode. Since each nucleotide carries a net negative charge, meaning the overall charge of a nucleic acid molecule is proportional to the total number of nucleotides (or its mass). In other words, nucleic acid molecules (DNA as well as RNA) carry a constant **charge-to-mass** ratio. Because the charge-to-mass ratio is nearly the same for all nucleic acid molecules, as a result, their mobility in gel electrophoresis is determined on the basis of **mass** (or **size**). For linear dsDNA fragments, migration distance is inversely proportional to the log of the molecular weight within a certain range.

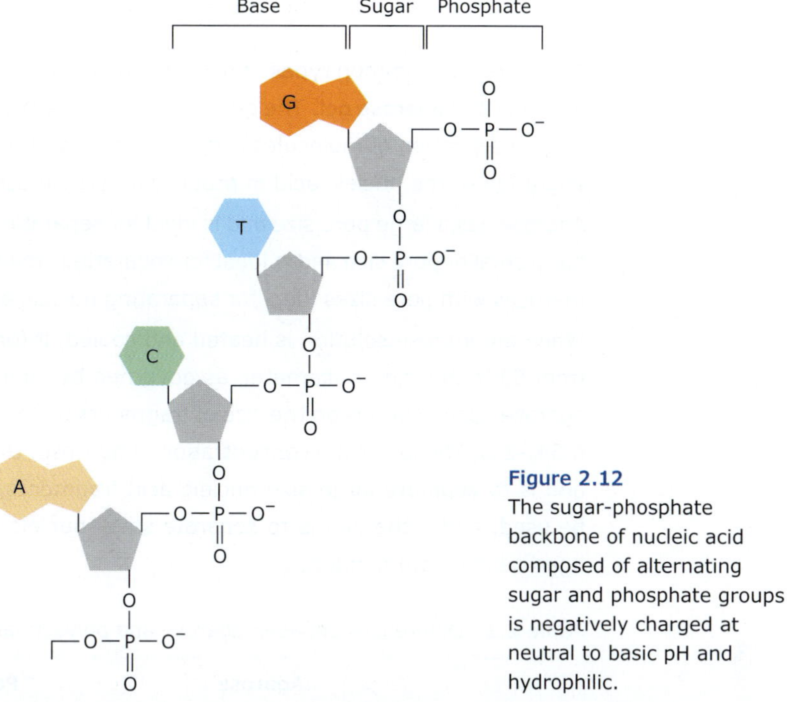

Figure 2.12 The sugar-phosphate backbone of nucleic acid composed of alternating sugar and phosphate groups is negatively charged at neutral to basic pH and hydrophilic.

Base composition and conformation of nucleic acid molecules

The mobility of nucleic acid molecules in gel electrophoresis is primarily determined by their *size*. Nevertheless, nucleic acids with an identical number of nucleotides, indicating the same size, may display different mobilities during electrophoresis due to variations in *base composition* and *conformation* (*shape*).

Effect of base composition: AT-rich DNA tends to migrate at a slower pace than GC-rich DNA of equivalent size, particularly evident in high-resolution electrophoresis.

Effect of conformation: Electrophoresis separates nucleic acid molecules not only based on their size but also on their conformation or shape. Nucleic acids can exist in various conformations, including circular and supercoiled, circular and relaxed, as well as linear forms. The migration of DNA molecules with the same sequence but different conformations is influenced by the compactness of each conformation. The more compact the DNA conformation, the more efficiently it migrates through the gel. Consequently, a relaxed circular DNA migrates more slowly than the same circular DNA in a highly supercoiled form.

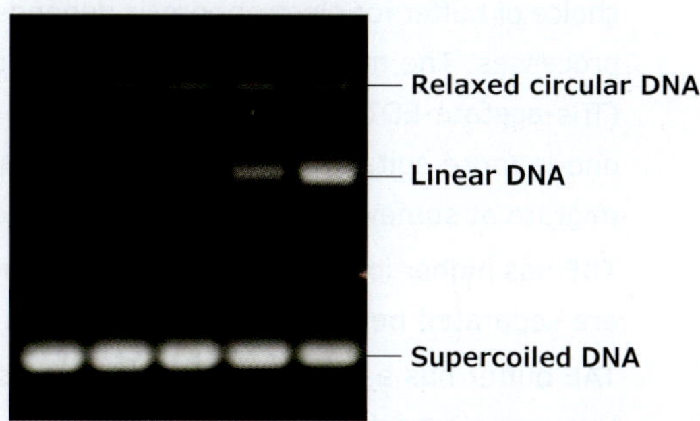

Figure 2.13 DNA molecules with different conformations move through the gel at different rates. Supercoiled, covalently closed circular DNA, due to its compact conformation, moves through the gel fastest, followed by a linear DNA fragment of the same size, with the relaxed circular form traveling the slowest. The basis of this separation lies in the fact that the greater the supercoiling, the more compact the conformation of a covalently closed circular DNA. The more compact the DNA, the more easily it can migrate through the gel.

It should also be noted that the mobility of dsDNA is greater than that of ssDNA at all ionic strengths due to the higher charge density of dsDNA. The mobilities of both ssDNA and dsDNA decrease with increasing ionic strength until reaching plateau values at ionic strengths greater than ~0.6 M.

Gel

There are two common types of gel used for separation of nucleic acid molecules – **polyacrylamide gel** and **agarose gel**. The gel acts as a porous support matrix and behaves like a molecular sieve. Separation of molecules is dependent upon the gel pore size. It selectively impedes the migration of the nucleic acid in proportion to its mass and conformation.

Agarose has a large pore size and is ideal for separating larger nucleic acids. Polyacrylamide gel has a smaller pore size and is ideal for separating smaller nucleic acid molecules. Agarose forms matrices with pore sizes ideal for separating nucleic acid molecules in the range of 0.1–25 kb.

When an agarose solution is heated and cooled, it forms a gel matrix with pore sizes ranging from 50 to 200 nm in diameter, as governed by agarose concentration. The concentration of agarose used depends on the size of fragments to be resolved, with most gels ranging between 0.5%-2%. The lower the concentration of agarose, the larger the pore sizes. In general, if the aim is to separate large size nucleic acid fragments, a low concentration of agarose should be used, and if the aim is to separate small nucleic acid fragments, a high concentration of agarose is recommended.

Table 2.1 Differences between agarose and polyacrylamide gels

	Agarose	**Polyacrylamide**
Chemical nature	Natural, polysaccharide	Synthetic, polymer of acrylamide crosslinked with bisacrylamide
Gel casting approach	Melt and solidify	Initiate chemical reactions
Nucleic acid recovery	Melt and extract	Dissolve and diffuse, or electroelute
DNA separation range	50–50,000 bp	5–3000 bp
Resolving power	5–10 nucleotides	Single nucleotides

Electrophoresis buffer

Effective separation of nucleic acids (DNA and RNA) by agarose or polyacrylamide gel electrophoresis depends upon the composition and ionic strength of the electrophoresis buffer. A *running buffer* is used to allow current flow while impeding pH changes that may occur. During electrophoresis, the negative electrode becomes more basic and the positive electrode more acidic because of electron flow, resulting in the electrolysis of water and shifts in pH. The choice of buffer for electrophoresis depends on sample sizes, run time, and post-electrophoresis processes. The most commonly used running buffer for nucleic acid electrophoresis are **TAE** (Tris-acetate-EDTA) and **TBE** (Tris-borate-EDTA). TAE and TBE have different properties, so one is more suitable than the other for any specific application. Nucleic acid fragments will migrate at somewhat different rates in these two buffers due to differences in ionic strength.

TBE has higher ionic strength and buffering capacity. Lower molecular weight samples (<1kb) are separated better with TBE buffer.

TAE buffer has a relatively low buffering capacity and is recommended for electrophoresis of larger nucleic acid fragments (>12 kb).

Agarose and polyacrylamide gels are prepared using an ionic solution (called **gel casting buffer**) with electrical conductivity to enable nucleic acid mobility during electrophoresis. The same buffer type is often used for both gel and running during electrophoresis to maintain the same pH and ionic strength.

Loading buffer

The nucleic acid samples to be loaded onto the gel are first mixed with the *gel loading buffer*, which usually comprises:

Density ingredient, such as glycerol or sucrose which increases the density of the sample, allowing it to sink into the wells of gel.

Salts such as Tris-HCl, create environment with favorable ionic strength and pH for the samples. Loading buffers with high salt concentrations may produce broader or distorted bands and smears.

Metal chelator, such as EDTA, prevents nucleases present in the sample from degrading nucleic acids.

Loading dyes provide color for easy monitoring of sample loading, progress of the electrophoretic run (i.e. serves as **tracking dye**) and often pH changes. The dyes move at standard rates through the gel, allowing the estimation of the distance that nucleic acid fragments have migrated. Typically, loading dyes are small and negatively charged molecules so that they migrate in the same direction as the nucleic acids. Some display pH-dependent colors, serving as pH indicators for samples during loading and running.

An ideal dye used in loading buffer, should have following characteristics:

- It should not interact with the DNA and gel.
- Like DNA, it should have negative charge and move towards the anode.
- It should impart color to DNA sample and its migration should easily be monitored visually.
- The size of the dye must be within the upper and lower resolution limit of the gel.

There are several dyes (e.g. xylene cyanol, bromophenol blue, bromocresol green) which are used as loading dyes. These dyes migrate at specific speeds in a given gel concentration and usually run ahead of the smallest fragments of DNA. They are not associated with the sample DNA, and thus they do not affect its separation. **Bromophenol blue** is used as an industrial dye, a laboratory indicator and a biological stain. Frequently present in loading dyes, it acts as a *tracking dye* because at moderate pH it is slightly negatively charged thus it moves in the same direction as that of DNA. The movement of the tracking dye is monitored, and when the dye approaches the end of the gel or the desired distance, electrophoresis is terminated. The length of the gel, the voltage used and the sizes of the molecules in the sample will determine the amount of time needed for electrophoresis. Run time should be monitored to ensure the smallest molecules in the samples or standards do not migrate off the gel. Run times shorter than necessary will not be sufficient to completely resolve the bands. Usually at specific time intervals during the gel run, the relative positions of the tracking dyes may be observed.

Denaturing buffers

The electrophoretic analysis of single-stranded nucleic acids like RNA is complicated because they tend to form secondary structures. Separation on the basis of molecular weight requires the inclusion of denaturing agents, which unfold the DNA or RNA strands and remove the influence of shape on their mobility. Denaturing conditions disrupt hydrogen bonds that may form between nucleic acids and thus reduce the formation of secondary structures such as hairpin loops. The most common denaturing agents used include *urea* and *formamide*. Denatured nucleic acid migrates through gels at a rate that is almost completely dependent on its size and base composition.

Denaturing electrophoresis is, therefore, more routine for RNA separation and analysis. Common denaturing buffers used with agarose and polyacrylamide gels in nucleic acid electrophoresis include:

Agarose: Glyoxal and DMSO in sodium phosphate buffer, NaOH-EDTA buffer, and formaldehyde or formamide in MOPS buffer.

Polyacrylamide: Urea in TBE buffer, since polyacrylamide gels are primarily prepared with TBE buffer supplemented with 7–8 M urea or a similar denaturant to maintain single-strandedness of the nucleic acids.

2.3.1 Samples visualization in the gel

After a gel run is complete, the samples need to be visualized. Since nucleic acids are not visible under ordinary ambient lighting, a detection method is required for visualization. Different methods are available for sample visualization or detection.

Gel staining

Fluorescent dyes are most widely utilized in sample detection due to their ease of use and high sensitivity. When excited with an appropriate wavelength, the dyes fluoresce. The intensity of fluorescence correlates to the amount of nucleic acid present, which is the basis for detection and quantitation of nucleic acids in electrophoresis.

Ethidium bromide (EtBr) is the most commonly used fluorescent dye that intercalates between bases of nucleic acids and allows very convenient detection of nucleic acid fragments in gels. It is commonly used in nucleic acid electrophoresis because of its short staining time (~30 min) and high sensitivity (detects ~1 ng of double-stranded DNA per band). EtBr works by intercalating itself in the nucleic acid molecule in a concentration-dependent manner. It detects both single- and double-stranded nucleic acids. This also gives an estimation of the DNA amount in any particular DNA band based on its intensity. It is important to note that ethidium bromide is a potent mutagen and moderately toxic after acute exposure. Therefore, it is highly recommended to handle it with considerable caution.

Two common approaches to staining nucleic acid samples are:

- In-gel, where the stain is incorporated into the gel (and running buffer).
- Post-electrophoresis, where the gel is stained after the run is complete.

In-gel staining is more convenient and requires less dye for visualization. The mode of binding of EtBr is intercalation between the base pairs. This binding changes the charge, weight, conformation, and flexibility of the DNA molecule. After running two identical gels, one without EtBr and one with EtBr in the running buffer, the mobilities of DNA fragments are always less in the gels with EtBr. Thus, dyes binding to nucleic acids alter the sample's migration, a phenomenon known as **gel shift**, where samples do not run true to size. Since post-electrophoresis staining does not affect samples during electrophoresis, it is the preferred method for accurate sizing of samples.

Alternative stains for DNA in agarose gels include SYBR Gold, SYBR green, Crystal Violet and Methyl Blue. Of these, Methyl Blue and Crystal Violet do not require exposure of the gel to UV light for visualization of DNA bands, thereby reducing the probability of mutation if recovery of the DNA fragment from the gel is desired. However, their sensitivities are lower than that of EtBr. SYBR gold and SYBR green are both highly sensitive, UV dependent dyes with lower toxicity than EtBr, but they are considerably more expensive. Moreover, all of the alternative dyes either cannot be or do not work well when added directly to the gel, therefore the gel will have to be post stained after electrophoresis.

UV shadowing

In place of staining with a dye, nucleic acids may be indirectly visualized by a method call **UV shadowing**, taking advantage of UV absorption by nucleic acids. For detection by UV shadowing, nanograms to micrograms of samples are needed, and a thin and transparent gel-like polyacrylamide should be used to ensure UV absorption and transmission. In a UV shadowing protocol, the gel is removed from the cassette after electrophoresis to maximize detection, wrapped in clear plastic film for protection, and then placed on a UV-fluorescent thin layer chromatography (TLC) plate. When the gel is exposed to UV radiation, absorption by the nucleic acid bands casts shadows on the TLC plate. The shadowy areas of the gel of desired sizes are cut out for further processing.

2.3.2 Quantification

Qualitative analysis of gels for the presence or absence of a band or relative mobilities of two bands can easily be performed by visual inspection. Answering 'How much?' and 'What size?' questions require additional work.

Amount: The amount of material in a band can be determined by a number of methods. The simplest is to visually compare the intensity of a band, either stained or autoradiographic, to standards of known quantity on the same gel. More accurate answers can be determined by using a densitometer to scan the stained gel or photograph/autoradiograph of the gel.

Size: The migration distances of nucleic acids in gel electrophoresis generally display a predictable correlation with their sizes, enabling calculation of the size of nucleic acids in a given sample. For linear double-stranded DNA fragments, migration distance is inversely proportional to the log of the molecular weight (expressed in base pairs), within a certain range. For approximate sizing, migration distances are compared to molecules of known sizes (called '**molecular weight standards**'). For a given gel concentration, usually, there will be some molecular weight range in which log of the molecular weight and migration distance are approximately linearly related. The DNA standard contains a mixture of DNA fragments of known sizes that can be compared against the unknown DNA samples. The exact sizes of separated DNA fragments can be determined by plotting the log of the molecular weight for the different bands of a DNA standard against the distance traveled by each band. Because shape affects the mobility of a molecule through a sieving gel, all the molecules in one gel must have similar shapes for valid comparisons. This does not present a problem for double-stranded DNA, because the shape of the molecules is virtually sequence-independent. Single-stranded nucleic acids, however, must be denatured to ensure similar shapes. For RNA or single-stranded DNA, denaturants added to the buffer may include formamide, urea, formaldehyde, or methylmercury hydroxide. Nucleic acids can also be denatured by treatment with glyoxal before electrophoresis.

2.4 Pulsed–field gel electrophoresis

The sizes of the DNAs that can be separated by conventional gel electrophoresis are limited to ~50 kb in size. Very large DNA fragments are unable to penetrate the pores in an agarose gel and thus cannot readily be resolved. However, larger DNA fragments can be resolved from one another by regular changes in the orientation of the electric field with respect to the gel. With each change in the electric-field orientation, the DNA must realign its axis prior to migrating in the new direction. This technique is known as *pulsed-field gel electrophoresis* (PFGE), invented by Schwartz and Cantor in 1984. With this technique DNA fragments up to 10 Mb can be separated. The method basically involves electrophoresis in agarose gel, where two electric fields are applied alternately at different angles for defined time periods. When the electrical field is applied to the gel, the DNA molecules elongate in the direction of the electrical field. The first electrical field is then switched to the second field. The DNA must change conformation and reorient before it can migrate in the direction of this field. As long as the alternating fields are equal with respect to the voltage and pulse duration, the DNA will migrate in a straight path down the gel.

There are several variants of PFGE, all of which share the essential characteristic that the orientation of the electric field is periodically changed in a manner that results in net migration of DNA. Currently, available advanced PFGE instrumentation utilizes multiple electrodes arranged in a hexagonal array. *Contour-clamped homogeneous electric fields* (**CHEF**) is one of the advanced variant of PFGE. In CHEF, some of the electrodes are clamped, or held to intermediate potentials, giving homogeneous electric fields necessary for straight, distortion-free lanes. This instrumentation allows the manipulation of pulse time, field strength and pulse angle, all of which influence the migration rate of DNA through an agarose gel and the resolution of the separation.

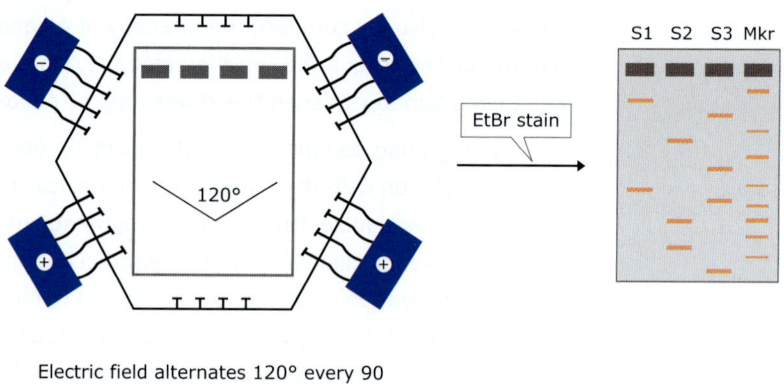

Figure 2.14 CHEF (Contour clamped homogeneous electric field) systems use a hexagonal gel box that alters the angle of the fields relative to the agarose gel. After PFGE, DNA fragments are visualized by staining with ethidium bromide.

2.5 Capillary electrophoresis

Capillary electrophoresis employs narrow-bore capillaries (typically 20-100 μm in internal diameter) to perform high-efficiency separations of both large and small molecules. These separations are facilitated by the use of high voltages, which may generate an electroosmotic and electrophoretic flow of buffer solutions and ionic species, respectively, within the capillary.

There are different modes of capillary electrophoretic separations. The distinct capillary electrophoresis modes include *capillary zone electrophoresis*, *capillary gel electrophoresis*, *capillary isoelectric focusing* and others. The basic instrumental configuration for capillary electrophoresis is relatively simple. It consists of a capillary, two electrode assemblies, two buffer reservoirs, a high-voltage power supply, a sample introduction system, a detector and an output device. The ends of the capillary are placed in the buffer reservoirs. After filling the capillary with buffer, the sample can be introduced by dipping the end of the capillary into the sample solution. In most frequently used injection mode, the tip of the capillary is dip into the sample vial. The solution in the vial is then pressurized to forced into the capillary.

Capillary zone electrophoresis, also known as *free solution capillary electrophoresis*, is the most commonly used technique. One of the fundamental processes that drive capillary electrophoresis is **electroosmosis**. The fused silica capillaries have silanol groups (–SiOH) that become ionized in the buffer. At pH levels greater than approximately 2 or 3, the silanol groups ionize to form negatively charged silanate ions ($-SiO^-$). Cations from the buffer are attracted to the silanate ions. As shown in the figure, some of these cations bind tightly to the silanate ions, forming a *fixed layer*. Because the cations in the fixed layer only partially neutralize the negative charge on the capillary walls, the solution adjacent to the fixed layer—what we call the *diffuse layer*—contains more cations than anions. Together these two layers are known as the *double layer*.

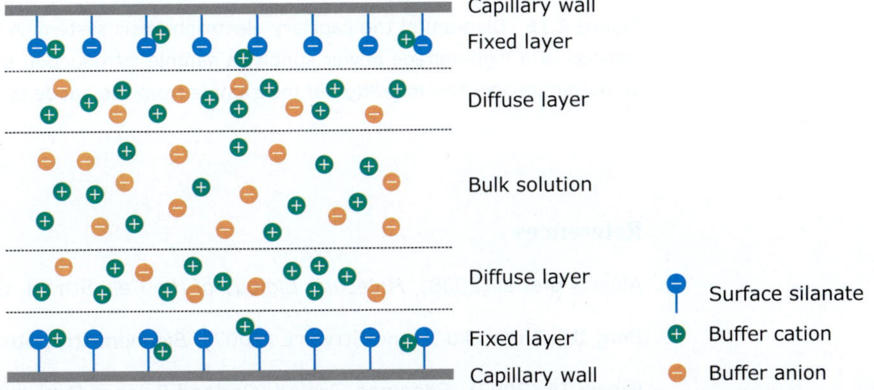

Figure 2.15 The production and effects of electroosmosis. The negatively-charged wall attracts positively-charged ions from the buffer, creating two layers of cations (called *diffuse double layers*) on the capillary wall. The first layer is referred to as the *fixed layer* because it is held tightly to the silanol groups. The outer, called the *mobile* or *diffuse layer*, is farther from the silanol groups. When a voltage is applied to the circuit, one electrode becomes net positive and the other net negative. The mobile cation layer is pulled in the direction of the negatively charged cathode. Since these cations are solvated, the bulk buffer solution migrates with the mobile layer, causing the *electroosmotic flow* of the buffer solution. The extent of electroosmosis, in this case, will depend on the pH of the running buffer, because this will affect the relative amount of the silanol groups that are present in their neutral acid form or charged conjugate base form.

In the presence of an applied electric field, cations in the diffuse layer migrate toward the cathode. Because these cations are solvated, the bulk solvent is also pulled along, producing the electroosmotic flow. Anions in solution are attracted to the positively charged anode, but get swept to the cathode as well. Cations with the largest charge-to-mass ratios separate out first, followed by cations with reduced ratios, neutral species, anions with smaller charge-to-mass ratios, and finally anions with greater ratios. The electroosmotic velocity can be adjusted by altering pH, the viscosity of the solvent, ionic strength, voltage, and the dielectric constant of the buffer.

Capillary electrophoresis of DNA

The separation of DNA fragments by capillary electrophoresis occurs within the walls of a fused-silica capillary. The capillaries are filled with a sieving matrix, and the DNA fragments are separated on the basis of size, analogously to standard slab gel separations. The matrix is either a chemically cross-linked gel, such as polyacrylamide, or a flowable polymer, such as modified cellulose or non-cross-linked polyacrylamide. The only difference between these separations is the separation matrix. Capillary electrophoresis offers a number of advantages over slab gel separations in terms of speed, resolution, sensitivity, and data handling.

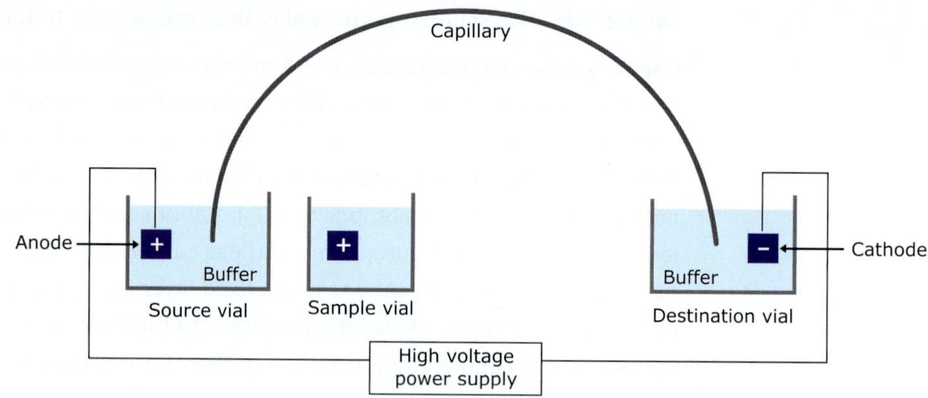

Figure 2.16 Diagram of the capillary electrophoresis system. A typical capillary electrophoresis system consists of a high-voltage power supply, a sample introduction system, a capillary tube, a detector and an output device. The migration of the sample from the anode to the cathode through the capillary tube.

References

Alberts B et al (2008), *Molecular Biology of the Cell*, 5th ed. Garland Science Publishing.

Berg JM, Tymoczko JL and Stryer L (2007), *Biochemistry*, 6th ed. W.H. Freeman and Company.

Brown TA (2007), *Genomes*, 3rd ed. Garland Science Publishing.

Chang R (2005), *Physical Chemistry for the Biosciences*, University Science Books.

Freifelder D (1982), *Physical Biochemistry*, 2nd ed. W.H. Freeman and Company.

Garfin DE (2003), Two-Dimensional Gel Electrophoresis: An Overview. *Trends in Analytical Chemistry* 22 263–272.

Hames BD (1998), *Gel Electrophoresis of Proteins: A Practical Approach*, 3rd ed. Oxford University Press, New York.

Holler FJ, Crouch SR (2014), *Fundamentals of analytical chemistry*, 9th ed. Cengage Learning.

Jorgenson JW (1986), Electrophoresis. *Analytical Chemistry* 58:743A–760A.

Lodish H, Berk A, Kaiser CA et al (2008), *Molecular Cell Biology*, 6th ed. W.H. Freeman and Company.

Manz A, Pamme N and Dimitri L (2004), *Bioanalytical Chemistry*, 3rd ed. Imperial College Press.

Pergande MR and Cologna SM (2017), *Isoelectric Point Separations of Peptides*, Proteomes.

Sambrook J and Russell DW (2001), *Molecular Cloning: A Laboratory Manual*, 3rd ed. Cold Spring Harbor Laboratory Press.

Thiel T, Bissen S and Lyons E (2002), *Biotechnology: DNA to Protein; A Laboratory Project*. McGraw Hill.

Westermeier R (2005), *Electrophoresis in Practice*, 4th ed. Wiley VCH.

3

Spectroscopy

Spectroscopy is the study of the interaction between electromagnetic radiation and matter as a function of the wavelength or frequency of the radiation. The matter can be atoms, molecules or ions. The nature of the interaction between radiation and matter may include – absorption, emission or scattering. It is the absorption, emission or scattering of radiation by matter that is used to quantitatively or qualitatively study the matter or a physical process. A study of the radiation absorbed or emitted by an atom or a molecule will give information about its identity and this technique is known as **qualitative spectroscopy**. Measurement of the total amount of radiation will give information about the number of absorbing or emitting atoms or molecules and is called **quantitative spectroscopy**.

3.1 Electromagnetic radiation

Electromagnetic radiation is a form of energy and has both electrical and magnetic characteristics. A representation of electromagnetic radiation with *electric field* (E) and the *magnetic field* (B) – at right angle to the direction of the wave – is depicted in the figure 3.1. The electric and magnetic fields in an electromagnetic wave oscillate along directions perpendicular to the propagation direction of the wave.

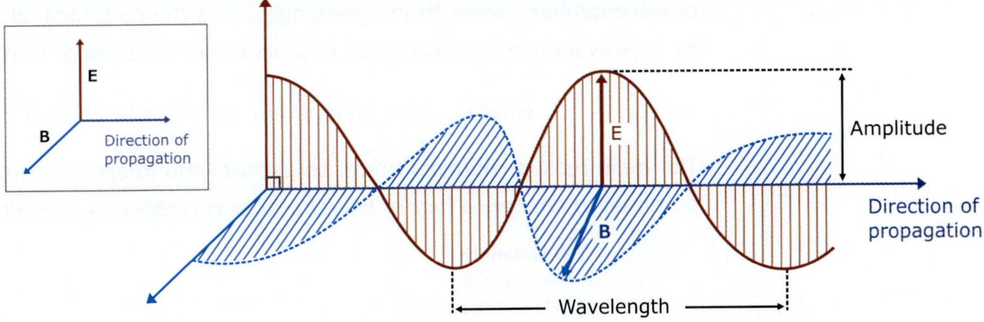

Figure 3.1
A representation of electromagnetic radiation with the electric field (**E**) and the magnetic field (**B**) at right angles to the direction of the wave movement. Both fields oscillate at the same frequency.

In a vacuum, all electromagnetic waves move at the same speed and differ from one another in their frequency. The classification of electromagnetic waves according to frequency or wavelengths is the **electromagnetic spectrum**. Different frequencies correspond to different wavelengths—waves of low frequencies have long wavelengths, and waves of high frequencies have short wavelengths. The electromagnetic spectrum ranges from very short wavelengths (such as gamma rays) to very long wavelengths (radio waves). The *visible region* of the spectrum extends approximately over the wavelength range 400–700 nm, the shorter wavelengths being the violet end of the spectrum and the longer wavelengths the red. The lowest frequency of light visible to our eyes appears **red**. The highest frequencies of visible light, which are nearly twice the frequency of red light, appear **violet**. The wavelengths between 400 and 200 nm comprise the near ultraviolet region of the spectrum, and wavelengths above 700 nm to approximately 2000 nm (2 µm) make up the infrared region. The boundaries describing the electromagnetic spectrum are not rigid, and overlap between spectral regions is possible.

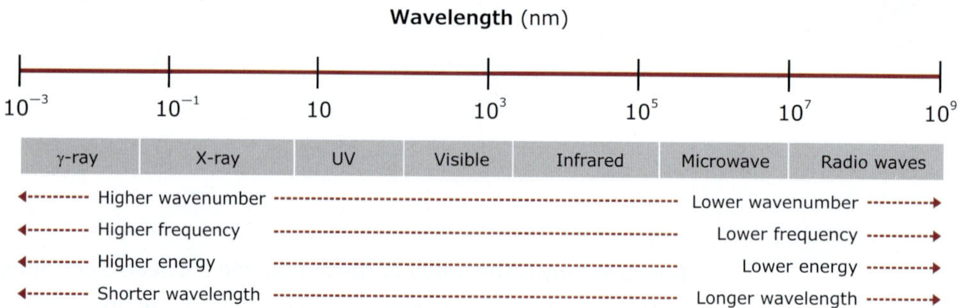

Figure 3.2 Range of electromagnetic radiation.

An electromagnetic radiation is characterized by several fundamental properties such as its velocity, amplitude, frequency, energy and polarization. The energy associated with a given segment of the spectrum is related to its frequency and wavelength. **Frequency** (ν) is the number of wave cycles that pass through a point in one second. It is measured in Hz, where 1 Hz = 1 cycle/sec. The **wavelength** (λ) is the length of one complete wave cycle. It is often measured in centimeters. It is inversely proportional to the frequency (ν) and is governed by the relationship;

$$\nu = \frac{c}{\lambda}$$ where, c = speed of light.

The energy is directly proportional to frequency and inversely proportional to wavelength. It is related to wavelength and frequency by the following equation:

$$E = h\nu = \frac{hc}{\lambda}$$ where, h is the Planck's constant (6.6×10^{-34} joules-sec).

The radiation in the infrared region of the electromagnetic spectrum is also expressed in terms of **wavenumber**, rather than wavelength. It is the reciprocal of the wavelength. It is denoted by $\bar{\nu}$. Wavenumber is expressed in units of per centimeter (cm^{-1}).

$$\bar{\nu} = \frac{1}{\lambda}$$ where, $\bar{\nu}$ is in units of cm^{-1} and λ is in units of cm.

The main reason chemists prefer to use wavenumbers as units is that they are directly proportional to energy (*a higher wavenumber corresponds to a higher energy*).

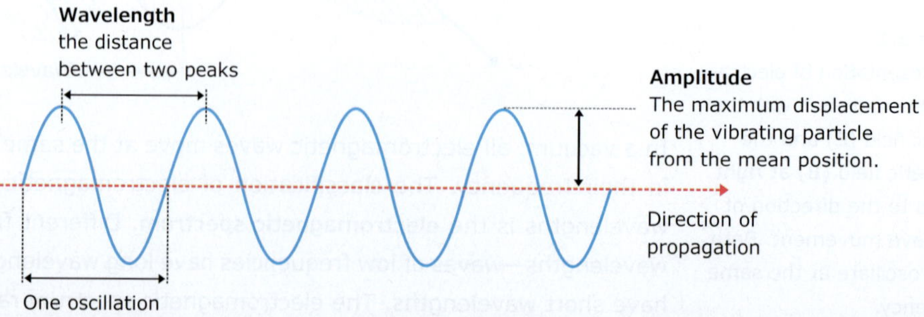

3.2 Types of spectroscopy

Spectroscopy is the collective term for a diverse group of techniques in which the interactions of electromagnetic radiation with a matter of interest are studied. When electromagnetic radiation meets matter, the radiation is either reflected off, scattered, emitted, transmitted or absorbed. This gives rise to three principal branches of spectroscopy: *absorption spectroscopy*, *emission spectroscopy* and *scattering spectroscopy*.

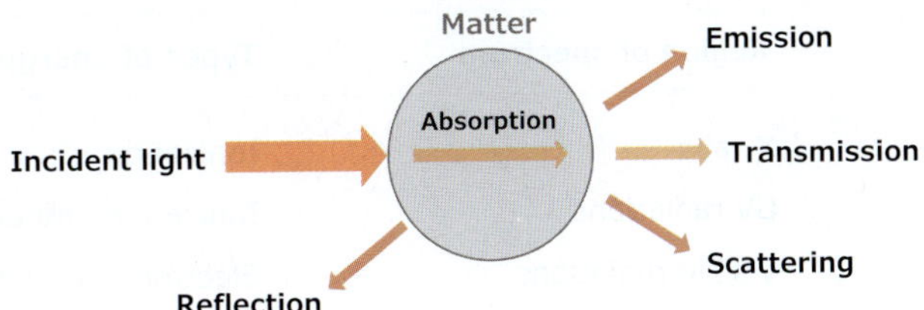

Absorption spectroscopy

Absorption spectroscopy studies radiation *absorbed* at various wavelengths. When a beam of electromagnetic radiation passes through a sample, much of the radiation passes through the sample without a loss in intensity. However, at selected wavelengths, the radiation's intensity is attenuated (decrease in the number of photons). This process of attenuation is called **absorption**. A plot of the amount of light absorbed by a sample versus the wavelength of the light is called an **absorption spectrum**. Through absorption spectroscopy, one can find both *qualitative* and *quantitative* information about the sample. In absorption spectroscopy, electromagnetic radiation is absorbed by an atom or molecule, which undergoes a transition from a lower-energy state to a higher energy or excited state. Absorption occurs only when the energy of radiation matches the difference in energy between two energy levels. The type of transition depends on the energy of electromagnetic radiation. Absorption spectroscopy can be categorized into different spectroscopic types (X-ray spectroscopy, UV-Vis spectroscopy, infrared spectroscopy) based on the region of the electromagnetic spectrum absorbed by an atom or molecule.

Table 3.1 Major types of absorption spectroscopy

Region of the electromagnetic spectrum	Spectroscopic type
X-ray	X-ray spectroscopy
UV–Vis	UV-Vis spectroscopy
Infrared (IR)	Infrared spectroscopy
	Raman spectroscopy
Microwave	Microwave spectroscopy
Radio wave	Electron spin resonance spectroscopy
	Nuclear magnetic resonance spectroscopy

Mass spectrometry is different from *spectroscopy*. It does not measure the absorption or emission of radiation. It determines the mass of the molecule.

Effect of interaction between electromagnetic radiation and matter

Atomic and molecular spectroscopies are mainly related to the absorption, emission, or scattering of electromagnetic radiation and the changes taking place in those systems due to the energy of the radiation. In absorption spectroscopy, the effect of electromagnetic radiation depends on the energy associated with the radiation. The absorption of very energetic radiations

A **photometer** is an instrument that measures the intensity of a beam of light.

A **spectrometer** is an apparatus used for recording and measuring spectra.

(such as UV and X-ray) may cause an ejection of the electron from the molecule (ionization). Radiations in the infrared region of the spectrum have much less energy than radiations in the X-ray and UV regions of the electromagnetic spectrum. Hence, IR radiation can't cause an ejection of the electron; it can cause a change in the vibrational energy of the chemical bond in a molecule. Similarly, microwave radiation is even less energetic than infrared radiation. It can neither induce electronic transition in molecules nor cause vibrations; it can only cause molecules to *rotate*.

Region of spectrum	Types of energy transitions
X-rays	Ionization
UV radiations	Ionization and electronic transition
Visible radiations	Electronic transition
Infrared	Molecular vibration
Microwaves	Molecular rotation
Radio waves	Nuclear spin (in case of nuclear magnetic resonance)
	Electronic spin (in case of electron spin resonance)

Spectrophotometer

Light absorption can be used in analytical chemistry for the characterization and quantitative determination of substances. An instrument used to measure the absorbance by measuring the amount of light of a given wavelength that is transmitted by a sample is termed a *spectrophotometer*. It measures the intensity of light passing through a sample solution in a cuvette and compares it to the intensity of the light before it passes through the sample (i.e. quantify the amount of light transmitted through a sample). It is an instrument that combines the function of both *spectrometer* and *photometer*. It measures the absorbance that uses a monochromator to select the wavelength. A method to measure how much a chemical substance absorbs light by measuring light intensity as a beam of light passes through a sample solution using a spectrophotometer is called **spectrophotometry**. All spectrophotometers contain four components: a source of light, an optical system (or monochromator), a sample holder (cuvette) and a light detector.

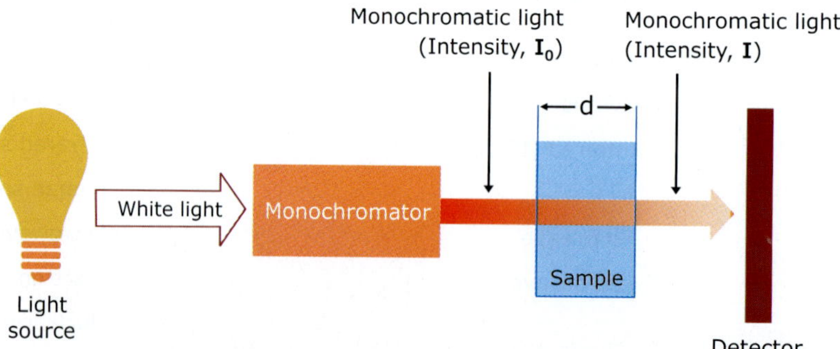

Figure 3.3 A spectrophotometer. Light from a light source passes through a monochromator for wavelength selection. A sample is contained in a cuvette in a cuvette holder. Light passes through a cuvette and is detected by a detector. The detector measures the intensity of light after passing through the sample solution. This fraction of light collected by the detector is called the transmitted intensity (I). The initial intensity of light before passing through the sample solution is I_0. The ratio between the two intensities I/I_0 is defined as transmittance.

Polychromatic Electromagnetic radiation of more than one wavelength.

Monochromatic Electromagnetic radiation of a single wavelength.

Emission spectroscopy

The absorption of light promotes an atom or molecule from ground state to an excited energy state. An excited-state molecule can return to a lower energy state by means of either a *nonradiative* or a *radiative transition*. In the nonradiative transition, the energy is ultimately dissipated into heat. In the radiative transition, the energy is released in the form of light. **Emission** (release of excitation energy in the form of light) following the absorption of light is also called **photoluminescence**. A graph of emission intensity versus wavelength is called an **emission spectrum**. Photoluminescence is divided into two categories: *fluorescence* and *phosphorescence*. **Emission spectroscopy** is a spectroscopic technique that examines the wavelengths of light emitted by atoms or molecules during their transition from an excited state to a lower energy state.

Fluorescence spectroscopy is a type of *emission spectroscopy*. **Fluorescence** is a radiative process that occurs during the transition of electrons from a singlet excited energy state to a singlet ground state (emission involving states of the same spin multiplicity). Fluorophores play a central role in fluorescence spectroscopy. A **fluorophore** is a component that causes a molecule to absorb the energy of a specific wavelength and then re-remit energy at a different but equally specific wavelength. The amount and wavelength of the emitted energy depend on both the fluorophore and the chemical environment of the fluorophore.

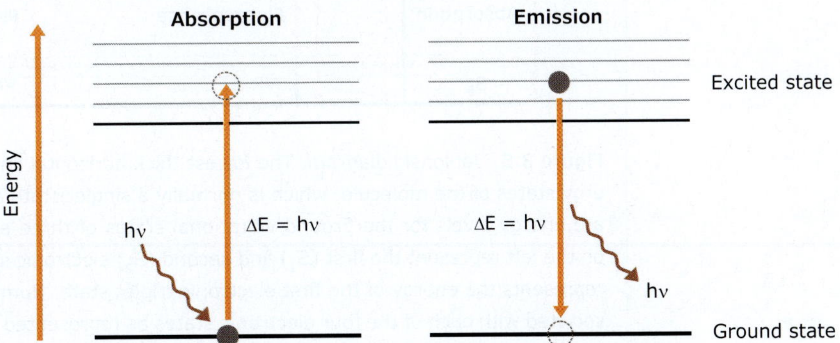

Figure 3.4 Simplified energy diagram showing the absorption and emission of a photon by an atom or a molecule. From the work of Bohr on atomic spectra, it could be established that absorption or emission of radiation is possible because of the quantization of atomic and molecular energy levels. When a photon of energy $h\nu$ strikes the atom or molecule, absorption may occur if the difference in energy, ΔE, between the ground state and the excited state is equal to the photon's energy. An atom or molecule in an excited state may emit a photon and return to the ground state. The photon's energy, $h\nu$, equals the difference in energy, ΔE, between the two states.

Scattering spectroscopy

When light interacts with matter, the light may be absorbed or scattered or may not interact with the matter and pass straight through it. **Scattering** is a physical process that causes light to deviate from a straight trajectory i.e. changes its direction. Scattering is different from absorption. In the case of absorption, specific wavelengths disappear when light encounters the matter. But both processes cause a light beam to be attenuated when passing through the solution of matter and the transmitted light intensity decreases. *Scattering spectroscopy* measures certain physical properties by measuring the amount of light that a substance scatters at certain wavelengths.

Jablonski diagram

The processes that occur between the absorption and emission of light are usually illustrated by the Jablonski diagram. A typical Jablonski diagram (basically an energy diagram) is shown in the figure. The singlet ground, first and second excited electronic energy states are depicted by S_0, S_1 and S_2, respectively. A molecular electronic energy state in which all electron spins are paired is called a *singlet state*. Each of the electronic energy states (ground or excited) has a number of vibrational energy levels, depicted by 0, 1, 2, etc. The lowest vibrational energy level for each electronic energy state is designated as 0, and the levels above it are successively 1, 2, etc. Each of these vibrational energy levels can be subdivided even further into rotational energy levels.

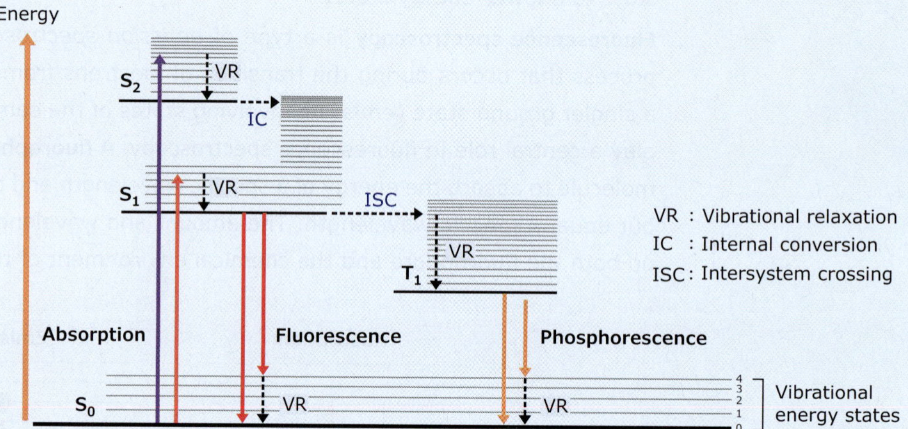

Figure 3.5 Jablonski diagram. The lowest thick horizontal line represents the ground electronic energy states of the molecule, which is normally a singlet state and is labeled S_0. The upper thick lines are energy levels for the ground vibrational states of three excited electronic states. The two lines on the left represent the first (S_1) and second (S_2) electronic singlet states. The one on the right (T_1) represents the energy of the first electronic triplet state. Numerous vibrational energy levels are associated with each of the four electronic states as represented by the thin horizontal lines. An excited molecule in a singlet or triplet excited energy state can undergo various modes of nonradiative as well as radiative decay. The light emitted by the radiative, singlet excited-to-ground state transition is called fluorescence (emission between states of same spin). The light emitted by the radiative, triplet-to-ground state transition is called phosphorescence (emission between states of different spin).

The absorbance of light of a particular wavelength by the molecule of interest causes transition of an electron from a lower electronic energy state to a higher electronic energy state. Only certain wavelengths of light are absorbed, that is, wavelengths that have energies that correspond to the energy difference between two different energy states of the particular molecule. The promotion of electrons to an excited energy state is called *excitation*. It is possible to excite the molecule to higher vibrational levels within the excited electronic states, so there are many possible absorption transitions.

Once an electron is in excited state, there are several ways that energy may be dissipated. The first is through a nonradiative process called **vibrational relaxation**. This is indicated on the Jablonski diagram as a dotted arrow between vibrational energy levels. This relaxation occurs between vibrational energy levels of one electronic energy state. Electrons will not change the electronic energy states.

However, if vibrational energy levels strongly overlap electronic energy state, a possibility exists that the excited electron can move from a vibration energy level of an excited electronic energy state to another vibration energy level in a lower energy state. This process is called **internal conversion**, a non-radiative process. It is mechanistically identical to vibrational relaxation but occurs between two vibrational energy levels in different electronic energy states. Internal conversion occurs because of the overlap of vibrational and electronic energy states.

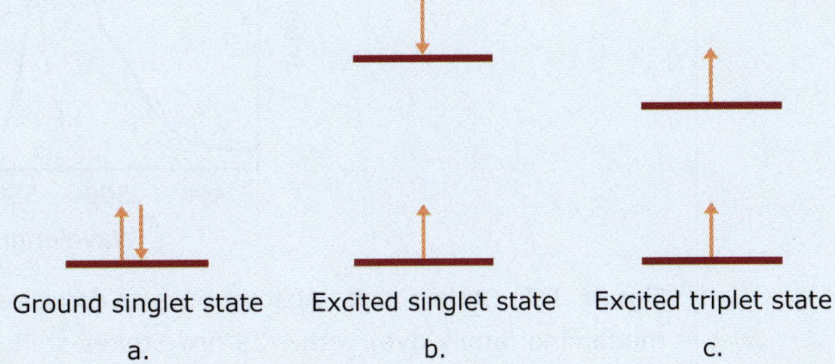

Figure 3.6 Electronic spin states of molecules. In **a**, the ground electronic state is shown. In the ground state, the spins are always paired, and the state is said to be a singlet state. In **b** and **c**, excited electronic states are shown. If the spins remain paired in the excited state, the molecule is in an excited singlet state **b**. If the spins become unpaired, the molecule is in an excited triplet state **c**.

An excited state may return to the ground state by emitting light. The emission of light may occur during the transition of electrons from a singlet first excited energy state (S_1) to a singlet ground state (S_0). This radiative process is called **fluorescence** (emission involving states of the same spin multiplicity). The energy of fluorescent light is always less than that of the exciting light. This difference is because in the excited state, some energy is always lost by non-radiative processes (such as transitions between vibrational states). Therefore, the energy of the emitted light is always less than that of the absorbed light, and hence, greater wavelengths than the absorbed light. The difference between the absorption (excitation) and emission wavelengths is called the **Stokes shift**.

Molecules in the singlet first excited energy state (S_1) can also undergo a spin conversion to the first triplet state, T_1. When one electron of a pair of electrons of a molecule is excited to a higher energy level, a singlet or a triplet state is formed. In the excited singlet state, the spin of the promoted electron is still paired with the ground state electron. In the triplet state, however, the spins of the two electrons have become unpaired and are thus parallel. Conversion of S_1 to T_1 or vice versa is called **intersystem crossing**, a nonradiative process. It is a spin-dependent internal conversion process where the electron changes spin multiplicity. Intersystem crossing leads to several interesting routes back to the singlet ground energy state (S_0). One direct transition is **phosphorescence** (emission involving states of different spin multiplicity), where a radiative transition from T_1 to S_0 occurs. This is also a very slow, forbidden transition. Another possibility is delayed fluorescence, the transition back to the S_1, leading to the radiative transition to the S_0.

Emission involving two states of the same spin is called fluorescence, while emission between states of different spin is phosphorescence. Emission from a triplet excited electronic state to a singlet ground state is thus phosphorescence.

Stokes shift

When a molecule absorbs a photon, it gains energy and enters an excited state. One way for the molecule to relax is to emit a photon, thus losing its energy. When the emitted photon has less energy than the absorbed photon, this energy difference is the Stokes shift. It is the difference (in wavelength units) between positions of the band maxima of the absorption and emission spectra of the same electronic transition.

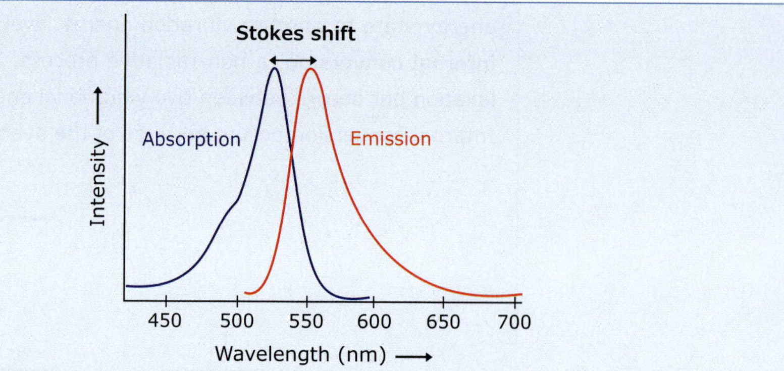

Figure 3.7 Stokes shift. Absorption and emission spectra of Rhodamine 6G (a highly fluorescent rhodamine family dye) with ~25 nm Stokes shift. It is the difference in wavelength or energy between the excitation and emission spectra of the same electronic transition. In practice, it is the difference between the excitation and emission maxima.

Principles of absorption spectroscopy

When electromagnetic radiation passes through a material, a portion of the electromagnetic radiation may be absorbed. If that occurs, the remaining radiation, when it is passed through a prism, yields a spectrum with a gap in it, called an *absorption spectrum*. The absorption spectrum is characteristic of a particular element or compound, and does not change with varying concentration. At a given wavelength, the measured absorbance has been shown to be proportional to the molar concentration of the absorbing species and the thickness of the sample the light passes through. This is known as the *Beer-Lambert law*.

Beer–Lambert law

When radiation falls on homogeneous medium, a portion of incident light is reflected, a portion is absorbed and the remainder is transmitted. The two laws governing the absorption of radiation are known as Lambert's law and Beer's law. In the combined form, they are referred to as *Beer-Lambert law.*

Lambert's law: It states that when monochromatic light passes through a transparent medium, the intensity of transmitted light decreases exponentially as the thickness of absorbing material increases.

Beer's law: It states that the intensity of transmitted monochromatic light decreases exponentially as the concentration of the absorbing substance increases.

Mathematical expression of Beer-Lambert law:

The relationship between concentration, length of the light path, and the light absorbed by a particular substance is expressed mathematically as shown below:

$$A = \log \frac{I_0}{I} = \varepsilon . c . l$$

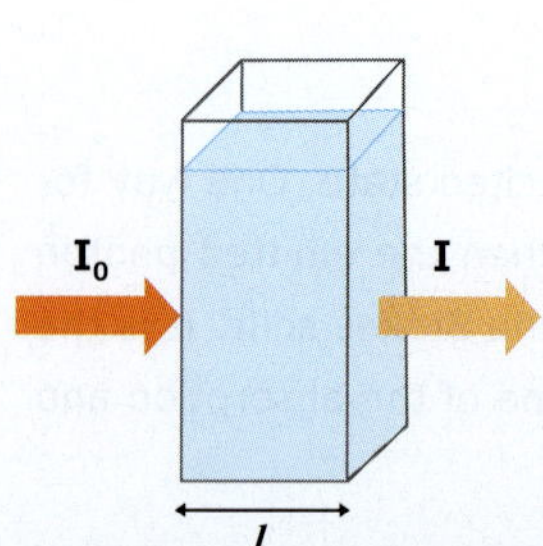

where, A = Absorbance

I_0 = Intensity of incident light

I = Intensity of light transmitted through the sample

ε = Extinction coefficient or absorption coefficient for an absorbing compound

c = Concentration of absorbing material in the sample

l = Path length (cm)

If the concentration is expressed in molarity, ε is termed as the **molar absorption coefficient** or **molar extinction coefficient.** Its unit is M^{-1} cm^{-1}. If the concentration is expressed in g/liter, ε becomes the **specific absorption coefficient**. The absorbance is a dimensionless quantity. Theoretically, absorbance (A) can have any positive value; in practice, for UV and visible spectrometers, 'A' normally varies between zero and one.

The ratio of the intensity of the transmitted light (I) to the intensity of the incident light (I_0) is called **transmittance** (T).

$$T = \frac{I}{I_0}$$

It measures the amount of light transmitted after passing through the medium. The smaller the transmittance, the greater the absorption of light. Transmittance is always a numerical value between zero (all light absorbed) and one (no light absorbed). It is common to convert transmittance into **percent transmittance** (%T). It is 100 × T.

$$\%T = \frac{I}{I_0} \times 100\% = T \times 100\%$$

Percent transmittance varies between 100% (all light transmitted) and 0% (no light transmitted). The transmittance and absorbance are inversely related. Moreover, the inverse relationship between transmittance and absorbance is not linear, it is logarithmic.

$$A = \log_{10}\frac{1}{T} = -\log_{10} T = -\log_{10}\frac{\%T}{100} = -\log_{10}\%T + \log_{10}100 = 2 - \log_{10}\%T$$

Molar extinction coefficients vs. Absorbances for 1% solutions

A *molar extinction coefficient* in the calculation gives an expression of concentration in terms of molarity:

$$\frac{A}{\varepsilon_{molar} \cdot l} = \text{Molar concentration}$$

However, many literatures do not provide molar extinction coefficients. Instead, they provide absorbance values for 1% (= 1g/100mL) solutions measured in a 1 cm cuvette. These values can be understood as *percent solution extinction coefficients* ($\varepsilon_{percent}$). Consequently, when these values are applied as extinction coefficients in the general formula, the units for concentration, c, are percent solution (i.e. 1% = 1g/100mL = 10mg/mL).

$$\frac{A}{\varepsilon_{percent} \cdot l} = \text{Percent concentration}$$

If one wishes to represent concentration in terms of mg/ml, then an adjustment factor of 10 must be made when using these percent solution extinction coefficients (i.e. one must convert from 10 mg/ml units to 1 mg/ml concentration units).

$$\frac{A}{\varepsilon_{percent} \cdot l} \times 10 = \text{Concentration in mg / ml}$$

The relationship between molar extinction coefficient (ε_{molar}) and percent extinction coefficient ($\varepsilon_{percent}$) is as follows:

(ε_{molar}) 10 = ($\varepsilon_{percent}$) × (molecular weight of protein)

Problem

A solution present in a 1 cm cuvette transmits 40% incident light. Calculate the concentration of the solution given that ε = 6000 M^{-1} cm^{-1}.

Solution

Transmittance (T) = 40% i.e. 0.40

Molar absorption coefficient (ε) = 6000 M^{-1} cm^{-1}

Path length (l) = 1 cm

Absorbance, A = –log T = –log 0.40 = 0.3980

According to Beer-Lambert law:

$A = \varepsilon.c.l$

$$c = \frac{A}{\varepsilon.l} = \frac{0.3980}{6000 \times 1} = 6.63 \times 10^{-5}\ M$$

Problem

A monochromatic radiation is incident on a solution of 0.06 molar concentration of an absorbing substance. The intensity of the radiation is reduced to one-fourth of the initial value after passing through 8 cm length of the solution. Calculate the molar extinction coefficient of the substance.

Solution

$$\text{Given, } \frac{I}{I_0} = \frac{1}{4} \text{ or } \frac{I_0}{I} = 4$$

According to Beer-Lambert law:

$$A = \log\frac{I_0}{I} = \varepsilon.c.l$$

$$\frac{\log 4}{0.06 \times 8} = \varepsilon, \text{ then } \varepsilon = \frac{0.60}{0.06 \times 8} = 1.254\ M^{-1}\ cm^{-1}$$

Problem

A solution containing NAD^+ and NADH had an optical density (i.e. absorbance) of 0.311 at 340 nm and 1.2 at 260 nm in a 1 cm cuvette. Calculate the concentrations of the NAD^+ in the solution. Both NAD^+ and NADH absorb at 260 nm, but only NADH absorbs at 340 nm. The extinction coefficients (ε) are given below.

	ε ($M^{-1} \times cm^{-1}$)	
Compound	260 nm	340 nm
NAD^+	18,000	~0
NADH	15,000	6220

Solution

The concentration of NADH from its absorbance at 340 nm,

$A = \varepsilon.c.l$

0.311 = 6220 × c × l; where, l = 1 cm

$c_{NADH} = 5 \times 10^{-5}$ M

The absorbance at 260 nm resulting from the NADH,

$A = \varepsilon.c.l$

$A = 15000 \times (5 \times 10^{-5}) \times 1 = 75 \times 10^{-2} = 0.75$

Now, the absorbance at 260 nm from the NAD^+,

= total absorbance at 260 nm − absorbance of NADH at 260 nm

$= 1.20 - 0.75 = 0.45$

Finally, from the absorbance of the NAD^+ at 260 nm, we can calculate the concentration of NAD^+.

$A = \varepsilon.c.l$

$0.45 = 18000 \times c \times 1$

$c = 2.5 \times 10^{-5}$ M

Problem

A protein has one tryptophan and one tyrosine in its sequence. Assume molar extinction coefficients at 280 nm of tryptophan and tyrosine as 3000 and 1500 $M^{-1}\ cm^{-1}$, respectively. What would be the molar concentration of that protein if its absorption at 280 nm is 0.90 in a 1 cm cuvette?

Solution

Molar extinction coefficient of tryptophan = 3000 $M^{-1}\ cm^{-1}$

Molar extinction coefficient of tyrosine = 1500 $M^{-1}\ cm^{-1}$

Total extinction coefficient = (number of tryptophan residue × 3000 $M^{-1}\ cm^{-1}$) + (number of tyrosine residue × 1500 $M^{-1}\ cm^{-1}$) = 4500 $M^{-1}\ cm^{-1}$

According to Beer-Lambert law:

As we know that, $A = \varepsilon.c.l$

$$c \text{ (in molarity)} = \frac{A}{\varepsilon \times l} = \frac{0.90}{4500 \times 1} = 0.2 \times 10^{-3}\ M$$

3.3 UV/VIS absorption spectroscopy

When electromagnetic radiation passes through a transparent material, a portion of radiation may be absorbed. As a result of energy absorption, atoms or molecules pass from a state of low energy (ground state) to a state of higher energy (excited state). The electromagnetic radiation that is absorbed has energy exactly equal to the energy difference between the excited and ground states.

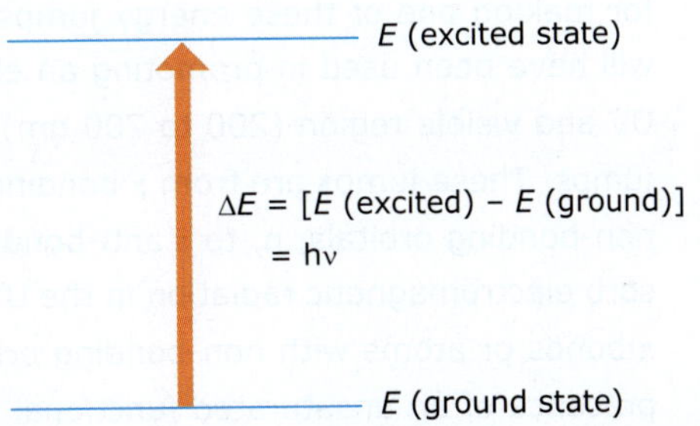

E (excited state)

$\Delta E = [E\text{ (excited)} - E\text{ (ground)}]$
$= h\nu$

E (ground state)

In an **electronic transition**, an electron moves from one orbital to another. Transition occurs between atomic orbitals in atoms and between molecular orbitals in molecules.

Ultraviolet and visible absorption spectroscopy is based on the transitions of electrons from one molecular orbital to another due to the absorption of electromagnetic radiation of UV and visible region. As a molecule absorbs energy, an electron is promoted from an occupied orbital

The **atomic spectroscopy** refers to the study of the electromagnetic radiation absorbed and emitted by atoms whereas the **molecular spectroscopy** refers to the study of the electromagnetic radiation absorbed and emitted by molecules.

to an unoccupied orbital of greater potential energy. Generally, the transition of electrons occurs from the *highest occupied molecular orbital* (**HOMO**) to the *lowest unoccupied molecular orbital* (**LUMO**).

Molecular orbitals with lowest energy are the σ-orbitals. The π-orbitals lie at high energy levels and **non-bonding orbitals** (*n*) lie at even higher energies. The non-bonding orbitals contain a lone pair of electrons and they are stable, filled orbitals. **Anti-bonding orbitals** (π* and σ*), are normally empty and have higher energy than bonding or non-bonding orbitals.

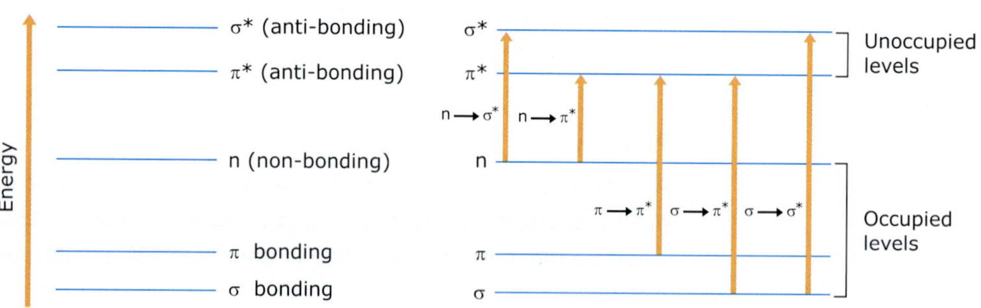

Figure 3.8 The possible electron jumps that light might cause are shown here. In each possible case, an electron is excited from a full orbital into an empty anti-bonding orbital. Each jump takes energy from the light, and a big jump obviously needs more energy than a small one. Each wavelength of light has a particular energy associated with it. If that particular amount of energy is just right for making one of these energy jumps, then that wavelength will be absorbed – its energy will be used in promoting an electron. The energy required to bring about transitions from the highest occupied energy level (HOMO) in the ground state to the lowest unoccupied energy level (LUMO) is less than the energy required to bring about a transition from a lower occupied energy level.

When electromagnetic radiation (light) passes through a compound, energy from the radiation is used to promote an electron from a bonding or non-bonding orbital into one of the empty anti-bonding orbitals. The possible electron jumps that electromagnetic radiation might cause are shown in figure 3.8. In each possible case, an electron is excited from a full orbital into an empty anti-bonding orbital. The energy gaps between these levels determine the wavelength of the electromagnetic radiation absorbed, and these gaps will be different in different compounds. The larger the gap between the energy levels, the greater the energy required to promote the electron to the higher energy level; resulting in light of higher frequency, and therefore shorter wavelength, being absorbed. Each wavelength of electromagnetic radiation has a particular energy associated with it. If that particular amount of energy is just right for making one of these energy jumps, then that wavelength will be absorbed – its energy will have been used in promoting an electron. An absorption of electromagnetic radiation in UV and visible region (200 to 700 nm) cause only a limited number of the possible electron jumps. These jumps are from π bonding orbitals to π anti-bonding orbitals (π to π*) and from non-bonding orbitals, n, to π anti-bonding orbitals (n to π*). This means that in order to absorb electromagnetic radiation in the UV and visible region, the molecule must contain either π bonds or atoms with non-bonding orbitals. Both *n* to π* and π to π* transitions require the presence of an unsaturated functional group to provide the p-orbitals. Molecules containing such functional groups and capable of absorbing UV and visible radiation are called **chromophores**, Greek words meaning 'to bear color'.

Chromophore
The bonds and functional groups that give rise to the absorption of ultraviolet and visible radiation.

Molecules that show increasing degrees of conjugation require less energy for excitation and as a result absorb radiation of longer wavelengths. Molecules that contain conjugated systems,

Electronic transitions

Transition	*Wavelength*
$\sigma \rightarrow \sigma^*$	<200 nm
$n \rightarrow \sigma^*$	160–260 nm
$\pi \rightarrow \pi^*$	200–500 nm
$n \rightarrow \pi^*$	250–600 nm

i.e. alternating single and double bonds, will have their electrons *delocalized* due to overlap of the p-orbitals in the double bonds. As the amount of delocalization in the molecule increases, the energy gap between the π bonding orbitals and π anti-bonding orbitals gets smaller; and therefore, light of lower energy, and longer wavelength, is absorbed.

3.4 IR absorption spectroscopy

The term **infrared** (IR) covers the range of the electromagnetic spectrum between 1 micrometer to 100 micrometers. It is commonly divided into three sub-regions – **near-IR, mid-IR** and **far-IR**. Each frequency (the number of wavelengths that pass a point per unit time) of infrared radiation has a specified amount of energy. If a particular frequency is being absorbed as it passes through the compound being investigated, it must mean that its energy is being transferred to the compound. As with other types of electromagnetic radiation absorption, molecules are excited to a higher energy state when they absorb IR radiation. A molecule absorbs only selected frequencies of IR radiation. The absorption of IR radiation corresponds to energy changes in the order of 8 to 40 kJ/mole. IR radiation does not have enough energy to induce electronic transitions as seen with UV and visible radiations. The energies of IR radiation correspond to the energies involved in bond vibrations.

Physical basis of IR spectroscopy

IR spectroscopy is one type of **vibrational spectroscopy**. At temperatures above absolute zero, the bonds within molecules all vibrate. There are two main types of bond vibrations – stretching and bending. A **stretching** vibration occurs along the line of the chemical bond whereas a **bending** vibration is any vibration that does not occur along the line of the chemical bond. Different bonds vibrate with characteristic frequencies.

Consider, for example, the C–H bonds in a typical organic compound. These bonds undergo various stretching and bending vibrations. Let us understand the situation by considering the C–H stretching bond vibration. This vibration takes place with a certain frequency ν; that is, it occurs a certain number of times per second. Suppose that a C–H bond has a stretching frequency of 9×10^{13} times per second. A wave of electromagnetic radiation can transfer its energy to the vibrational wave motion of the C–H bond only if *there is an exact match between the frequency of the radiation and the frequency of the vibration*. Thus, if a C–H vibration has a frequency of 9×10^{13} times per second, then it will absorb energy from radiation with the same frequency. When radiation of this frequency interacts with a vibrating C–H bond, energy is absorbed and both the frequency and the intensity of the bond vibration increase. That is, after absorbing energy, the bond vibrates with a greater frequency and with a larger amplitude (a larger stretch and tighter compression). This absorption gives rise to the peak in the IR spectrum. Eventually, the bond returns to its normal vibration frequency, energy is released in the form of heat.

A normal mode of vibration that gives rise to an oscillating dipole in the IR spectral range is said to be **IR-active**. Conversely, a normal mode of vibration that does not give rise to an oscillating dipole moment (e.g. the stretching of a homonuclear diatomic molecule) cannot lead to IR absorption and is said to be **IR-inactive**.

Thus, when the natural frequency of bond vibration in a molecule is equal to the frequency of the IR radiation directed on the molecule, the molecule absorbs the radiation. Different bonds have different vibrational frequencies and therefore different IR absorption frequencies. When IR radiation is absorbed, both the frequency and the intensity of the bond vibration increase. That is, after absorbing energy, the bond vibrates with a greater frequency and with a larger amplitude.

Dipole moment depends on the variation in distribution of electrons along the bond and also its length. For bonds between unlike atoms, the larger the difference in electronegativity, the greater the dipole moment.

Not all bonds of a molecule are capable of absorbing IR radiation, even if the frequency of the IR radiation exactly matches that of the bond vibrational frequency. To absorb IR radiation, the molecule must have a **dipole moment** and that must change during vibration. A polar chemical bond has a bond dipole. If a polar bond vibrates with a particular frequency, its bond dipole vibrates with the same frequency. The magnitude of the bond dipole is proportional not only to the amount of charge on each bonded atom but also to the distance between the atoms, that is, the *bond length*. Therefore, as the bond stretches, the bond dipole increases, and, as the bond compresses, the bond dipole decreases. When the frequencies of the IR radiation and the vibrating bond dipole match, the bond dipole absorbs energy from the IR radiation.

Thus, a bond must have a dipole moment that is changing at the same frequency as the incoming radiation for energy to be transferred. Molecular vibrations that occur but do not give rise to IR absorptions are said to be **infrared-inactive**. In contrast, any vibration that gives rise to an IR absorption is said to be **infrared-active**.

Diatomic molecules such as H_2, N_2 and O_2 (having two identical atoms) have zero dipole moment and vibration of the bonds also do not produce it. These molecules are said to be infrared-inactive. For diatomic molecules, having two different atoms, usually have a permanent dipole moment due to a difference in the electronegativity of the bonded atoms. A greater electronegativity difference between the two atoms usually leads to a larger dipole moment for that bond. However, all molecules having a permanent dipole moment are also not infrared-active. In order to be infrared-active, the bond vibration (stretching or bending) must cause a change in the dipole moment of the molecule.

It is also possible for a molecule that has a zero dipole moment but a specific molecular vibration creates a temporary dipole moment in the molecule. Consider the CO_2 molecule as an example. It is a linear symmetric molecule in which the three atoms are arranged linearly with a partial positive charge on carbon and partial negative charge on oxygens. Its equilibrium dipole moment is zero. In the case of **symmetrical stretching**, both bonds are compressed or stretched to the same extent. As a result dipole moment remains zero. Thus, symmetric stretching vibrations do not absorb IR radiation and said to be *infrared-inactive*. But in **asymmetrical stretching**, one of the bonds is compressed while the other is stretched. Hence, there is change in bond length and dipole moment. So it is *infrared-active*. Bending vibration is also allowed to this molecule. This also makes molecule infrared-active.

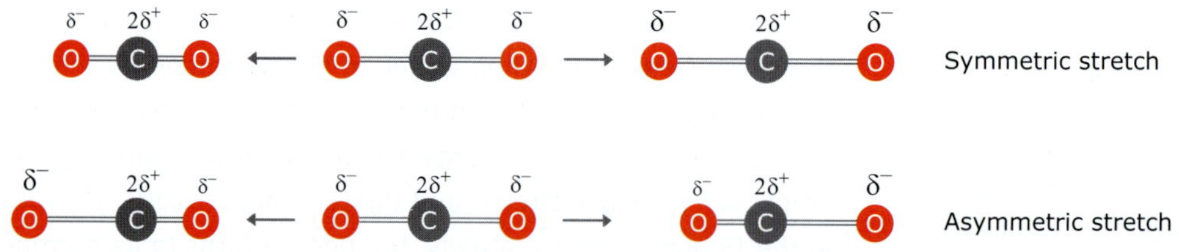

Figure 3.9 Symmetric and asymmetric stretching vibration of the carbon dioxide molecule. In the symmetrical stretch, the two C=O bonds stretched (or compressed) at the same time so that the molecule maintains its symmetry. In the asymmetrical stretch, one C=O bond shortens when the other lengthens. In case of the symmetrical stretch, both bonds are increased; but, because they are exactly equal and oppose each other, the dipole moment remains zero. Hence, the symmetrical stretching vibration is infrared-inactive. In an asymmetrical stretch, one C=O bond is reduced in length while the other is increased. Because the 'long' C=O bond has a greater bond dipole than the 'short' C=O bond, the two bond dipoles no longer cancel. Thus, the asymmetrical stretch imparts a temporary dipole moment to the CO_2 molecule. Consequently, this vibration is infrared-active—it gives rise to an IR absorption.

Fundamental vibrational motion falls into the two main categories – *stretching* and *bending*. **Stretching** is a vibrational motion in which the bond length alters and the two nuclei move harmonically relative to each other. **Bending** vibrations are characterized by a change in the angle between two bonds and are of four types: **scissoring**, **rocking**, **wagging** and **twisting**. The only possible type of vibration in a diatomic molecule (for example, H–F) is a stretching vibration. However, when a molecule contains more than two atoms, both stretching and bending vibrations are possible. Furthermore, stretching and bending vibrations can be *symmetrical* or *asymmetrical*. The allowed vibrations of a molecule are called its **normal vibrational modes.**

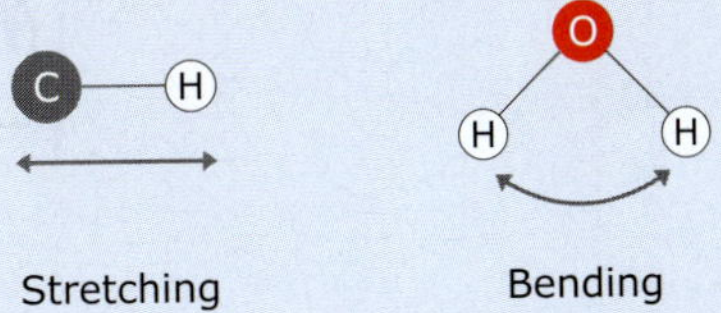

Stretching and bending vibrations in water and carbon dioxide molecules

Water molecule has a nonlinear structure. It has three fundamental vibrational modes – symmetrical stretching vibration, asymmetrical stretching vibration and bending vibration. On the other hand, carbon dioxide molecule has a linear structure. It has four fundamental vibrational modes: symmetric stretching vibration, asymmetric stretching vibration and two bending vibrations – *in-plane* bending vibration and *out of plane* bending vibration.

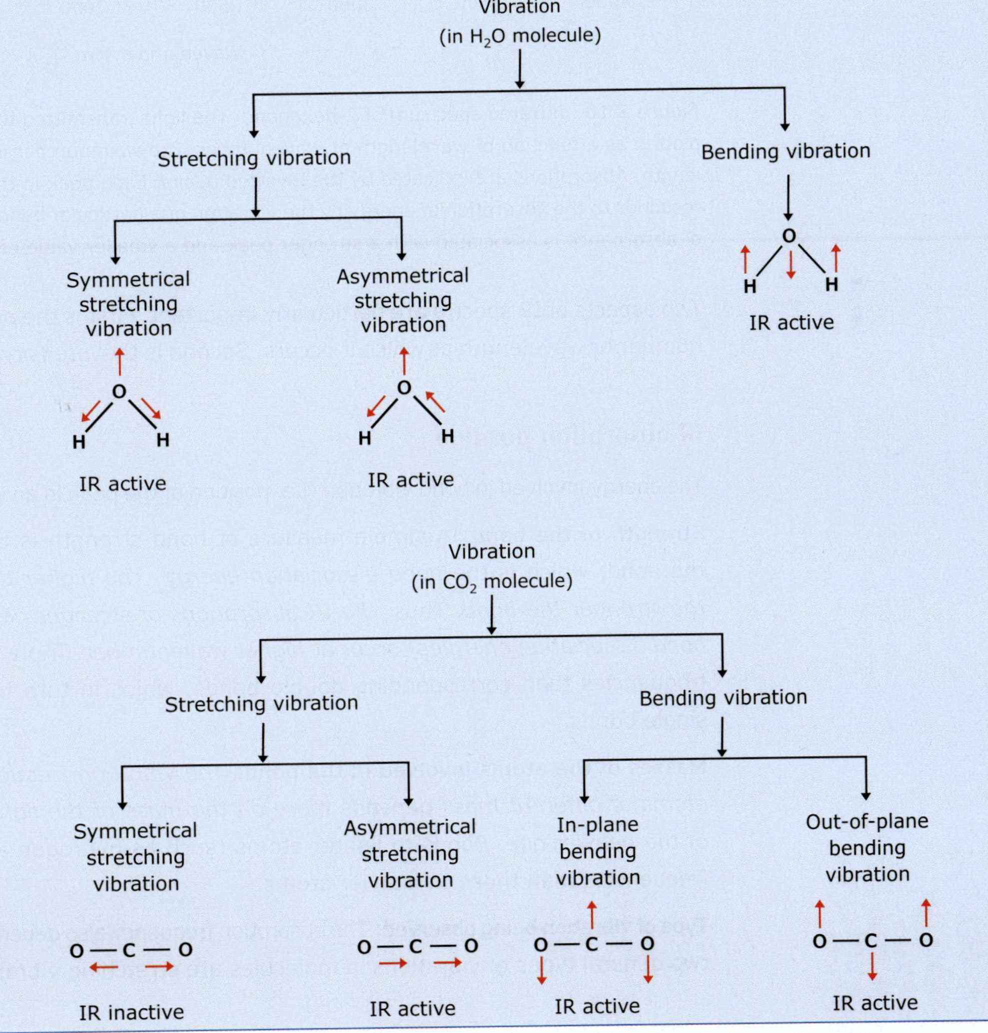

A **stretching vibration** involves changes in the length of an interatomic bond along its axis. **Bending vibrations** are characterized by a change in the angle between two bonds.

Infrared spectra

An infrared spectrum, like any absorption spectrum, is a record of the IR radiation absorbed by a compound as a function of wavelength. The instrument that determines the IR absorption spectrum for a compound is called an **infrared spectrometer**. An infrared spectrum provides information about what functional groups are present in a compound.

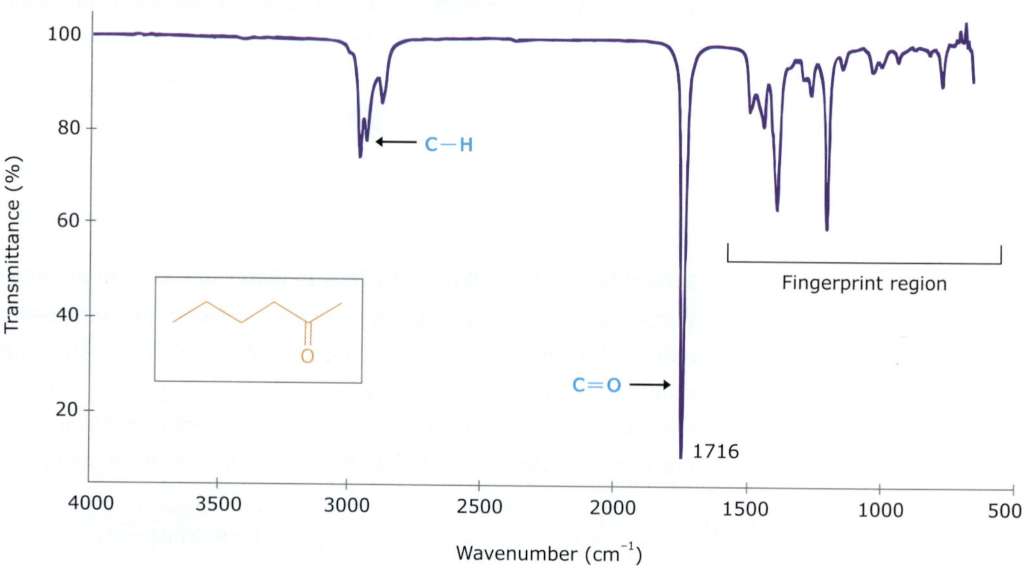

Figure 3.10 Infrared spectrum of 2-hexanone. The light transmitted through a sample of 2-hexanone is plotted as a function of wavelength or wavenumber. The wavenumber is simply the inverse of the wavelength. Absorptions are indicated by the inverted peaks. Each peak in the IR spectrum of a molecule corresponds to the absorption of energy by the vibration of a particular bond or group of bonds. A larger value of absorbance is associated with a stronger peak and a smaller value of absorbance with a weaker peak.

Two aspects of IR spectra are particularly important. First is the *position* of the peak—the wavenumber or wavelength at which it occurs. Second is the *intensity* of the peak—how strong it is.

IR absorption position

The energy involved in bond vibration (i.e. position of the peak in an infrared spectrum) depends on:

Strength of the bond: A simple measure of bond strength is the energy required to break the bond, which is the *bond dissociation energy*. *The higher the bond dissociation energy, the stronger the bond*. Thus, *the IR absorptions of stronger bonds (e.g. bonds with greater bond dissociation energies) occur at higher wavenumber*. Triple bonds have higher stretching frequencies than corresponding double bonds, which in turn have higher frequencies than single bonds.

Masses of the atoms involved in the bond: *The vibration frequency for a bond between two atoms of different mass depends more on the mass of the lighter object than on the mass of the heavier one.* Bonds to lighter atoms (such as hydrogen atom) have higher stretching frequencies than those to heavier atoms.

Type of vibration being observed: The absorption frequency also depends on the type of vibration. The two general types of vibrations in molecules are stretching vibrations and bending vibrations.

In general, bending vibrations occur at lower frequencies (higher wavelengths) than stretching vibrations of the same groups. It is easier to bend a bond than to stretch or compress it.

Values chiefly affected by mass of atoms (*lighter atom, higher frequency*)			
C–H	C–D	C–O	C–Cl
3000 cm^{-1}	2200 cm^{-1}	1100 cm^{-1}	700 cm^{-1}
Values chiefly affected by bond strength (*stronger atom, higher frequency*)			
C≡O	C=O	C–O	
2143 cm^{-1}	1715 cm^{-1}	1100 cm^{-1}	

D, deuterium, has twice the mass of H, and Cl has about twice the mass of O.

This means that each different bond will vibrate in a different way, involving different amounts of energy. The amount of energy it needs to do this will vary from bond to bond, and so each different bond will absorb a different frequency (and hence energy) of IR radiation.

IR absorption intensity

The different peaks in an IR spectrum typically have very different intensities. The intensity of an IR absorption depends on:

Concentration of molecules in the sample: The greater the number of molecules per unit volume, the greater is the intensity of IR absorption.

Magnitude of dipole moment: The intensity of IR absorption varies with the change of dipole moment. *The greater the change in the dipole moment, the stronger will be the absorption.* If the bond is perfectly symmetrical, there is no change in dipole moment and there is no IR absorption. As we know, the magnitude of the bond dipole is proportional not only to the amount of charge on each bonded atom but also to the distance between the atoms—that is, the bond length.

In IR spectroscopy, spectroscopists normally use **wavenumbers** instead of frequency. A wavenumber is the inverse of the wavelength. It has units of cm^{-1}. It is directly proportional to the frequency and the energy of the radiation– IR radiation with a high wavenumber has higher frequency and energy than IR radiation with a lower wavenumber. Because wavenumber and frequency are directly proportional to one another, it is common for the two terms to be used almost interchangeably.

Infrared absorption and chemical structure

The most common application of IR spectroscopy is perhaps to identify the functional groups. This is possible because different functional groups vibrate at different frequencies allowing their identification. *In all compounds, a given type of functional group absorbs IR radiation in the same general region of the IR spectrum*. The frequency of vibration, however, depends on additional factors such as delocalization of electrons, H-bonding, and substitutions at the nearby groups.

An IR spectrum usually extends from radiation around 4000 cm^{-1} to 650 cm^{-1} and can be split into the **functional group region** and the **fingerprint region**. Many functional groups absorb infrared radiation at about the same frequency, regardless of the structure of the rest of the molecule. The absorption bands in the 4000 – 1500 cm^{-1} region allow identification of functional groups; this region therefore is also termed the *functional group region* of the IR spectrum. For example, C–H stretching vibrations usually appear between 3200 and 2800 cm^{-1}; and carbonyl (C=O) stretching vibrations usually appear between 1800 and 1600 cm^{-1}. This makes these bands diagnostic markers for the presence of a functional group in a sample. These types of infrared bands are called *group frequencies* because they tell us about the presence or absence of specific functional groups in a sample.

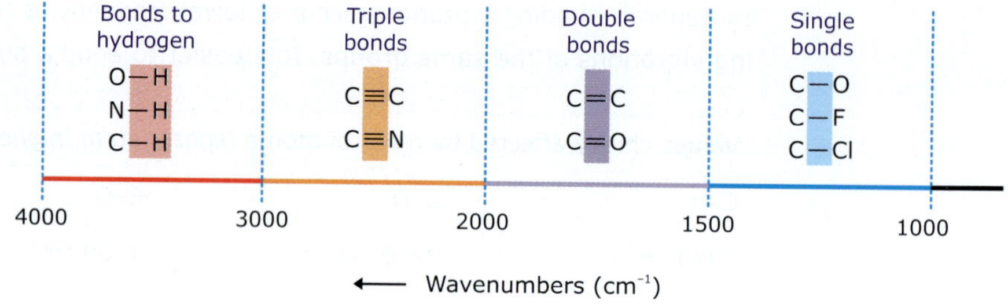

Figure 3.11 Infrared spectrum. The approximate regions where various common types of bonds absorb for stretching vibrations only. The first region, from 4000 to 2500 cm^{-1} is the region for C-H, N-H, and O-H bond stretching. The triple bond region from about 2500 to 2000 cm^{-1}. Double bonds appear about 2000–1500 cm^{-1} and single bonds come below 1500 cm^{-1}.

The region of the infrared spectrum from 1200 to 700 cm^{-1} is called the **fingerprint region**. This region usually contains a very complicated series of absorptions. Many different vibrations, including C—O, C—C and C—N single bond stretches, C—H bending vibrations, and some bands due to benzene rings are found in this region. The fingerprint region is often the most complex and confusing region to interpret. This region is different for each molecule just like a fingerprint is different for each person. Two different molecules may have similar functional group regions because they have similar functional groups, but they will always have a different fingerprint region. Since every type of bond has a different natural frequency of vibration, and since two of the same types of bonds in two different compounds are in two slightly different environments, no two molecules of different structure have exactly the same infrared absorption pattern or infrared spectrum. By comparing the infrared spectra of two substances, one can establish whether they are identical or non-identical. A second use of the infrared spectrum is to determine structural information about a molecule. The absorptions of each type of bond (N—H, C—H, O—H, C—O, C—C and other) are found only in certain small portions of the infrared region. Thus, a small range of absorption can be defined for each type of bond.

In **infrared spectra:**

Position of band depends on	Mass of atoms – **Light** atoms give **high** frequency. Bond strength – **Strong** bonds give **high** frequency.
Strength of band depends on	Change in dipole moment A **large** change in dipole moment gives **strong** absorption.
Width of band depends on	Hydrogen bonding **Strong** H-bond gives **wide** peak.

3.5 Nuclear magnetic resonance

Nuclear magnetic resonance (NMR) is a spectroscopic technique that involves a change in nuclear spin energy in the presence of an external magnetic field. It is based on the magnetic properties of nuclei that result from a property called *nuclear spin*. NMR allows us to detect atomic nuclei and say what sort of environment they are in, within their molecule. For example, the hydrogen of the hydroxyl group in propanol is different from the hydrogens of its carbon skeleton. With the help of a proton NMR, one can easily distinguish between these two sorts of hydrogens. '^{1}H NMR' and 'proton NMR' are interchangeable terms. The hydrogen atom is denoted by ^{1}H. Likewise, **carbon NMR** (^{13}C NMR) can easily distinguish between the three different carbon atoms in propanol.

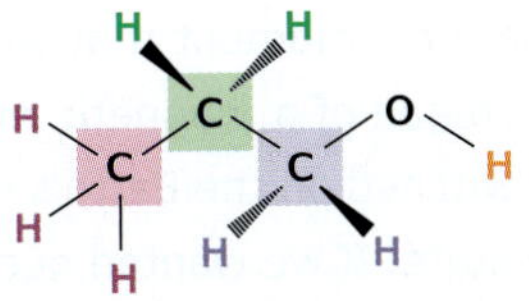

Figure 3.12 ^{1}H NMR distinguishes the coloured hydrogens and ^{13}C NMR distinguishes the boxed carbons.

NMR spectroscopy is used to detect nuclei, but only those nuclei that have a magnetic property as a result of nuclear spin. Spin comes in multiples of 1/2 and can be + or –. Not all atomic nuclei have *nuclear spin*. The rules for determining the net spin of a nucleus are as follows;

- If the number of neutrons and the number of protons are both *even*, then individual spins are paired and the overall spin becomes zero (i.e. nuclei have no spin).
- If the number of neutrons and the number of protons are both *odd*, then the nucleus has an integer spin (i.e. 1, 2, 3).
- If the number of neutrons plus the number of protons is *odd*, then the nucleus has a half-integer spin (i.e. 1/2, 3/2, 5/2).

Table 3.2 Number of protons, number of neutrons and nuclear spin quantum number of some elements

Number of protons	Number of neutrons	Spin quantum number (*I*)	Element
Even	Even	0	^{12}C, ^{16}O
Odd	Odd	Integer (1,2,...)	^{14}N
Even	Odd	Half-integer (1/2, 3/2,...)	^{13}C
Odd	Even	Half-integer (1/2, 3/2,...)	^{19}F, ^{15}N

For each nucleus with spin, the number of allowed spin states – it may adopt – is determined by its **nuclear spin quantum number** (*I*). A nucleus of spin quantum number *I* has 2*I* + 1 allowed spin states. For example, ^{1}H has the nuclear spin quantum number *I* = 1/2 and has two allowed spin states [2(1/2) + 1 = 2] for its nucleus, –1/2 and +1/2. For the chlorine nucleus, *I* = 3/2 and there are four allowed spin states [2(3/2) + 1 = 4] for its nucleus, –3/2, –1/2, +1/2 and +3/2.

Table 3.3 Spin quantum numbers and number of spin states of some common element

Element	$^1_1$H	$^2_1$H	$^{12}_6$C	$^{13}_6$C	$^{14}_7$N	$^{16}_8$O	$^{17}_8$O	$^{19}_9$F
Spin quantum number	½	1	0	½	1	0	5/2	½
Number of spin states	2	3	0	2	3	0	6	2

The hydrogen nucleus ^{1}H has a nuclear spin that can assume either of two spin states, –1/2 and +1/2. The ^{12}C nucleus has no spin whereas ^{13}C nucleus has spin that can assume either of two spin states, –1/2 and +1/2.

Physical basis of NMR

NMR spectroscopy is based on the magnetic properties of nuclei that result from a property called *nuclear spin.* All nucleus with spin act like a tiny magnet. We can compare the behaviour of a nucleus acting like a tiny magnet with magnetic compass placed in an external magnetic field. Imagine for a moment that we were able to 'switch off' the Earth's magnetic field. Compass needle (made of a magnetic material) will point randomly in any direction. However, as soon as we switched on the Earth's magnetic field back, the needle would point North—their *lowest* energy state. If we wanted needle to point South, we would have to apply force (energy). But after removal of external force, the needle would return to its lowest energy state, pointing North. How hard it is to turn the compass needle depends on how strong the magnetic field is and also on how well the needle is magnetized. If the needle is not magnetized at all, it is free to rotate.

Nevertheless, there is an important difference between a compass needle and the atomic nucleus acting like a tiny magnet. A real compass needle can rotate through 360° and have a virtually infinite number of different energy levels, all higher in energy than the lowest energy state (pointing North). When atomic nuclei acting like tiny magnets are placed in an external magnetic field, they have different energy levels. Fortunately, the number of energy levels an atomic nucleus can adopt is very less. For example, a ^{1}H or C^{13} nucleus in a magnetic field can have two energy levels.

Let's take an example of the hydrogen nuclei in a chemical sample. In the discussion, the word 'proton' is used for '^{1}H nuclei'. In the absence of an external magnetic field, the nuclear magnetic poles are oriented randomly. When an external magnetic field is applied to ^{1}H nuclei, they can either align themselves parallel to the field, which would be the lowest energy state, or they can align themselves antiparallel to the field, which is higher in energy. The magnetic poles of nuclei with a spin of +½ are oriented *parallel* to the applied field, and those nuclei with a spin of -½ are oriented *antiparallel* to the applied field.

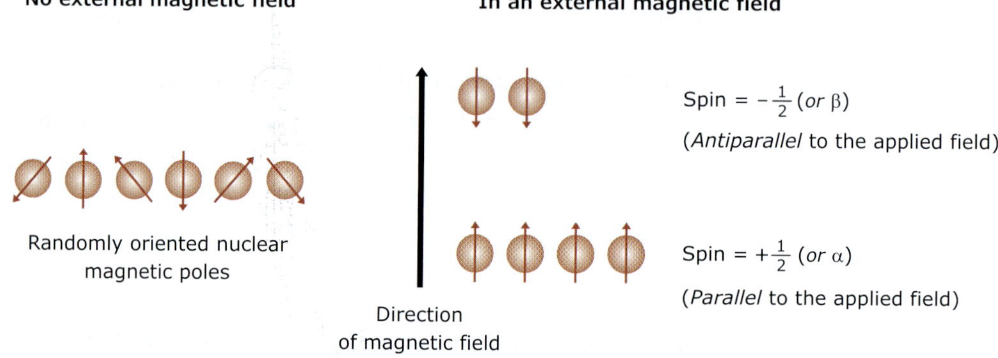

Figure 3.13 In the absence of an external magnetic field, the nuclear magnetic poles are oriented randomly (with arrows indicating their magnetic, north-south, polarity). In an external magnetic field or applied magnetic field, the spin state +½ is aligned parallel to the field, while the spin state −½ is aligned antiparallel to the applied field. More nuclei are oriented along with the applied field because this arrangement is lower in energy.

In the absence of an external magnetic field, the two spin states have the *same* energy. But when an external magnetic field is applied, the two spin states have *different* energies: the +½ spin state has lower energy than the -½ spin state. The energy difference between two spin states is a function of the strength of the applied magnetic field. The stronger the applied magnetic field, the higher the energy difference between the possible spin states. However, the energy difference between the two nuclear spin states is very small—so small that a very, very strong magnetic field is required to see any difference at all.

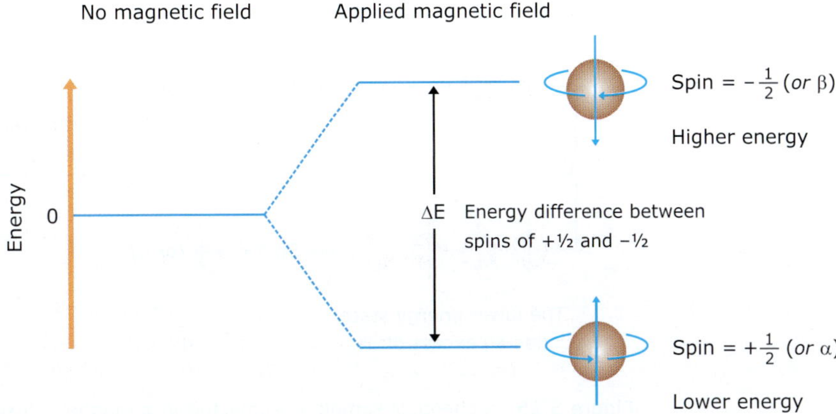

Figure 3.14 Effect of increasing magnetic field at a ^1H nucleus on the energy difference between its +½ and -½ spin states. The two spin states have identical energies when the external magnetic field is absent. The difference in energy between the two spin states depends on how strong the applied external magnetic field is, and also on the properties of the nucleus itself. The energy difference between the two spin states grows with increasing field. The stronger the magnetic field, the higher the energy difference between the two states.

Thus, for a nucleus in an external magnetic field, the difference in energy between two spin states in an applied magnetic field depends on:

- How strong the magnetic field is, and
- The magnetic properties of the nucleus itself.

The energy difference between the two spin states is given by the fundamental equation of NMR:

$$\Delta E = \frac{h\gamma_H}{2\pi} B_p$$

where, h is Planck's constant,

B_p is the magnitude of the magnetic field at the H^1 nucleus (proton), in gauss; and

γ_H is a fundamental constant of the proton, called the **gyromagnetic ratio**. The value of this constant is 26,753 radians gauss^{-1} s^{-1}.

This equation shows that when the external magnetic field is zero, there is no energy difference between the spin states and as the magnetic field is increased, the energy difference between the two spin states increases.

When the ^1H nuclei in a chemical sample are subjected in a magnetic field; each ^1H nucleus is in one of two spin states that differ in energy by an amount ΔE, and; a small excess of ^1H nuclei have spin +½. Even at a very strong magnetic field, the difference in the populations of the two spin energy states is very small. If the sample is now subjected to electromagnetic radiation with energy *exactly equal* to ΔE, this energy is absorbed by some of the ^1H nuclei in the +½ spin state. The absorbed energy causes these protons to 'flip' their spins and assume a more energetic state with a spin −½. Since the energy difference between the two states even in a very very strong external magnetic field is so small, the amount of energy needed to flip the nuclei can be provided by electromagnetic radiation of radio wave frequency. Radio waves flip the nucleus from the lower energy state to the higher state. This absorption phenomenon, called **nuclear magnetic resonance**, can be detected in a type of absorption spectrometer called an **NMR spectrometer**.

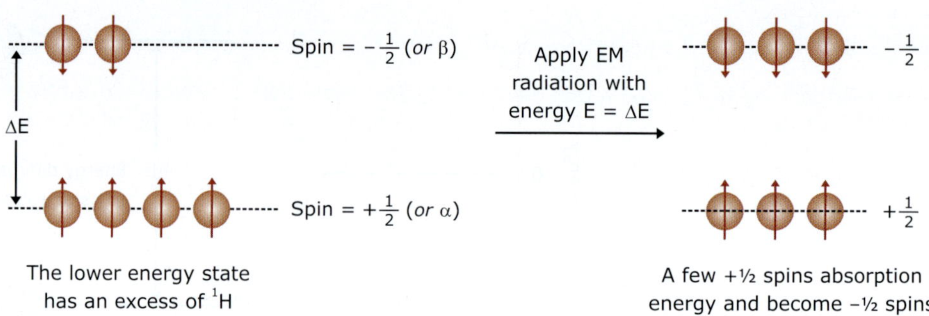

Figure 3.15 A chemical sample is subjected in a magnetic field; each ^{1}H nucleus is in one of two spin states that differ in energy by an amount ΔE, and; a small excess of ^{1}H nuclei have spin +½. The lower energy level contains slightly more nuclei than the higher level. The amount of energy needed to flip the nucleus can be provided by electromagnetic radiation (EM) of radio wave frequency. Radio waves flip the nuclei from the lower energy state to the higher state.

Chemical shift

To understand the chemical shift, let us analyze the **proton NMR** (^{1}H NMR) spectrum of **dimethoxymethane**. There are two absorption peaks in the protons NMR spectrum. Each peak represents a different kind of protons (hydrogen nuclei): each one absorbs energy at a different frequency. But why should protons present in the same molecule be different? We might expect all protons to resonate at one particular frequency. But they don't. In an applied magnetic field, not all protons in a molecule resonate at exactly the same frequency. This variability is due to the fact that the hydrogens in a molecule exist in slightly different electronic environments from one another.

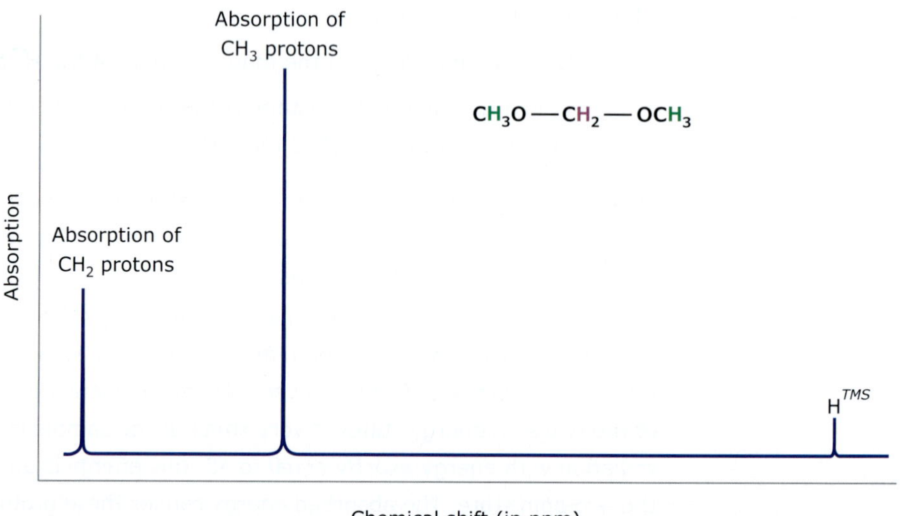

Figure 3.16 The proton NMR (^{1}H NMR) spectrum of dimethoxymethane. This spectrum is a plot of energy absorption on the y-axis versus relative frequency of the radiation from which energy is absorbed on the x-axis. Peaks in an NMR spectrum are called *absorptions*, or *lines*. The position of absorption on the horizontal axis is called its *chemical shift*. We usually express the peak positions in ppm. The small peak at the far right is from the protons of tetramethylsilane (TMS), a reference standard.

Each nucleus is surrounded by electrons. The valence-shell electron densities vary from one hydrogen to another. In an applied magnetic field, the valence-shell electrons are caused to circulate. This circulation, called a **local diamagnetic current**, generates own magnetic field that opposes the applied magnetic field (B_0). Circulation of electrons around a nucleus can be viewed as being similar to the flow of an electric current in an electric wire. In physics, we know that the flow of a current through a wire induces a magnetic field. In an atom, the local

Changes in the distribution of electrons around a nucleus affect:

the local magnetic field that the nucleus experiences

the frequency at which the nucleus resonates

the chemistry of the molecule at that atom

diamagnetic current generates an induced magnetic field that has a direction opposite to that of the applied magnetic field. This causes a reduction in the *effective* or *local magnetic* field. The effective or local magnetic field ($\mathbf{B_p}$) sensed by a proton is less than external magnetic field ($\mathbf{B_0}$). Thus, the electrons are said to *shield* the nucleus from the external magnetic field. The reduction in the local field due to the circulation of nearby electrons is called **shielding**.

As a result, each proton in a molecule is shielded from the applied magnetic field to an extent that depends on the electron density surrounding it. The greater the electron density around a nucleus, the greater the shielding. Therefore, atoms of relatively low electronegativity near a proton increase the electron density at the proton and thus increase the shielding of the proton from the applied magnetic field. Conversely, more electronegative atoms near a proton reduce the electron density at the proton and decrease the shielding of the proton. In summary, the local or effective magnetic field at a more-shielded proton is smaller than the local field at a less-shielded proton.

We can understand the magnetic shielding by the following example. If you use an umbrella on a rainy day, the umbrella shields you from the rain. If you don't use an umbrella, you get soaked. If you use a very small umbrella, you get partially wet, but perhaps not soaked. Likewise, we can think of electrons as providing a 'magnetic umbrella' that shields nuclei from the external applied magnetic field. If electron density is high in the vicinity of a proton, the 'magnetic umbrella' is relatively large, and the effective local field at the proton is smaller. In this case, the chemical shift is small. If electron density is smaller at a proton, the 'magnetic umbrella' is smaller, and the local field at the proton is higher. In this case, the chemical shift is larger.

Let's apply these ideas to the chemical shifts in the NMR spectrum of dimethoxymethane.

$$\overset{a}{\mathbf{CH_3O}} - \overset{b}{\mathbf{CH_2}} - \overset{a}{\mathbf{OCH_3}}$$

Dimethoxymethane

Let $\mathbf{B_a}$ be the local magnetic field at protons a and $\mathbf{B_b}$ be the local magnetic field at protons *b*. Protons *b* are adjacent to *two* oxygens and protons *a* are adjacent to only one oxygen. If the electron distribution varies from protons *a* to protons *b*, so does the local magnetic field. Protons *b* are less shielded than protons *a*. Therefore, the effective or local magnetic field at protons *b* is higher as compared to protons *a*.

$$B_b > B_a$$

The less shielded a proton is, the greater is its chemical shift. Thus, protons *b* undergo resonance at a greater frequency than protons *a*. These frequency differences are due to the different chemical environments of protons *a* and *b*. Hence, there are two separate absorption peaks for protons *a* and *b* in ^{1}H NMR spectrum. This variation in frequency is known as the **chemical shift** and it is denoted by δ. It is defined as the difference between its resonance frequency and that of the protons in the reference compound *tetramethylsilane* (**TMS**). The absorption position of TMS defines the δ = 0 position on the *x*-axis of each spectrum. TMS is used as a standard because it has a single strong absorption, it is chemically inert, and its chemical shift is smaller than that of most common organic compounds.

The **chemical shift** in absolute terms is defined by the frequency of the resonance expressed with reference to a standard compound which is defined to be at 0 ppm. The scale is made more manageable by expressing it in parts per million (ppm) and is independent of the spectrometer frequency.

The chemical shift of protons *b* is greater than that of protons *a*. *The size of a peak* (actually, the area under the peak) *is proportional to the number of protons contributing to the absorption.* The less shielded a proton is, the greater is its chemical shift. Protons *b* are less shielded and thus have a greater chemical shift, than protons *a*. This reduction in shielding (sometimes called **deshielding**) is a consequence of decreased electron density at the *b* protons.

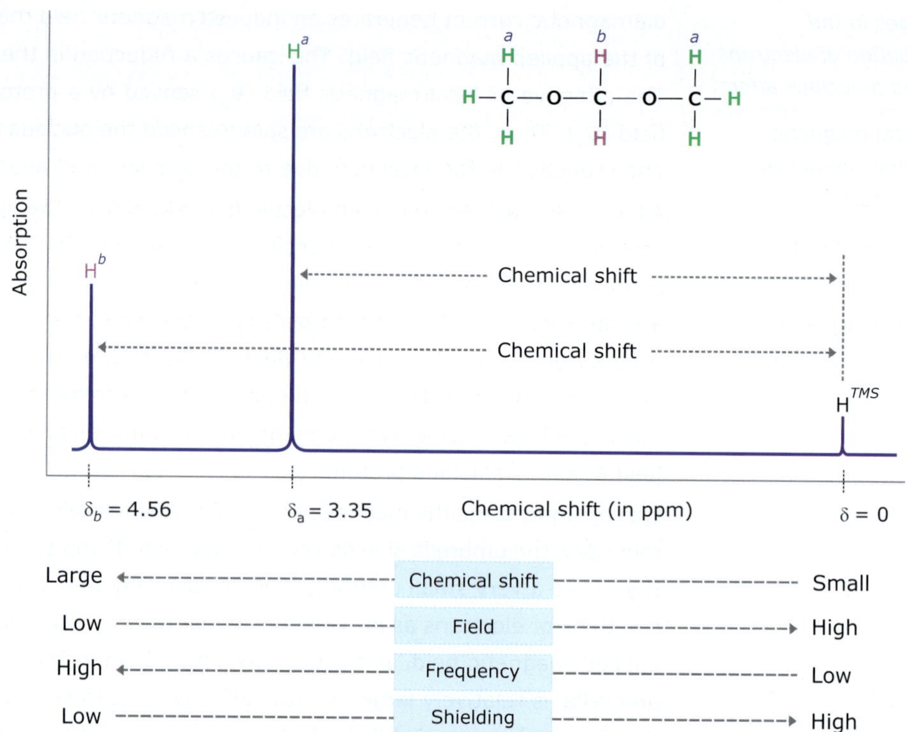

Figure 3.17 The proton NMR spectrum of dimethoxymethane. This spectrum is a plot of energy absorption on the y-axis versus relative frequency of the radiation from which energy is absorbed on the x-axis. Peaks in an NMR spectrum are called absorptions or lines. The position of an absorption peak on the horizontal axis is called its chemical shift. We express the peak positions in parts per million or ppm. The chemical shift of an absorption is written with a δ followed by the numerical value of the peak position. The absorption position of TMS (a reference standard) defines the $\delta = 0$ position on the x-axis of the spectrum. The chemical shift scale runs to the left from zero (where TMS resonates). The other two peaks—the ones at $\delta = 3.35$ and $\delta = 4.56$—are the NMR absorptions of the protons of dimethoxymethane. We can say that the spectrum contains peaks at 0, 3.35 and 4.56 ppm. Two absorptions ($\delta = 3.35$ and $\delta = 4.56$) indicate that there are two chemically distinguishable sets of protons in dimethoxymethane, the CH_2 protons and the CH_3 protons. The two peaks have different sizes because different numbers of protons contribute to each absorption. The resonance at $\delta = 3.35$ is larger because more protons (six) contribute to this resonance than to the resonance at $\delta = 4.56$ (two).

In summary, proton NMR provides four types of information:

1. the number of sets of chemically nonequivalent protons
2. the chemical environments of each set of protons (chemical shift)
3. the number of protons within each set
4. the number of protons in adjacent sets

The reference sample — Tetramethylsilane

$Si(CH_3)_4$

Tetramethylsilane

Unlike IR and UV/Vis spectroscopy, where the signals are fixed at particular frequencies or wavelengths, in NMR, the signal is dependent on the field strength. Since no two magnets have the same field, the frequency at which signals are obtained would vary correspondingly. Therefore, there was a need to characterize and specify the location of the signals. In order to avoid this situations, it has been decided to eliminate the problems by keeping a standard reference with respect to which a numerical value can be assigned. The compound we use as a reference sample in ^{1}H and ^{13}C NMR is usually **tetramethylsilane**, TMS. This is silane (SiH_4)

with each of the hydrogen atoms replaced by methyl groups to give $Si(CH_3)_4$. Because of molecular symmetry, all 12 protons of TMS absorb at the same frequency and all 4 carbons absorb at the same frequency. The frequency of absorption for a nucleus of interest relative to the frequency of absorption of a standard is called the *chemical shift* of the nucleus. The chemical shift of the ^1H nuclei in the ^1H NMR spectrum or ^{13}C nuclei in the ^{13}C NMR spectrum of TMS appear at 0 ppm. Typically, it increases from 0 on the right-hand side of the spectrum to 10 ppm on the left hand side of an ^1H NMR spectrum or from 0 on the right hand side to 200 ppm on the left hand side of a ^{13}C NMR spectrum. The reason frequencies of absorption are recorded on the δ-scale relative to those of a standard molecule. It makes the position of absorption independent of the spectrometer used to record the spectrum, in particular independent of the strength of the magnetic field of the spectrometer.

3.6 Circular Dichroism

Light is a transverse electromagnetic wave. The electric and magnetic fields in an electromagnetic wave oscillate along directions perpendicular to the propagation direction of the wave. Light can be *unpolarized* or *polarized* depending on how the electric field is oriented. In **unpolarized light,** the electric field vector oscillates in more than one plane; all perpendicular to the direction of propagation. If the electric field vector oscillates in single plane perpendicular to the direction of propagation, it is called **linearly** or **plane polarized light**. The orientation of a linearly polarized light is defined by the direction of the electric field vector. Linear polarization is obtained by passing unpolarized light through a polarizer that absorbs all electric field vectors not lying along a particular plane.

If the electric field vector rotates around the propagation axis maintaining a constant magnitude, it is called **circularly polarized light**. Circularly polarized light is the antithesis of linearly polarized light. In linear polarized light, the direction of the vector stays constant and the amplitude oscillates whereas in circularly polarized light, the amplitude stays constant while the direction oscillates. Circularly polarized light is obtained by superimposing two plane polarized light of same wavelengths and amplitudes but having a phase difference of 90°. A phase difference of 90° means that when one wave is at its peak, then the other one is just crossing the zero line.

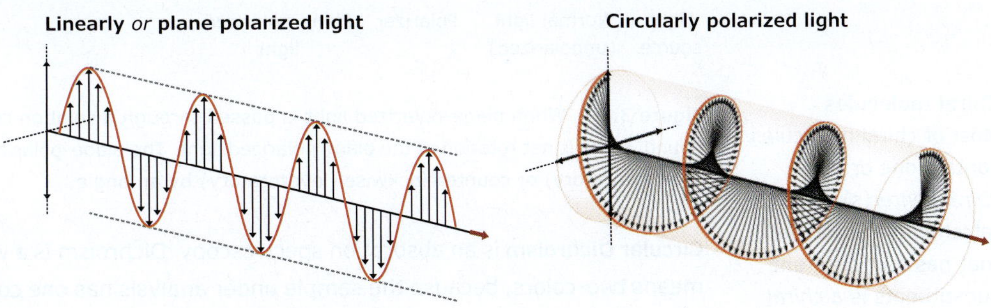

Figure 3.18 Linearly polarized light results when the direction of the electric field vector is restricted to a plane perpendicular to the direction of propagation while its amplitude oscillates. In circularly polarized light, the amplitude of the oscillation is constant and the direction oscillates.

A circularly plane polarized light may be right or left-handed circularly polarized depending on the rotation direction. For **left circularly polarized light** with propagation towards the observer,

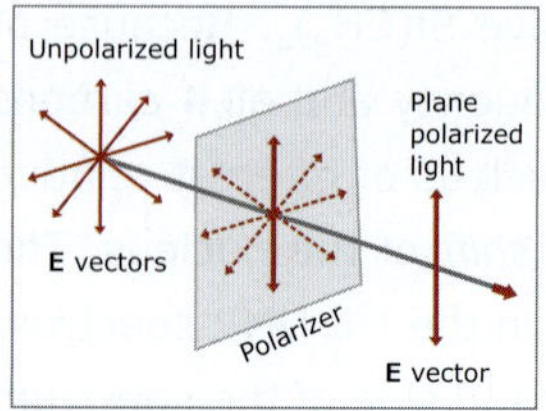

Generation of plane polarized light.

the electric field vector rotates counter-clockwise. For **right circularly polarized light**, the electric field vector rotates clockwise. The superposition of left and right circularly polarized light beams of equal amplitudes can result in linearly polarized light. Thus, linear polarized light can be viewed as a superposition of opposite circular polarized light of equal amplitude and phase.

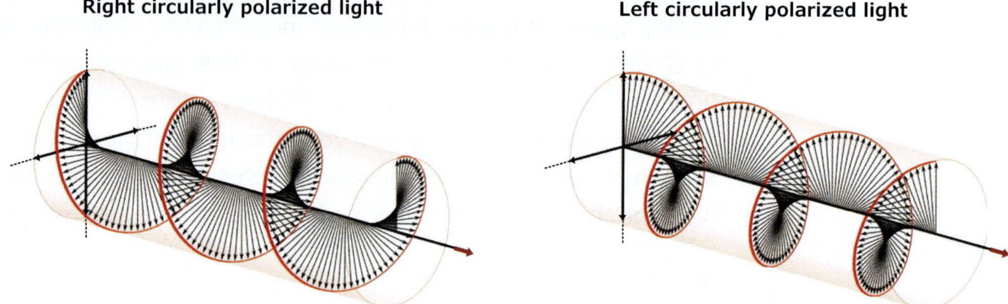

Figure 3.19 Schematic representation of right circularly polarized and left circularly polarized light. In both cases, the length of the vector remains constant. In linearly polarized light, the electric vector stays in the same plane but its length changes.

Optical rotatory dispersion (ORD) and **circular dichroism** (CD) are two phenomena that result when *chiral molecules* (molecules that have no plane and center of symmetry) interact with polarized light. Most of biological molecules are chiral molecules and thus they are **optically active**. An optically active compound shows a phenomenon called **optical rotation**. It means that the plane of polarization of a linearly polarized light rotates as it passes through an optically active medium. The technique of ORD measures the ability of optically active compounds to rotate plane polarized light as a function of the wavelength (i.e. wavelength dependence of optical rotation).

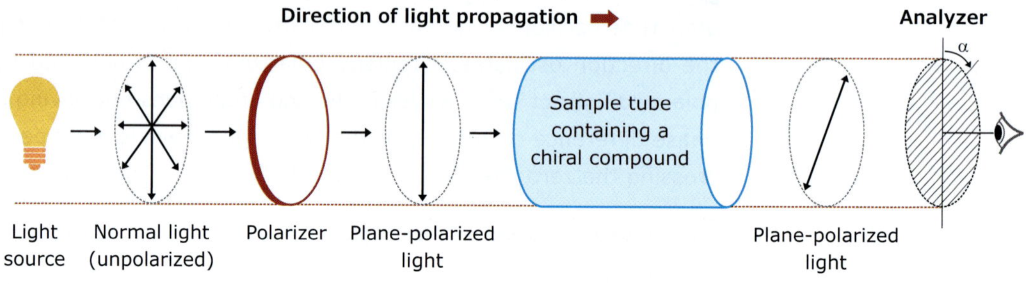

Figure 3.20 When plane-polarized light is passed through a solution containing an optically active compound, there is net rotation of the plane polarized light. The plane-polarized light is rotated either clockwise (dextrorotatory) or counterclockwise (levorotatory) by an angle.

Chiral molecules
Most of chiral molecules contain one or more *chiral center*(s). Any tetrahedral carbon atom that has four different substituents is a *chiral center*. A molecule is not chiral (even if it has chiral centers) if it has a plane or center of symmetry. To be a chiral molecule there should be no plane and center of symmetry.

Circular Dichroism is an absorption spectroscopy. Dichroism is a word derived from Greek which means two-colors, because the sample under analysis has one color if illuminated with the right polarized light and a different color if illuminated with the left one. The color, in fact, depends on light absorption. Circular dichroism is observed when optically active matter absorbs left and right circularly polarized light slightly different. In fact, circular dichroism is the absorption difference between left and right circularly polarized light at a given wavelength. Absorption is quantitated by the **molar extinction coefficient** (ε). Optically active samples have distinct molar extinction coefficients for left (ε_L) and right (ε_R) circularly polarized light. The difference in absorbance of left and right circularly polarized light is a measure of circular dichroism.

An **elliptically polarized** light is obtained by superimposing two plane polarized light vibrating at right angle to each other having same wavelengths and unequal amplitudes but there is a phase difference of 90° between them.

Thus, **CD** = $A_L - A_R = \Delta A$

From Beer-Lambert law the difference in the absorbance of left and right circularly polarized light (i.e. ΔA) can be given by,

$\Delta A = (\varepsilon_L - \varepsilon_R).c.l$ where, c is the molar concentration and l is the path length (cm)

$\Delta A = \Delta\varepsilon.c.l$ where, Δε is the *molar circular dichroism*.

The differential absorption of the left and right circularly polarized light means that the amplitudes of the right and left circularly polarized components of the transmitted beam will differ. The superposition of the two components is no longer a linearly polarized wave but it rotates along an ellipsoid path. Such a light wave is called an **elliptically polarized light**. How elliptical the plane-polarized wave becomes after traversing the medium is determined by the difference between the absorptions of the two circularly polarized components. In the most extreme case, the material almost completely extincts one left or right component and then the transmitted wave almost becomes a perfect circularly polarized light because the other circular component disappears.

The ellipticity is proportional to the difference in the absorbance of the two components, $A_L - A_R$. Thus, the CD is equivalent to ellipticity. The relationship between CD and ellipticity (θ) is given by:

$$\theta = 2.303\,(A_L - A_R)\,\frac{180}{4\pi} = 33\,(A_L - A_R) = 33\,\Delta A \text{ degree}$$

Units of CD data

CD data are presented in terms of either ellipticity [θ] or differential absorbance (ΔA). The data are normalised by scaling to molar concentrations of either the whole molecule or the repeating unit of a polymer. For far-UV CD of proteins, the repeating unit is the peptide bond. The *mean residue weight* (MRW) for the peptide bond is calculated from MRW=M/(N−1), where M is the molecular mass of the polypeptide chain (in Da), and N is the number of amino acid residues in the chain; the number of peptide bonds is N−1.

The **mean residue ellipticity** (MRE) at wavelength λ is given by,

$$[\theta]_{MRE} = \frac{MRW \times \theta_\lambda}{10 \times l \times c}$$

Where, θ_λ = observed ellipticity (in degrees),
l = path length (in cm) and
c = concentration (in g/ml)

If we know the molar concentration (c) of a solute, the molar ellipticity at wavelength λ is given by,

$$[\theta]_{molar} = \frac{100 \times \theta_\lambda}{c \times l}$$

The units of mean residue ellipticity and molar ellipticity are degrees cm^2 decimol^{-1}.

Applications of circular dichroism

Circular dichroism is an excellent method for the study of the conformations adopted by proteins and nucleic acids in solution. Although not able to provide the detailed residue-specific information as obtained from NMR and X-ray crystallography, analysis by CD has a number of advantages. CD analysis can be made on molecules of any size and also, even in very small concentrations. It allows to study dynamic systems and kinetics. It provides one of the best methods for monitoring any structural alterations that might result from changes in environmental conditions, such as pH, temperature, and ionic strength. However, it only provides qualitative analysis of data. It does not provide atomic-level structural analysis. Also, the observed spectrum is not enough for claiming one and only possible structure.

CD has a wide range of potential applications in the study of structure and function of biomolecules such as proteins and nucleic acids. Following are some important applications.

1. Determination of protein's secondary and tertiary structure.
2. Comparison of the secondary and tertiary structure of wild-type and mutant proteins.
3. Nucleic acid structure and changes upon protein binding or melting.
4. Determination of conformational changes due to protein-protein interactions, protein-DNA interactions and protein-ligand interactions.

Analysis of protein's structural and conformational changes

Circular dichroism relies on the differential absorption of left and right circularly polarized light by chromophores. Depending on the wave-length of the light used for the generation of circularly polarized light, there are **far-UV CD** (190-250 nm), **near-UV CD** (250-320 nm) and **visible CD.**

Proteins possess a number of chromophores which can give rise to CD signals. In proteins, the chromophores of interest include the *peptide bond*, *aromatic amino acid side chains* and *disulphide bonds*. The *peptide bonds* have an absorption range of 240 nm to 190 nm. This region is called the 'far-UV' region. The *aromatic side chains* of tyrosine, tryptophan and phenylalanine absorb light in the 260 nm to 320 nm region. This region is the 'near-UV' region. The three residues, phenylalanine, tyrosine and tryptophan, exhibit fine structure peaks between 255-270 nm, 275-285 nm, and 290-305 nm, respectively. The *disulphide bonds* absorb light near 260 nm and are generally quite weak.

In addition, various non-protein *cofactors* can also absorb light over a wide spectral range, such as *pyridoxal phosphate* around 330 nm, *flavins* in the range 300 nm to 500 nm (depending on oxidation state), *heme* groups strongly around 410 nm with other bands in the range from 350 nm to 650 nm (depending on spin state and coordination of the central Fe ion) and *chlorophyll* moieties in the visible and near IR regions.

The CD spectrum in the **far-UV region** (240 nm-190 nm) gives quantitative information about the overall secondary structure of the protein. CD spectra can be readily used to estimate the fraction of a molecule that is in the α-helix, the β-sheet, or random coil conformation.

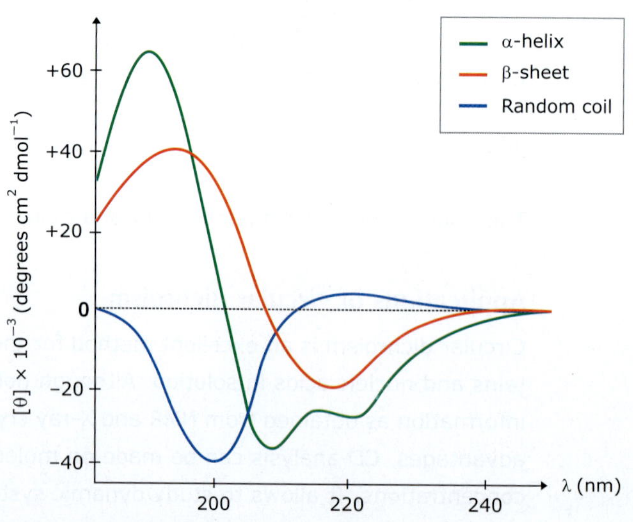

Figure 3.21 Circular dichroism spectra in far-UV region and protein secondary structure. The absorption spectrum of synthetic homopolypeptide, poly-L-lysine, in the far UV region between 180 nm and 250 nm. The α-helix, β-sheet and random coil structures each give rise to a characteristic shape and magnitude in the CD spectrum.

Each of the three basic secondary structures of a polypeptide chain shows a characteristic CD spectrum. It should be noted that in CD, only the relative fractions of residues in each conformation can be determined but not specifically where each structural feature lies in the protein molecule.

The CD spectrum in the **near-UV region** (320–260 nm) gives information about the *tertiary structure* of the protein. The CD signals obtained in this region are due to the absorption by aromatic amino acids and disulfide bonds. The CD signals in the near-UV region are sensitive to the overall tertiary structure of the protein. If a protein retains a secondary structure but no defined three-dimensional structure (e.g. molten-globule structure), the signals will be nearly zero. On the other hand, the presence of significant near-UV signals is a good indication that the protein is folded into a well-defined structure. Near-UV CD can also be used to provide structural information about the binding of prosthetic groups in proteins.

3.7 Raman spectroscopy

When light interacts with matter, the photons which make up the light may be absorbed or scattered, or may not interact with the material and may pass straight through it. *Raman spectroscopy* involves neither absorption nor emission, but scattering of radiation by the sample. A molecule can react to incoming light via the processes of absorption and scattering. If the energy of a photon matches a difference between two energy levels in a molecule, absorption can occur causing a transition from the lower to higher energy state. Rotational transitions occur at low energies (microwave region of the electromagnetic spectrum), while vibrational transitions occur in the infrared, and electronic transitions occur in the visible and ultraviolet region of the electromagnetic spectrum.

Raman spectroscopy is based on *inelastic scattering* of light. **Inelastic scattering** means that light is scattered at frequencies that differ from the incident light. When light is scattered at the same frequency as that of the incident light, it is termed **elastic scattering** (**Rayleigh scattering**). Unlike the elastic scattering, inelastic scattering involves a net energy transfer between the incident photons and a material. *Raman scattering* is an example of *inelastic* process.

Raman spectroscopy is a nondestructive, reagentless, vibrational spectroscopic technique. It differs from IR spectroscopy in that information is derived from light scattering whereas in IR spectroscopy information is obtained from an absorption process. Raman scattering is associated with normal modes that produce a change in the polarisability of the molecule, while IR absorptions are observed for vibrational modes that change the dipole moment of the molecule.

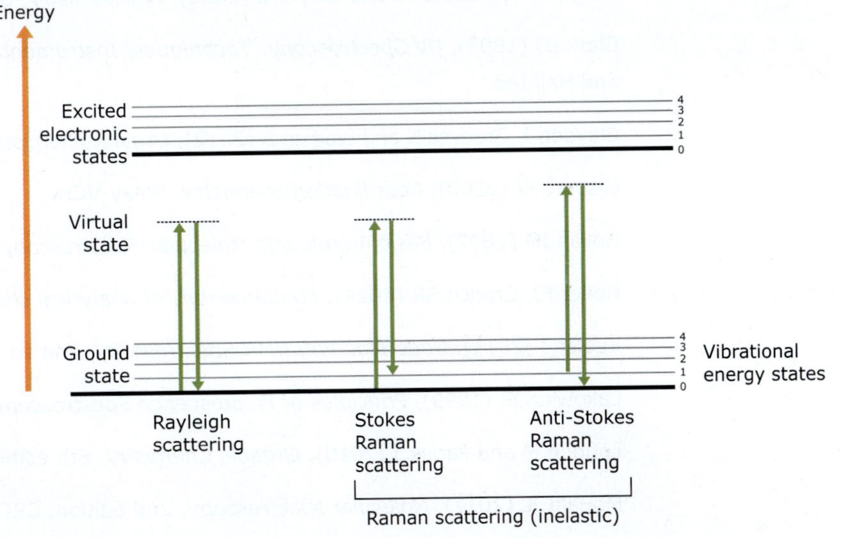

Figure 3.22 Energy level diagram to explain Raman scattering.

The change in the energy of the scattered photon corresponds exactly to the photon energy. As a result, the wavelength of the scattered photons can be longer (**Stokes Raman scattering**) or shorter (**anti-Stokes Raman scattering**). The energy of the scattered radiation is less than the incident radiation for the Stokes line and the energy of the scattered radiation is more than the incident radiation for the anti-Stokes line. The difference in energy between the incoming and scattered photon (**Raman shift**) corresponds to the energy difference between vibrational energy levels of the molecule. The different vibrational modes of a molecule can therefore be identified by recognizing Raman shifts (or 'bands') in the inelastically scattered light spectrum.

Both Raman and infrared spectroscopy is based on molecular vibrations associated with chemical bonds in a sample to obtain information on molecular structure, composition, and intermolecular interactions. IR spectroscopy monitors the net absorption of incident radiation by a sample in the IR region of the electromagnetic spectrum. Absorption occurs where the frequency of the incident radiation matches that of a vibration.

The wavelength of IR absorption bands is characteristic of vibrational modes of specific bond types in a sample, whereas Raman spectroscopy profiles vibrational and rotational motion of molecules that arise from an inelastic scattering event that depends on nuclear vibrations that create a change in polarizability (the ability to form instantaneous dipoles) of a molecule as it vibrates and rotates. Thus, Raman spectroscopy provides intensity profiles of scattered light as a function of frequency. The frequency difference between the incident and scattered light is the frequency of vibration. Unlike infrared absorption, Raman scattering does not require matching of the incident radiation to the energy difference between the ground and excited states. In Raman scattering, the light interacts with the molecule and distorts (polarizes) the cloud of electrons round the nuclei to form a short-lived state called a 'virtual state'. This state is not stable and the photon is quickly reradiated.

References

Banwell CN, McCash EM (2013), *Fundamentals of Molecular Spectroscopy*, 5th ed. McGraw Hill Education.

Berg JM, Tymoczko JL and Stryer L (2006), *Biochemistry*, 6th ed. W.H. Freeman and company.

Clark BJ (1993), *UV Spectroscopy: Techniques, Instrumentation and Data Handling,* Chapman and Hall Inc.

Clayden J, Greeves N and Warren S (2012), *Organic chemistry*, 2nd edition, Oxford University Press.

Gispert JR (2008), *Coordination chemistry*. Wiley-VCH.

Hollas JM (2002), *Basic Atomic and Molecular Spectroscopy*, RSC Publishing.

Holler FJ, Crouch SR (2014), *Fundamentals of analytical chemistry*, 9th ed. Cengage Learning.

Keeler J (2011), *Understanding NMR Spectroscopy*, 2nd ed. John Wiley and Sons.

Lakowicz JR (1999), *Principles of Fluorescence Spectroscopy*, 2nd ed. Kluwer, New York.

Loudon M and Parise J (2016), *Organic Chemistry*, 6th edition, W.H. Freeman and company.

McHale JL (2017), *Molecular spectroscopy*, 2nd edition. CRC press.

Pavia DL, Lampman GM, Kriz GS and Vyvyan JR (2009), *Introduction to spectroscopy*, 4th ed. Brooks/Cole Cengage Learning.

4

Mass spectrometry

Mass spectrometry is an analytical technique that determines the mass of an ionic species by measuring the mass-to-charge (m/z) ratio. In mass spectrometry, an instrument called a **mass spectrometer** is used to determines the m/z ratio by performing four essential steps – *ionization*, *acceleration*, *deflection* or *separation*, and *detection*. In the process of *ionization*, a sample is first converted into gas-phase ions. Once the gas-phase ions are formed, the ions are accelerated so that they all have the same kinetic energy (*acceleration*). A magnetic field then deflects the ions according to their masses (*deflection*). Finally, the beam of ions passing through the machine is detected electrically (*detection*). A mass spectrometer has three basic components: the **ionization source** (generates gas-phase ions), the **mass analyzer** (causes separation of ions) and the **detector** (detects the relative number of ions of each mass).

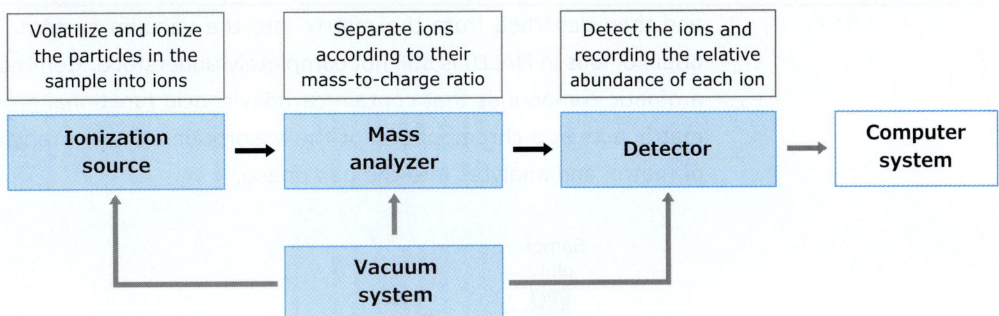

The operation of the *mass spectrometer* involves the following steps:

1. The particles in the sample (atoms or molecules) are converted in to gas-phase ions.
2. The ions are *accelerated* so that they all have the same kinetic energy.
3. Different ions are *deflected* or *separated* by the magnetic field. The amount of deflection depends on: the *mass* of the ions (lighter ions are deflected more than heavier ones) and the *charge* on the ions. These two factors are combined into the mass-to-charge ratio.
4. Detection of the relative number of ions of each mass.

Ionization process

The particles in the sample (atoms or molecules) studied by mass spectrometer must be converted to gas phase-charged particles by the ionization process before they can be analyzed and detected. However, the term *ionization* is misleading because most ionization processes do not perform the ionization of sample particles per se. Instead, the term ionization relates to the transfer to the gas-phase of a sample while maintaining its charge and/or acquiring a charge from the sample environment, typically in the form of a proton. Several methods are used for converting the sample to ions. These methods include - electron ionization, chemical ionization, desorption ionization and electrospray ionization. All these ionization processes are fundamentally different techniques, but they achieve essentially the same end result — the generation of gas-phase ions via non-destructive vaporization and ionization.

Electron ionization

In electron ionization, a beam of high-energy electrons strikes the molecules. The high-energy electrons cause the ejection of an electron from the molecule, creating a radical cation. For example, if CH_4 is treated with high electrons, it loses an electron from one of the C–H bonds.

Chemical ionization

In chemical ionization, the sample molecules are combined with an ionized reagent gas. When the sample molecules collide with the ionized reagent gas, some of the sample molecules are ionized by various mechanisms like proton transfer, electron transfer and adduct formation.

Desorption ionization

In desorption ionization, the sample to be analyzed is dissolved in a matrix and placed in the path of high energy beam of ions or high intensity photons. In case of high intensity photons, it is termed as **MALDI** (matrix-assisted laser desorption/ionization). In MALDI, analytes are placed in a light-absorbing solid matrix. A matrix is used to protect the analytes from being destroyed by direct laser beam. Solid matrix strongly absorbs the laser radiation and acts as a receptacle for energy deposition. With a short pulse of laser light, the analytes are ionized and then desorbed from the matrix into the vacuum system. The exact mechanism of the origin of ions in MALDI is still not completely understood. Commonly used matrix materials are aromatic compounds that contain carboxylic acid functional groups. The aromatic ring of the matrix acts as a chromophore for the absorption of laser irradiation leading to the desorption of matrix and analytes into the gas phase.

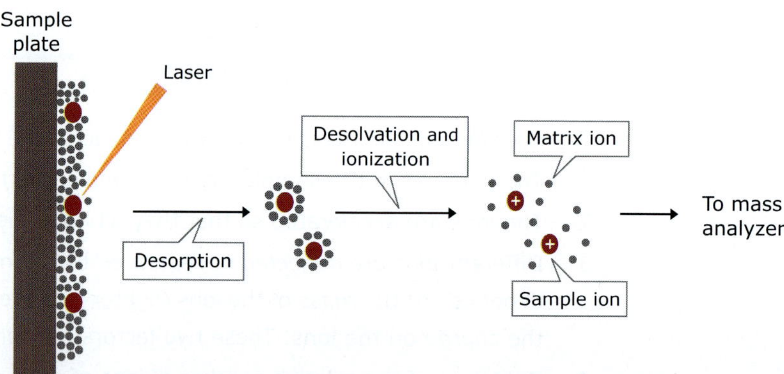

Figure 4.1 Schematic representation of MALDI. It is a mass spectrometry ionization method for directly vaporizing and ionizing non-volatile samples. This method involves three steps. First, the heat-sensitive sample is mixed with the matrix and applied to the metal plate. Secondly, a pulsed laser is used to evaporate the matrix and the sample into the gas phase. Finally, ionization takes place in the gas phase. This is achieved by proton transfer, leading to predominantly $[M + H]^+$ species being produced.

Electrospray ionization

During standard electrospray ionization, the sample is dissolved in a polar, volatile solvent and pumped through a narrow, stainless steel capillary. A high voltage of 3 or 4 kV is applied to the tip of the capillary. As a consequence of this strong electric field, the sample emerging from the tip is dispersed into an aerosol of highly charged droplets. The charged droplets diminish in size by solvent evaporation, assisted by a flow of drying gas. Eventually charged sample ions, free from solvent, are released from the droplets.

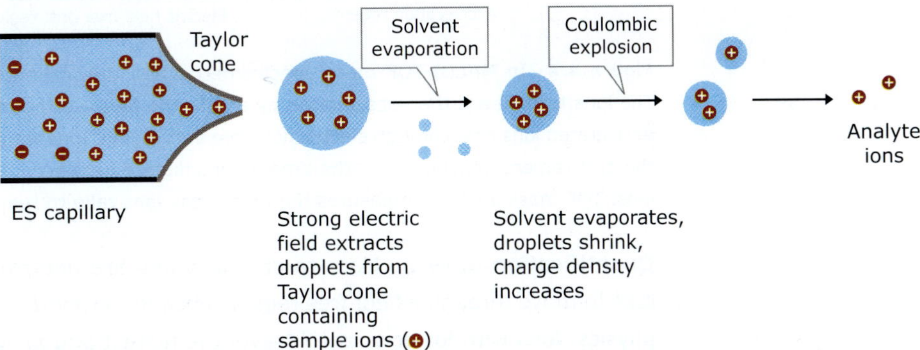

Figure 4.2 Electrospray ionization: The liquid effluent containing the analytes is electrostatically dispersed. This generates highly charged droplets, which are normally positively charged. Once the droplets are airborne, the solvent evaporates, which decreases the size and increases the charge density of the droplets. Desolvated ions are generated by the desorption of sample ions from the droplet surface due to high electrical fields and/or the formation of very small droplets due to repetitive droplet fission until each droplet contains, on average, only one sample ion.

Mass analysis

Once the sample has been ionized, the beam of ions is *accelerated* by an electric field and then passes into the mass analyzer, the region of the mass spectrometer where the ions are separated according to their mass-to-charge ratio. Just like ionization methods, there are several types of mass analyzer. Some examples include **quadrupoles, ion traps, orbitraps, Fourier transform** and **time-of-flight**. The most common type of mass analyzer is **time-of-flight** (TOF) mass analyzer. The TOF mass analyzer measures ion flight time. It is based on the simple idea that velocities of two ions with the same kinetic energy will vary depending on the mass of an ion – the lighter ion will have higher velocity. Because the ions have different velocities, the ions reach the detector at different times. The smaller ions reach the detector first because of their greater velocity as compared to the larger ions. Hence, the analyzer is called **TOF** because the mass is determined from the ion's time-of-flight.

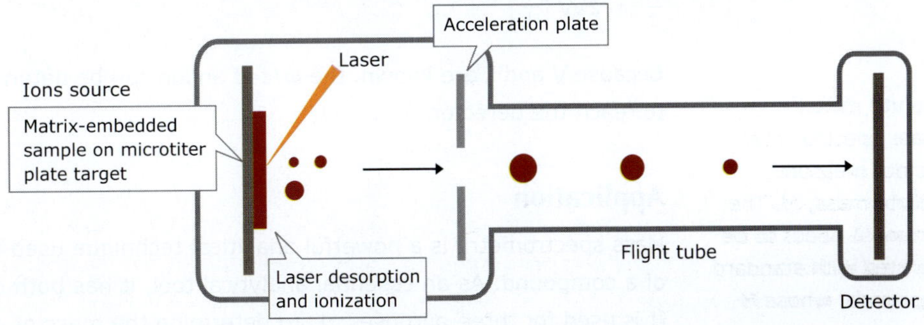

Figure 4.3 Schematic representation of a MALDI/TOF mass spectrometer.

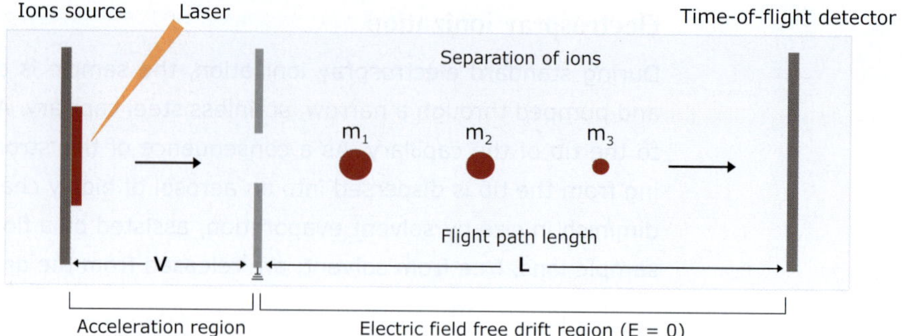

Figure 4.4 In MALDI-TOF mass spectrometry, first desorption and ionization of the sample occurs. The ions formed are then accelerated by an electric field, so that they all have the same kinetic energy. Accelerated ions pass through electric field free drift region. Ions with smaller m/z value move faster through the drift region. Consequently, the time of ions flight differs according to the mass-to-charge ratio of the ions. TOF mass analyzer measures the time these ions take to reach a detector.

Quantification of m/z: Mass-to-charge ratios are determined by measuring the time ions take to move through a field-free region between the source and the detector. Based on classical physics, ions with lower m/z will travel the fastest and arrive at the detector first, while ions with larger m/z will travel the slowest and arrive at the detector last. Once the ion source generates gas-phase ions, these ions are accelerated so that they all have the same kinetic energy. Thus, the potential energy given to ions in the accelerated regions is converted to kinetic energy for all ions.

Potential energy of a charged particle in the electric field = zeV

where, z is the number of charges, e is the amount of charge on an electron and V is the voltage

When an ion is accelerated into a flight tube (or TOF tube) by the voltage V, its potential energy is converted to kinetic energy.

Kinetic energy = $\frac{1}{2}mv^2$

where, v = velocity of sample ion and m = mass of sample ion

The potential energy is converted to kinetic energy.

Hence, $zeV = \frac{1}{2}mv^2$

The velocity of the sample ion, $v = \frac{\text{Flight path length (L)}}{\text{Time required to reach the detector (t)}}$

$$zeV = \frac{1}{2}mv^2 = \frac{1}{2}m(L/t)^2$$

$$\frac{m}{z} = 2eV\left[\frac{t}{L}\right]^2$$

Because V and L are known, the m/z of an ion can be determined from the time the ion takes to reach the detector.

Absolute masses

A mass spectrometer does not measure absolute mass, *M*. The instrument needs to be calibrated with standard compounds, whose *M* values are known very accurately.

Application

Mass spectrometry is a powerful analytical technique used to determine the molecular mass of a compound. As an essential analytical tool, it has both qualitative and quantitative uses. It is used for three purposes: 1. to determine the mass of an unknown compound, 2. to determine the structure (or a partial structure) of an unknown compound and 3. to confirm the structures of compounds with known or suspected structures. It has a wide range of applications in the diverse field such as **metabolomics** (e.g. global metabolic fingerprinting analysis, biomarker discovery and profiling, lipidomics studies and metabolic disorder profiling), **proteomics** (such as characterization of proteins and protein complexes, sequencing of peptides, and identification of post-translational modifications), **pharmaceutical** (e.g. drug discovery, pharmacokinetic and pharmacodynamic analyses, **forensic analysis** (such as identification of explosive residues) **environmental analysis** (such as pesticide screening and quantitation, soil contamination assessment).

Polypeptide sequencing

Sequence information of a polypeptide can be extracted using **tandem MS** or **MS-MS** (two mass spectrometers coupled in series). It is a fast and very sensitive method and can sequence not only 20 standard amino acids but also modified amino acids according to their mass. In this method, a solution containing the protein under investigation is first treated with a protease or chemical reagent to hydrolyze it to a mixture of shorter peptides. After proteolytic hydrolysis, the mixture is injected into the first mass spectrometer (MS-1). The function of MS-1 is to separate peptide ions of different masses. The peptide ion is further fragmented in a **collision cell** between the two mass spectrometers due to the high-energy impact with a collision gas. Finally, the m/z for each fragment is measured in the second mass spectrometer (MS-2).

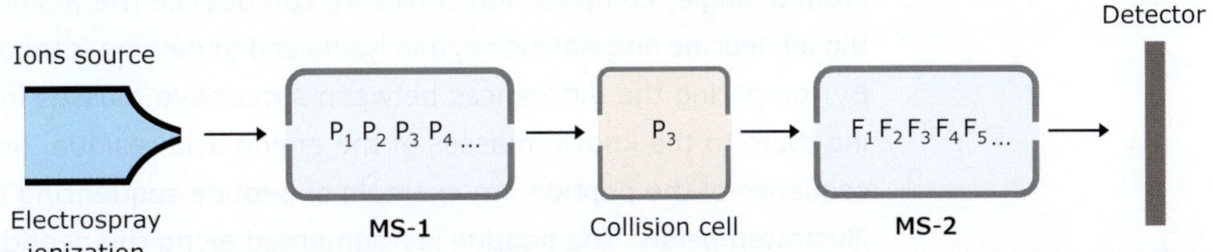

Figure 4.5 Tandem mass spectrometry in peptide sequencing. The ion source (ESI) generates gas-phase peptide ions, labeled P_1, P_2, etc., from the hydrolyzed protein. These peptides are separated by the MS-1 according to their m/z values, and one of them is directed into the collision cell, where it collides with helium atoms. This treatment breaks the peptide into fragments (F_1, F_2, etc.), which are directed into the MS-2 for determination of their m/z values. This latter process is carried out for each of the P_n peptide ions.

Three different types of bonds can break along the amino acid backbone: the NH–CH, CH–CO, and CO–NH bonds. Each bond breakage gives rise to two species – one neutral and the other one charged, and only the charged species is monitored by the mass spectrometer. The fragmentation of a peptide is shown in the figure below.

a_1 b_1 c_1 a_2 b_2 c_2 a_3 b_3 c_3

N-terminus $H_2N-CH(R_1)-C(=O)-NH-CH(R_2)-C(=O)-NH-CH(R_3)-C(=O)-NH-CH(R_4)-COOH$ C-terminus

x_3 y_3 z_3 x_2 y_2 z_2 x_1 y_1 z_1

The most common cleavage sites are at the CO–NH bonds (peptide bonds). However, bonds other than the peptide bond can be broken in the fragmentation process. The breakage of the peptide bonds gives two prominent sets of charged fragments. Those fragments that appear to extend from the amino terminus are termed 'b ions'. The fragment containing only the amino terminal amino acid is termed b_1. The fragment containing the first two amino terminal amino acids is termed b_2 ion, and so forth. Similarly, groups of peptide fragment ions appear to extend from the C-terminus are termed 'y ions'. Because the bond-breaking in the collision cells does not yield full carboxyl and amino groups at the sites of the breaks, the only intact α-amino and α-carboxyl groups on the peptide fragments are those at the very ends. The two sets of fragments can, thereby, be identified by the resulting slight differences in mass.

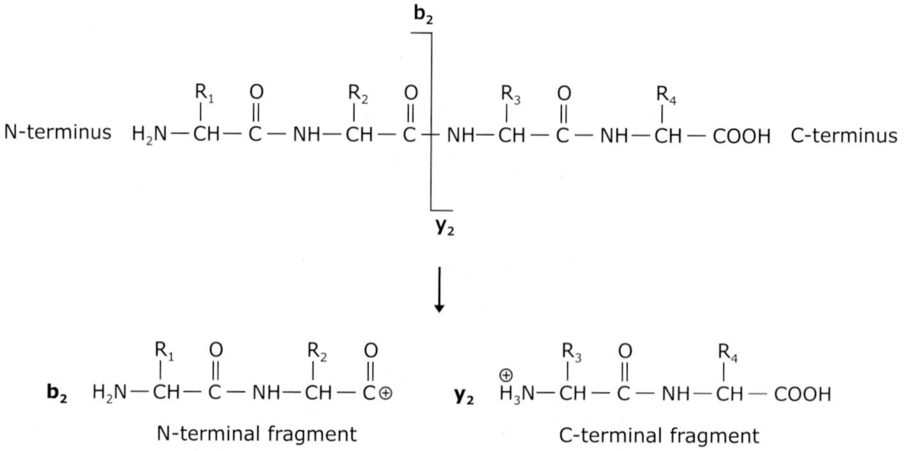

b_2

N-terminus $H_2N-CH(R_1)-C(=O)-NH-CH(R_2)-C(=O)$ | $NH-CH(R_3)-C(=O)-NH-CH(R_4)-COOH$ C-terminus

y_2

↓

b_2 $H_2N-CH(R_1)-C(=O)-NH-CH(R_2)-C^{\oplus}(=O)$ **y_2** $H_3N^{\oplus}-CH(R_3)-C(=O)-NH-CH(R_4)-COOH$

N-terminal fragment   C-terminal fragment

From a single, complete ion series we can deduce the amino acid sequence of a peptide, although leucine and isoleucine, and lysine and glutamine, cannot be differentiated in this manner. By comparing the differences between successive masses in each of the series and comparing them to the known masses of the amino acid residue, one can determine the amino acid sequence of the peptide. An example of peptide sequencing by tandem mass spectrometry is illustrated below. The peptide is fragmented along the peptide backbone to form **b** and **y-ions** and resulting fragment ions are measured to produce the MS-MS spectrum. Let us consider a following peptide to understand the concept.

Ala – Val – Ala – Gly – Cys – Ala – Gly – Ala – Arg

If one can identify either the y-ion or b-ion series in spectrum, the peptide sequence can be determined. The main idea during sequencing is to use the mass difference between two fragment ions to calculate the mass of an amino acid residue on the peptide backbone. The predicted b- and y-ion fragmentations for above mentioned peptide Ala-Val-Ala-Gly-Cys-Ala-Gly-Ala-Arg (AVAGCAGAR) is illustrated below. Starting from the N-terminus, cleavages are generating an ascending series of fragment ion corresponding to the b-series and a descending order of y-series fragments. For example, the b_4- and y_5-fragments are formed by the cleavage of peptide bond between Gly and Cys.

b_1 A VAGCAGAR **y_8**
b_2 AV AGCAGAR **y_7**
b_3 AVA GCAGAR **y_6**
b_4 AVAG CAGAR **y_5**
b_5 AVAGC AGAR **y_4**
b_6 AVAGCA GAR **y_3**
b_7 AVAGCAG AR **y_2**
b_8 AVAGCAGA R **y_1**

The mass difference between adjacent members of a series can be calculated. For example, the mass differences between y_7 – y_6 = 605.3 – 534.3 = 71 Da which is equivalent to amino acid residue alanine; and similarly, y_6 – y_5 = 534.3 – 477.0 = 57 Da which is equivalent to a glycine residue. The complete y-ions from y_8 to y_1 establishes the VAGCAGAR motif. The complete b-ions from b_8 to b_1 corresponds to the AVAGCAGA motif. Combining the results of the y- and b-ion series, the MS-MS spectrum provides definitive confirmation of the sequence AVAGCAGAR.

b_1 (72.08) b_2 (171.2) b_3 (242.0) b_4 (299.3) b_5 (401.4) b_6 (472.5) b_7 (529.6) b_8 (600.6)

Ala, A — Val, V — Ala, A — Gly, G — Cys, C — Ala, A — Gly, G — Ala, A — Arg, R

^+H_2N–C–C–N–C–C–N–C–C–N–C–C–N–C–C–N–C–C–N–C–C–N–C–C–N–C–COOH

y_8 (703.8) y_7 (604.6) y_6 (533.6) y_5 (476.5) y_4 (374.42) y_3 (303.34) y_2 (246.2) y_1 (175.2)

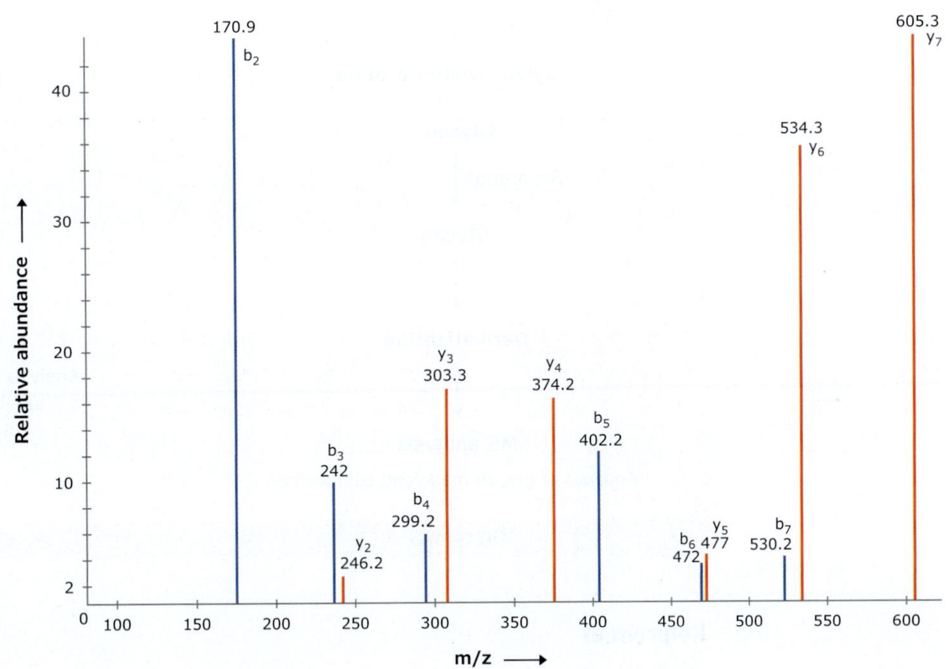

Figure 4.6 In an MS/MS, many copies of the same peptide are fragmented at the peptide backbone to form b- and y-ions. The spectrum consists of peaks at the m/z (mass-to-charge) values of the corresponding fragment ions. The b-series ions have been labeled with blue vertical lines and the y-series ions have been labeled with red vertical lines.

Characterization of glycoproteins

A **mass spectrum** is a graph of the relative amount of each ion (called the *relative abundance*) as a function of the ionic mass (or m/z). The peak in the spectrum with highest intensity is called the base peak.

A glycoprotein is a protein that contains covalently linked carbohydrates (or glycan). The nature of carbohydrates in glycoproteins may be monosaccharides, oligosaccharides, or polysaccharides. *N*- and *O*-linked glycosylated proteins are the most common forms of glycoproteins present in cells. Mass spectrometry is one of the most sensitive and powerful techniques for glycoprotein analysis – to identify the glycoproteins, evaluate glycosylation sites, and elucidate the glycan structures. The complete glycan structural elucidation includes the determination of the composition, sequence, branching, glycosidic linkages and anomeric configuration.

Glycan
According to IUPAC, the term glycan is used for polysaccharides meaning a compound consists of a large number of covalently linked monosaccharides. However, in general use, this term may also be used to refer to the carbohydrate portion of a glycoconjugate, such as a glycoprotein or a glycolipid, even if the carbohydrate is only an oligosaccharide.

Numerous approaches are used for the characterization of glycoprotein through mass spectrometry. However, two approaches are most common:

- Removal of carbohydrates from glycoprotein by chemical or enzymatic treatment and analysis of the deglycosylated protein and the released carbohydrates separately. The differences in protein–sugar linkage chemistry of *N*- and *O*-glycans require different approaches for the release of covalently-linked carbohydrates.
- Proteolytic digestion of the glycoprotein and analysis of the resulting glycopeptides.

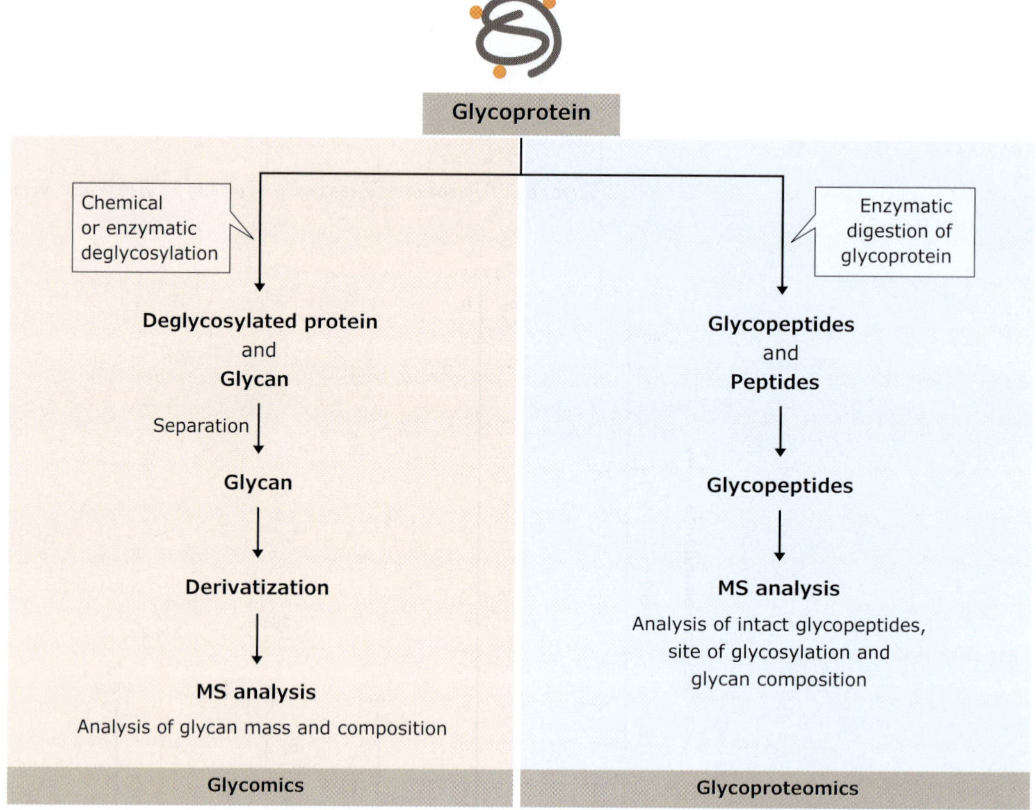

References

Banwell CN, McCASH EM (2013), *Fundamentals of Molecular Spectroscopy*. McGraw Hill Education.

Berg JM, Tymoczko JL and Stryer L (2007), *Biochemistry*, 6th ed. W.H. Freeman and Company.

Branden C and Tooze J (1999), *Introduction to Protein Structure*, 2nd ed. Garland Publishing Inc.

Domon B and Aebersold R (2006), *Mass Spectrometry and Protein Analysis*. Science 312: 212–217.

Hoffmann E and Stroobant V (2007), *Mass Spectrometry: Principles and Applications*, 3rd ed. John Wiley and sons.

Liebler DC (2002), *Introduction to Proteomics: Tools for the New Biology*. Humana Press Inc.

Morelle W and Michalski JC (2007), Analysis of protein glycosylation by mass spectrometry, Nature Protocols volume 2.

Pan S, Chen R, Aebersold R, and Brentnall TA (2011), Mass Spectrometry Based Glycoproteomics—From a Proteomics Perspective, by The American Society for Biochemistry and Molecular Biology, Inc.

White FA and Wood GM (1986), Mass Spectroscopy, Applications and Engineering, John Wiley.

5

Centrifugation

Centrifugation is a technique used to separate or concentrate materials suspended in a liquid medium by applying centrifugal force. The separation of particles is based on their differences in sedimentation rate under the centrifugal field. The sedimentation rate of particles depends on a number of different factors, including size, density, and shape. A device that generates centrifugal force by spinning the fluid at high speed and separate particles is called a **centrifuge**. In addition to the separation of molecules, it is also used to measure the physical properties (such as molecular weight, density, and shape). If centrifugation is used to separate one type of material from others, it is termed **preparative centrifugation**; whereas if it is used for measurement of physical properties of materials, then termed **analytical centrifugation.**

Principle of centrifugation

Particles suspended in a solution are pulled downward by Earth's force due to gravity (*or* simply, gravity). The force due to gravity depends only on the mass of the particles and not on charge, shape, and chemical composition. Because the Earth's gravitational field is weak, a solution containing particles of very small masses usually remains suspended due to random thermal motion. Hence, forces much larger than Earth's force due to gravity are required to cause appreciable sedimentation of such small masses. Such forces can be obtained by subjecting particles to centrifugation. By applying centrifugal forces, the sedimentation of these particles can be enhanced. A centrifuge does the same thing. It increases the sedimentation by generating centrifugal forces as great as 10,00,000 times the force of gravity.

Let us understand this principle by considering a solution being spun in a centrifuge tube. The centrifugal force acting on a solute particle of mass m,

Centrifugal force = $m\omega^2 r$

where, ω is the angular velocity in radians per second,

r is the distance from the center of rotation to the particle, and

$\omega^2 r$ is the centrifugal acceleration.

A particle will move through a liquid medium when subjected to a centrifugal force. Hence, we must also consider the particle's buoyancy due to the displacement of the solvent molecules by the particle. This buoyancy reduces the force on the particle by $\omega^2 r$ times the mass of the displaced solvent.

Buoyant force = $m_0\omega^2 r$

where, m_0 is the mass of fluid displaced by the particle

$m_0 = m\bar{v}\rho$

Where, ρ is the density of the solution (g/mL) and $\bar{v}$ is the *partial specific volume* of the particle. As measuring the volume of a very small particle is difficult, for convenience we use a term called partial specific volume. It is the volume in mL that each gram of the particle occupies in solution. Thus, the net force acting on the particle downward is given by,

$$\begin{aligned} \text{Net force} &= \text{Centrifugal force} - \text{Buoyant force} \\ &= m\omega^2 r - m_0\omega^2 r \\ &= m\omega^2 r - m\bar{v}\rho\,\omega^2 r \end{aligned}$$

When particles move downward through the solution, the motion is also opposed by the frictional force. The frictional force is equal to the product of the *frictional coefficient*, f ($Nm^{-1}s$) and the sedimentation velocity, v. It acts in the opposite direction to the net force.

$F_{friction} = f.v$

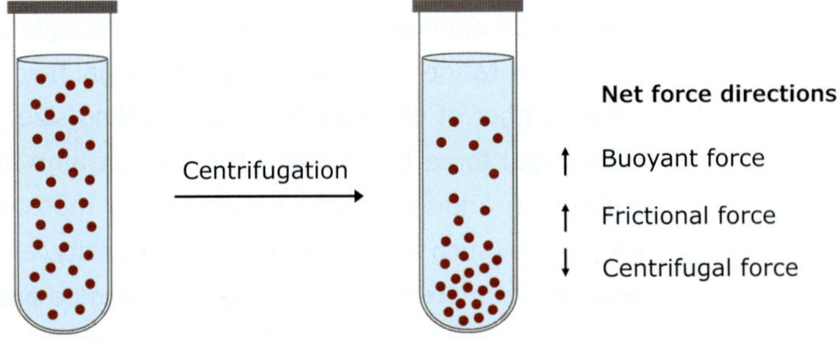

Figure 5.1 Direction of different forces and distribution of solute particles during centrifugation.

At steady state, then, the *frictional force* is equal to the net force and the molecule moves with velocity v downward:

$fv = m\,\omega^2 r - m\bar{v}\rho\,\omega^2 r$

$fv = m\,\omega^2 r\,(1 - \bar{v}\rho)$

Since, m (mass of a single particle in grams) = M/N.

Where, M is the molar weight of the solute in g/mol and N is the Avogadro's number.

$$fv = \frac{M}{N}\,\omega^2 r\,(1 - \bar{v}\rho)$$

$$v = \frac{M\,\omega^2 r\,(1 - \bar{v}\rho)}{Nf}$$

Now, the *sedimentation coefficient*,

$$s = \frac{v}{\omega^2 r} = \frac{M}{N}\,\frac{(1 - \bar{v}\rho)}{Nf}$$

The **sedimentation coefficient** is the ratio of a velocity to the centrifugal acceleration. The sedimentation coefficient (s) has units of second. A sedimentation coefficient of 1×10^{-13} second is defined as one **Svedberg**, S, (1S = 10^{-13} second). This unit is named for The Svedberg, a pioneer in the field of centrifugation.

Important conclusions drawn from the equation:

1. The sedimentation velocity of particle is proportional to the particle size (mass).
2. The sedimentation velocity is proportional to the difference in density between the particle and the medium. It is zero when the particle density is the same as the medium density.
3. The sedimentation velocity depends on the density of the medium (ρ). Particle sinks when $\bar{v}\rho < 1$ and floats when $\bar{v}\rho > 1$.
4. The sedimentation velocity also depends on the shape of particles since the particle's shape affects the frictional coefficient.

Sedimentation coefficients of biological macromolecules are normally obtained in buffered solutions whose viscosity and density may differ from those of water. It is also measured at different temperatures. Hence, we standardize the sedimentation coefficient value in standard conditions i.e. pure water at 20°C (denoted by $s_{20,w}$).

$$s_{20,w} = s_{exp} \frac{1 - \bar{v}\rho_{20,w}}{1 - \bar{v}\rho_{exp}} \frac{\eta_{exp}}{\eta_{20,w}}$$

Where,

$s_{20,w}$ = the sedimentation coefficient of the molecule in pure water at 20°C,

s_{exp} = the experimentally measured sedimentation coefficient of the molecule,

η_{exp} = the viscosity of the solvent at the experimental temperature T (°C),

$\eta_{20,w}$ = the viscosity of water at 20°C,

$\rho_{20,w}$ = the density of water at 20°C,

ρ_{exp} = the density of the solvent at given temperature T (°C) and

$\bar{v}$ = the partial specific volume of the molecule.

Relative centrifugal field

Particles suspended in a fluid move, under the influence of gravity, towards the bottom of a vessel at a rate that depends, in general, on their size and density. Centrifugation utilizes centrifugal forces which are greater than the Earth's gravitational force to increase the sedimentation rate of particles. This is achieved by spinning the vessel containing the fluid and particles about an axis of rotation so that the particles experience a centrifugal force acting away from the axis. The force acting on particles during centrifugation is called **relative centrifugal force** (RCF) or more commonly, the **g force** and it is expressed as multiples of the Earth's gravitational force. For example, an RCF of 500 ×g indicates that the centrifugal force applied is 500 times greater than Earth's gravitational force.

Relationship between RCF, revolutions per minute and radius of rotation

The RCF generated by a rotor depends on the speed of the rotor in *revolutions per minute* (rpm) and the *radius of rotation* (i.e. the distance from the axis of rotation). The equations that permit calculation of the RCF from a known rpm and radius of rotation is:

$$RCF = r\omega^2/g$$

where, r is the radius in cm, g is the gravitational acceleration (cm/sec^2) and ω is the angular velocity.

$$\text{Angular velocity } (\omega) = \frac{2\pi \times rpm}{60} \text{ radians/sec}$$

$$RCF = \frac{r}{g} \times \left[\frac{2\pi \times rpm}{60}\right]^2 = r \times (rpm)^2 \times \left[\frac{4 \times 3.14 \times 3.14}{60 \times 60 \times 980}\right] = 1.12 \times r \times (rpm)^2 \times 10^{-5}$$

RCF describes and compares the strength of the fields generated by different-sized rotors and different operating speeds. A centrifuge with a higher radius of rotation will spin samples at a higher RCF. For example, when revolving at 2000 rpm, a larger centrifuge with a higher radius of rotation will spin samples at a higher RCF than a smaller centrifuge with a shorter radius of rotation.

Fixed-angle or **swing-bucket rotors**

Two types of rotors are commonly used in the laboratory – *fixed-angle* and *swing-bucket* rotors. However, these rotors are fundamentally different in their geometry and consequently in their sedimentation properties. In the following diagram, major differences between these rotors have been shown.

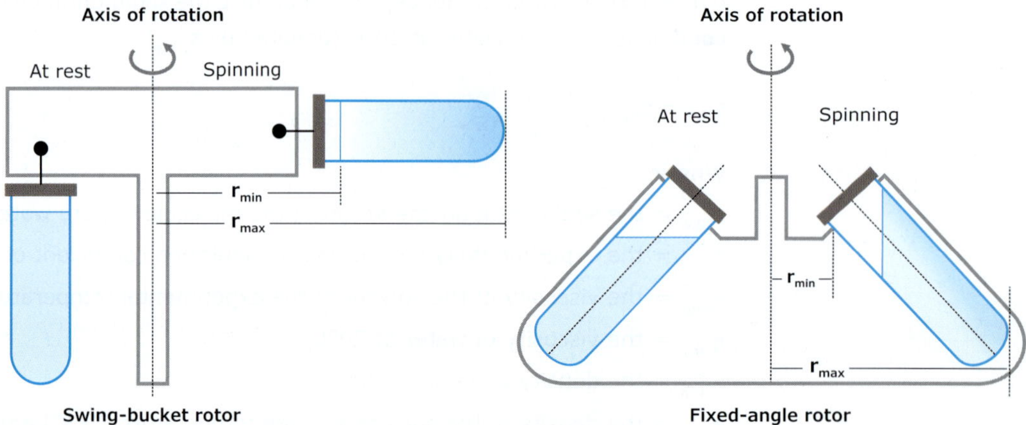

Figure 5.2 Schematic sketches of the two types of rotors: *swinging bucket* and *fixed angle* rotors.

Differential centrifugation

Differential centrifugation is the simplest form of centrifugation. Particles of different densities or sizes present in a suspension will sediment at different rates, with the largest and most dense particles sedimenting fast followed by less dense and smaller particles. These sedimentation rates can be increased by using centrifugal force. This centrifugation process is used mainly for the separation of sub-cellular components.

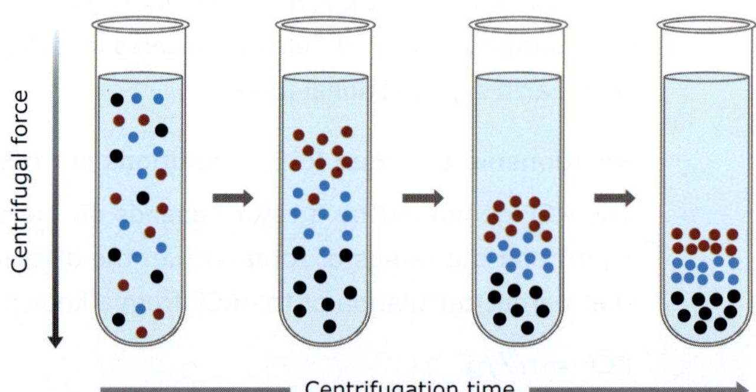

Figure 5.3 Differential centrifugation separates particles based on differences in sedimentation rate, which reflect differences in size or density. Particles which are large or dense sediment rapidly, those that are intermediate in size or density sediment less rapidly, and the smallest or least dense particles sediment very slowly. Eventually, all of the particles reach the bottom of the tube.

Separation of sub-cellular components by differential centrifugation: In the process of separation of sub-cellular components, the preparation of broken cells is poured into a centrifuge tube and is initially centrifuged at low centrifugal force long enough to completely sediment the largest and heaviest sub-cellular component. The supernatant obtained is carefully decanted and is again centrifuged at a higher centrifugal force for sedimenting the next heavier entity in the extract. This process is continued and at each ensuring step the *centrifugal force* as well as the *centrifugal time* is increased to successively sediment the lighter components and particles. In this method, the separation occurs due to the differential sedimentation rates of the sub-cellular organelles because of the differences in their sizes and densities.

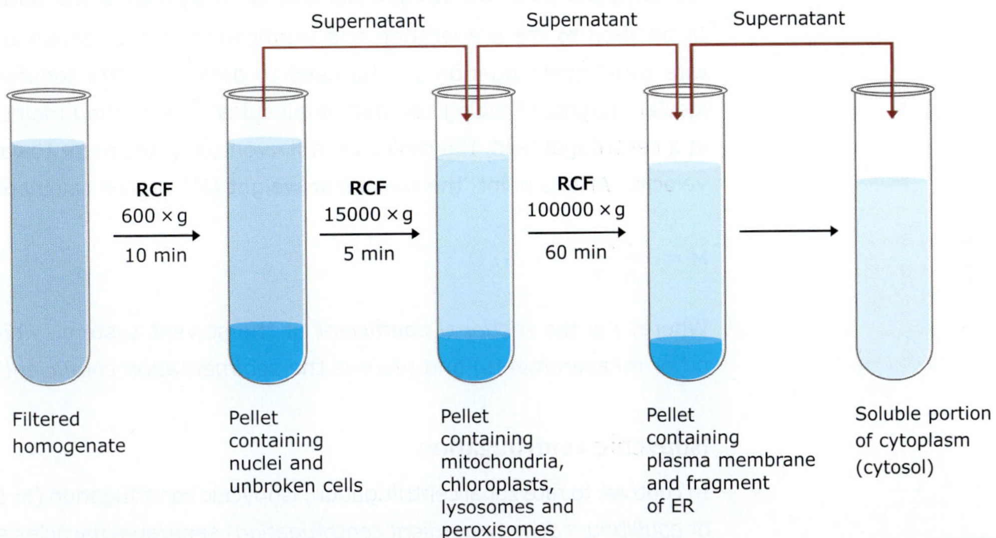

Figure 5.4 Differential centrifugation and separation of cell organelles. The different sedimentation rate of various cellular components make their separation possible. The tissue of interest is first homogenized. Subcellular fractions are then isolated by subjecting the homogenate and subsequent supernatant fractions to successively higher centrifugal forces and longer centrifugation times. During each step of the centrifugation process, particles of a given size and density are removed as a pellet from suspension. Each time, the supernatant from one step is decanted into a new centrifuge tube and subjected to greater centrifugal force to obtain the next pellet.

Density gradient centrifugation

In differential centrifugation, the particles about to be separated were uniformly distributed throughout the solution prior to centrifugation. Density gradient centrifugation is a variation of differential centrifugation in which the sample is centrifuged in a medium that gradually increases in density from top to bottom. The gradient consists of an increasing concentration of solute (and therefore density) from the top of the tube to the bottom. It is of two types: **Rate zonal centrifugation**, in which the sample is centrifuged in a preformed gradient and **isopycnic centrifugation**, in which a self-generating gradient forms during centrifugation.

Rate zonal centrifugation

Ultracentrifuges are capable of speeds in excess of 30,000 rpm and RCFs of over 600,000 × g.

Rate zonal centrifugation (also termed *velocity centrifugation*) is used to separate particles on the basis of differences in their sedimentation rate. The separation of particles occurs according to their size or density. Although sedimentation rate is strongly influenced by the size (mass) and density of particles, even slight variation in shape also affects sedimentation rate. In the rate zonal centrifugation method, materials used for the preparation of density gradients are sucrose, glycerol, ficoll, etc. A 5–20% sucrose solution is commonly used to form a density gradient. The density range is chosen so that the density of the particles is greater than the density of the medium at all points during the separation. In this centrifugation process, the function of the density gradient is to stabilize the bands of particles and improves particle resolution by inhibiting convection currents.

In this centrifugation process, samples are centrifuged just long enough to separate the molecules of interest into discrete zones. However, since the density of the particles is greater than the density of the gradient, hence if a sample is centrifuged much longer than necessary,all the components of the sample will end up in a pellet at the bottom of the tube.

In addition to the preparation and purification of macromolecules and cellular components, rate zonal centrifugation can be used to determine the sedimentation coefficients and molecular weights of biological macromolecules. If a purified molecule such as a protein is spun in a centrifugal field, the molecule will eventually sediment towards the bottom at a constant velocity. At this point, the molecular weight (M) can be calculated as:

$$M = \frac{f \times v}{\omega^2 r}$$

Where, f is the *frictional coefficient* of the solvent system (which has been calculated from other measurements) and $v/\omega^2 r$ is the *sedimentation coefficient*.

Isopycnic centrifugation

In contrast to rate zonal centrifugation, isopycnic centrifugation (or *buoyant density centrifugation* or *equilibrium density-gradient centrifugation*) separates particles solely on the basis of buoyant density and is independent of the shape and size of particles. It is also independent of the time of centrifugation. This technique is used to separate particles of similar size but different densities. Sedimentation of particles in density gradient occurs until the buoyant density of the particle is equal to the density of the gradient. A steep density gradient containing sucrose (20–70%) or CsCl is generally used.

Table 5.1 Density gradient media

	CsCl	Sucrose
Molecular weight	168.36	342.30
Solubility (g ml^{-1} in water, 20°C)	1.2	2.0
Maximum density (g ml^{-1} in water, 20°C)	1.91	1.32

The *zonal* and *isopycnic centrifugation* methods are often confused because both utilize gradients through which the sedimenting molecules move. However, the density gradient in rate zonal centrifugation plays no direct role in the separation of the sedimenting molecules while the density gradient performs the separation in isopycnic centrifugation. Another distinction is that when the centrifugation time of a zonal centrifugation is too long, all of the sedimenting molecules will be found on the bottom of the centrifuge tube. In contrast, lengthening the centrifugation time in isopycnic centrifugation will have no effect.

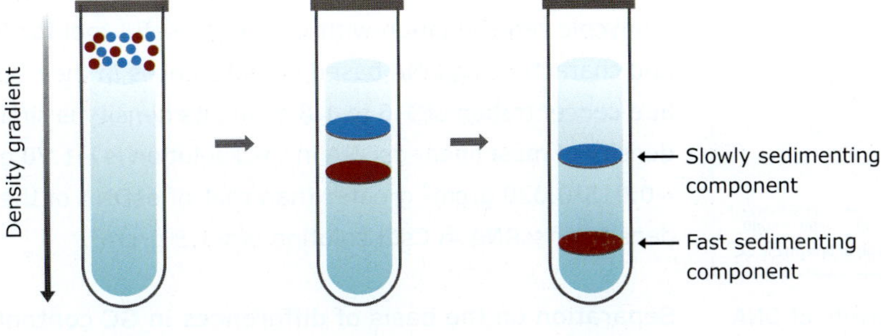

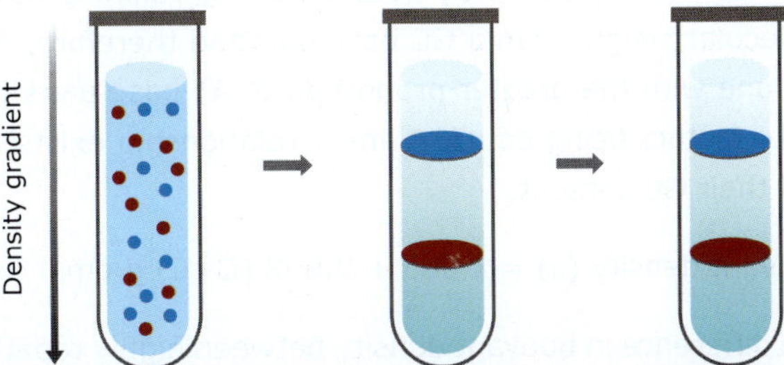

Figure 5.5 (a) Rate-zonal centrifugation is a variation of differential centrifugation in which the sample for fractionation is placed as a thin layer on top of a gradient of solute. The gradient consists of an increasing concentration of solute—and therefore density—from the top of the tube to the bottom. When subjected to a centrifugal force, particles differing in size or density move downward as discrete bands, that migrate at different rates. Because of the gradient of solute in the tube, the particles at the leading edge of each zone continually encounter a slightly denser solution and are, therefore, slightly impeded. As a result, each zone remains very compact, maximizing the resolution of different particles.
(b) Isopycnic centrifugation also includes a gradient of solute that increases in concentration and density, but in this case the solute is concentrated so that the density gradient spans the range of densities of the particles about to be separated. During centrifugation, the particles move into the gradient until each reaches its equilibrium (or buoyant) density—the point in the gradient at which the density of the particle is exactly equal to the density of the gradient.

When macromolecules present in the CsCl solution are centrifuged, they form bands at distinct points in the gradient and exactly where a particular macromolecule forms band will depend on its buoyant density. DNA has a buoyant density of about 1.70 g/cm^3, and therefore it will migrate to the point in the gradient where the CsCl density is also 1.70 g/cm^3. In contrast, protein molecules have much lower buoyant densities and so float at the top of the tube, whereas RNA forms a pellet at the bottom of the tube due to high buoyant density. Isopycnic centrifugation can therefore separate DNA, RNA and protein.

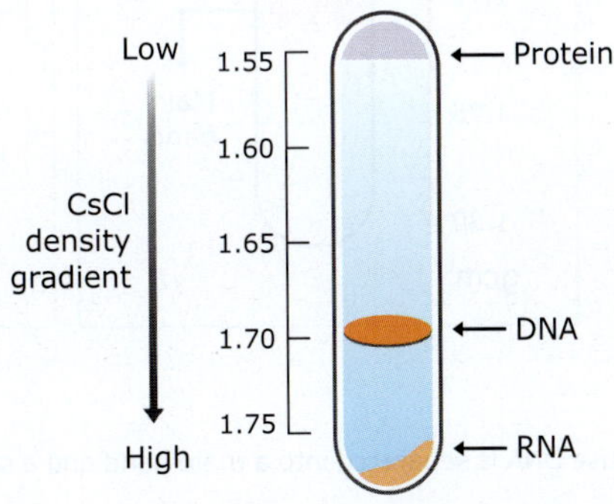

Figure 5.6 CsCl equilibrium density gradient centrifugation. Separation of protein, DNA and RNA in a density gradient.

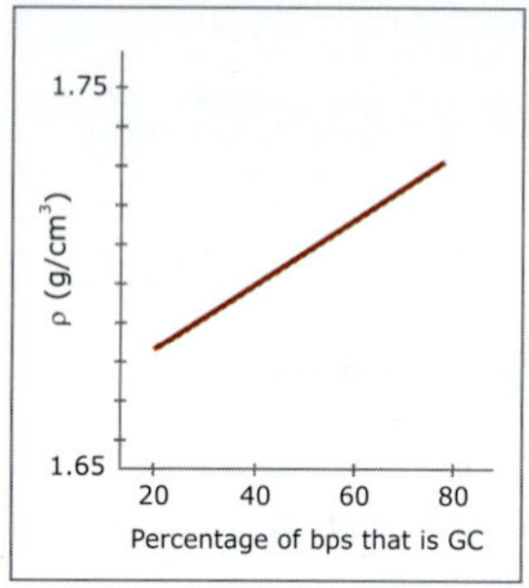

Buoyant density of DNA as a function of GC content.

DNA separation by isopycnic centrifugation

Isopycnic centrifugation with CsCl is a useful tool for fractionating, quantitatively separating and characterizing DNA based on differences in their buoyant densities. CsCl is used because, at a concentration of 1.6 to 1.8 g/mL, its density is similar to the density of DNA. The buoyant density of most linear dsDNA in CsCl solution is ~1.70 g/cm^3. Whereas the density of ssDNA is ~0.015-0.020 g/cm^3 greater than that of dsDNA of the same base composition. The buoyant density of ssRNA in CsCl solution is >1.8 g/cm^3.

Separation on the basis of differences in GC content

The DNA molecules having differences in the relative proportions of AT base pairs to GC base pairs can be separated by CsCl density gradient centrifugation. An AT base pair has a lower molecular weight than a GC base pair and therefore, for two DNA molecules of equal length, the one with the greater proportion of AT base pairs will have a lower buoyant density if all other factors being equal. A linear relationship exists between the buoyant densities of DNA and their GC content.

Buoyant density (ρ) = 1.660 + 0.098 (G+C) g/cm^3

The difference in buoyant density between highly repetitive satellite DNA from the rest of DNA is used to separate these two by CsCl density gradient centrifugation. DNA that consists of very large numbers of short tandem repeats (termed **satellite DNA**) may have a base composition (and thus GC content) different from that of the genome as a whole. If so, the satellite DNA will have a different buoyant density from the rest of the DNA, as this property depends on the base composition. DNA may be fractionated according to density by CsCl equilibrium density gradient centrifugation. Each fraction of DNA forms a band at the position corresponding to its own density. If the GC content varies by 5% or more, separate bands are obtained. For example, when fragments of mouse nuclear DNA are centrifuged on a CsCl density gradient, two DNA bands are seen. One contains 92% of the DNA with a density of 1.701 gm/cm^3 (GC content ~42%) and the thin satellite band contains 8% of the DNA with a density of 1.690 gm/cm^3 (GC content ~30%). Satellite DNA was originally defined by this density separation. However, in cases where the average satellite DNA base composition is close to that of the genome as a whole, the satellite DNA cannot be physically separated using a density gradient.

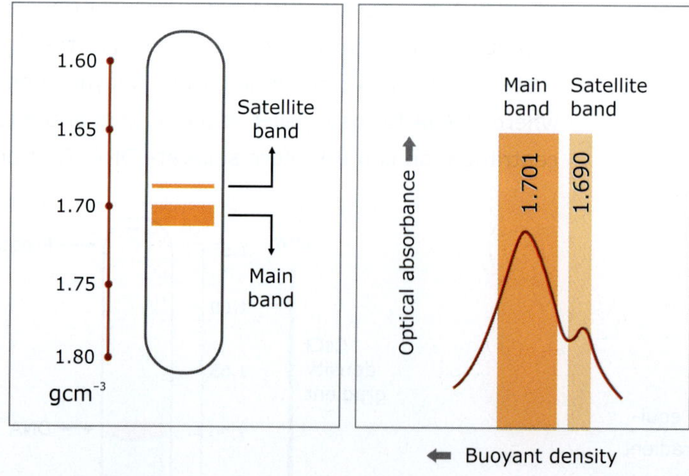

Figure 5.7 Mouse DNA is separated into a main band and a satellite band by CsCl density gradient centrifugation.

EtBr is a planar compound which intercalates between base pairs in the DNA double helix. Each molecule of EtBr which intercalates causes the double helix to unwind, decreasing Tw and increasing Wr. When one molecule of EtBr intercalates in the DNA double helix, the helix untwists by 26°. Addition of 14 EtBr molecules to a DNA molecule results in the unwinding of one full turn. Thus, Tw decreases by one.

Separation on the basis of conformation

In the laboratory, CsCl equilibrium density gradient centrifugation is also used to separate linear DNA (relaxed) molecules from circular supercoiled DNA. It is carried out in the presence of DNA intercalating dye EtBr. DNA-EtBr intercalation causes the unwinding of the DNA helix, which reduces the buoyant density of DNA by as much as 0.125 g/cm^3 for linear DNA. However, covalently closed circular supercoiled DNA, with no free ends, has very little freedom to unwind and can only bind a limited amount of EtBr. As a result, the decrease in buoyant density of a circular supercoiled DNA is much less, only about 0.085 g/cm^3. In contrast, linear DNA molecules with free ends are not as topologically constrained and can bind more EtBr molecules, resulting in a more decrease in buoyant density. Consequently, circular supercoiled DNA form a band in an EtBr-CsCl gradient at a different position to linear DNA.

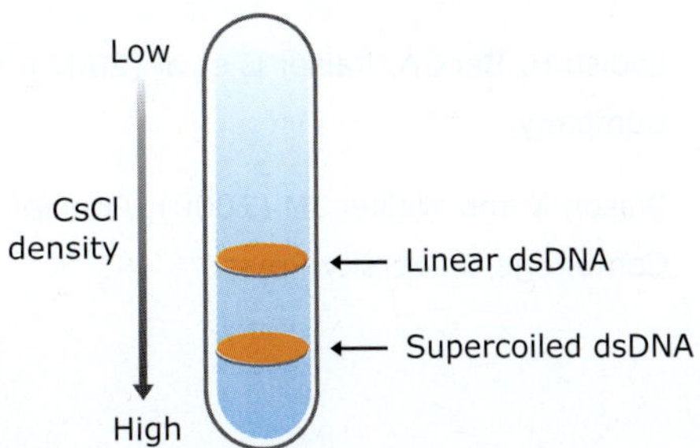

Figure 5.8 The separation of covalently closed supercoiled DNA from non-supercoiled linear DNA molecules by equilibrium density gradient centrifugation in the presence of ethidium bromide (EtBr) is shown here. EtBr binds to DNA molecules by intercalating between adjacent base pairs, causing partial unwinding of the double helix. This unwinding results in a decrease in the buoyant density. However, supercoiled circular DNA, with no free ends, has very little freedom to unwind and can only bind a less amount of EtBr as compared to linear dsDNA. The decrease in buoyant density of a supercoiled molecule is, therefore, much less. As a consequence, supercoiled molecules form a band in an EtBr-CsCl gradient at a different position to linear DNA.

References

Becker WM, Kleinsmith LJ and Hardin J (2006), *The world of the cell*, 6th ed. Pearson Education Inc.

Berg JM, Tymoczko JL and Stryer L (2006), *Biochemistry*, 6th ed. W.H. Freeman and Company.

Chang R (2005), *Physical Chemistry for the Biosciences*, University Science Books.

Freifelder D (1982), *Physical Biochemistry*, 2nd ed. W.H. Freeman and Company.

Harrison RG, Todd, Paul, Rudge, Scott R and Petrides DP (2003), *Bioseparations - Science and Engineering*. Oxford University Press, 2003.

Howlett GJ, Minton AP and Rivas G (2006), Analytical ultracentrifugation for the study of protein association and assembly. *Current Opinion in Chemical Biology* Vol. 10, pp. 430–436.

Lodish H, Berk A, Kaiser C et al (2007), *Molecular Biology of the Cell*, 4th ed. W.H. Freeman and Company.

Wilson K and Walker JM (2000), *Principles and Techniques of Practical Biochemistry*, 5th ed. Cambridge University Press.

6

Microscopy

Microscopy is a technique for making very small things visible to the unaided eye. An instrument used to make the small things visible to the naked (unaided) eye is called a **microscope**. There are two fundamentally different types of microscopes based on the source of illumination: the light microscope and the electron microscope.

6.1 Light microscope

Light or *optical microscope* uses visible light as a source of illumination. Because the light travels through the specimen, this instrument can also be called a *transmission light microscope*.

The simplest form of light microscope consists of a single lens, a magnifying glass. Microscope made up of more than one glass lens in combination is termed **compound microscope**. Compound microscope includes the *condenser lens*, the *objective lens* and the *eyepiece lens*. **Condenser lens** focuses the light from the light source at the specimen. The one facing the object is called the **objective** and the one close to the eye is called the **eyepiece**. The objective has a smaller aperture and smaller focal length than those of the eyepiece (also referred to as the **ocular**). A compound microscope with a single eyepiece is said to be **monocular** and one with two eyepieces is said to be **binocular**. The word 'compound' refers to the fact that two lenses, the objective lens and the eyepiece, work together to produce the final magnified image.

The objective lens is responsible for producing the magnified image. It is available in different varieties (4x, 10x, 20x, 40x, 60x, 100x). The power of a lens is described with a number followed by the letter 'x'. For example, if through a microscope one can see something 25 times larger than actual size, its magnification power is 25x. The eyepiece works in combination with the objective lens to further magnify the image. Eyepieces usually magnify by 10x, since an eyepiece of higher magnification merely enlarges the image, with no improvement in resolution.

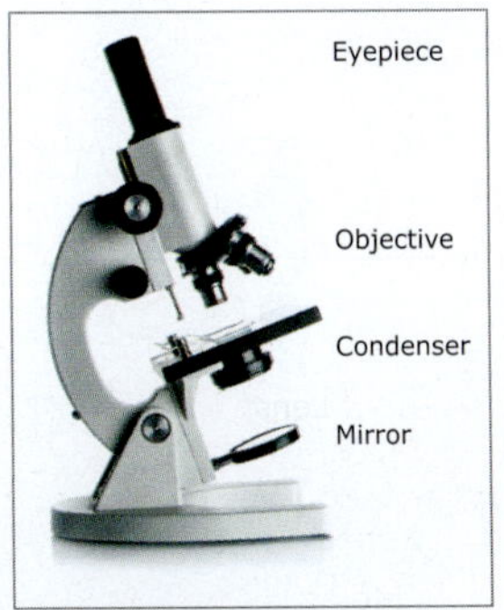

Credit: Triff / Shutterstock

Compound microscope

In *compound microscope*, the objective lens forms a real and inverted magnified image of the object (called the *real intermediate image*) in the focal plane of eyepiece. This image works as an object for the eyepiece. It magnifies the intermediate image. The image produced by the eyepiece is a magnified virtual image. It is erect with respect to the first image and hence, inverted with respect to the object. The eye views the virtual image created by the eyepiece, which serves as the object for the lens in the eye. The virtual image formed by the eyepiece is well outside the focal length of the eye, so the eye forms a real image on the retina.

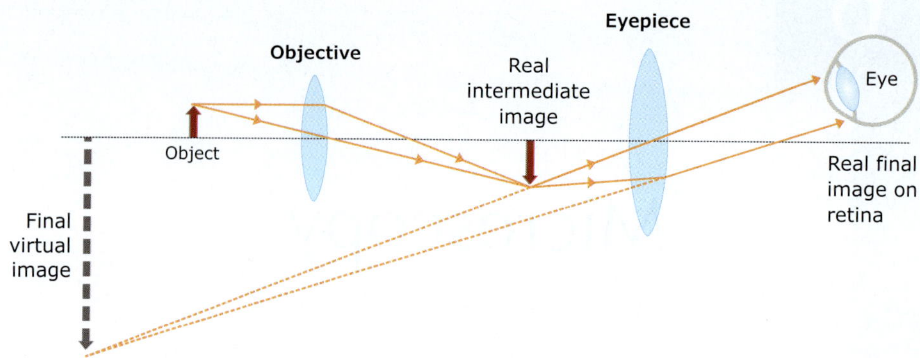

Figure 6.1 Compound microscope and the ray diagram for image formation. It consists of two converging lenses arranged coaxially. The one facing the object is called the objective and the other close to eye is called the eyepiece.

Magnification

The magnification of a compound microscope is the product of the magnification of the objective and the eyepiece. The magnification produced by the *objective lens* is called the **linear magnification** of the compound microscope, because it is measured in linear dimensions. It is the ratio of the image size to the object size.

Linear magnification by objective lens

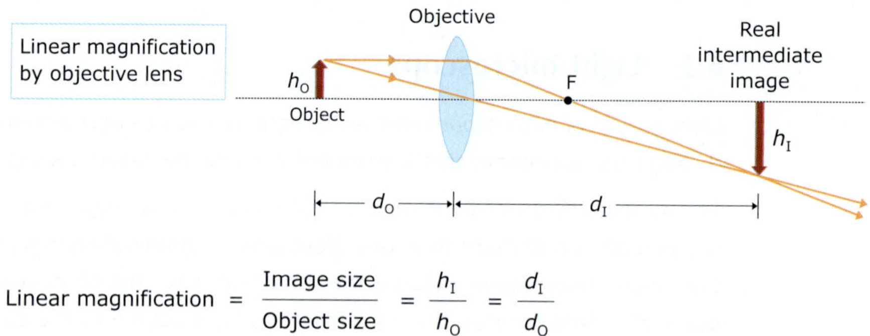

$$\text{Linear magnification} = \frac{\text{Image size}}{\text{Object size}} = \frac{h_I}{h_O} = \frac{d_I}{d_O}$$

The magnification produced by the *eyepiece* is called the **angular magnification**. This is also known as the **magnifying power**. It is the ratio of the angle subtended by the image on the eye when the microscope is used and the angle subtended by the object without microscope (i.e. naked eye).

Magnification means how large the image is as compared to the actual specimen size. In contrast, the **resolution** is the smallest distance below, which we will see two discrete objects as one. Magnification is a function of the number of lenses. Resolution is a function of the ability of a lens to gather light.

The *maximum magnification* of the compound light microscopes is usually 1500x and has a limit of resolution of about 0.2 μm.

Angular magnification by eyepiece

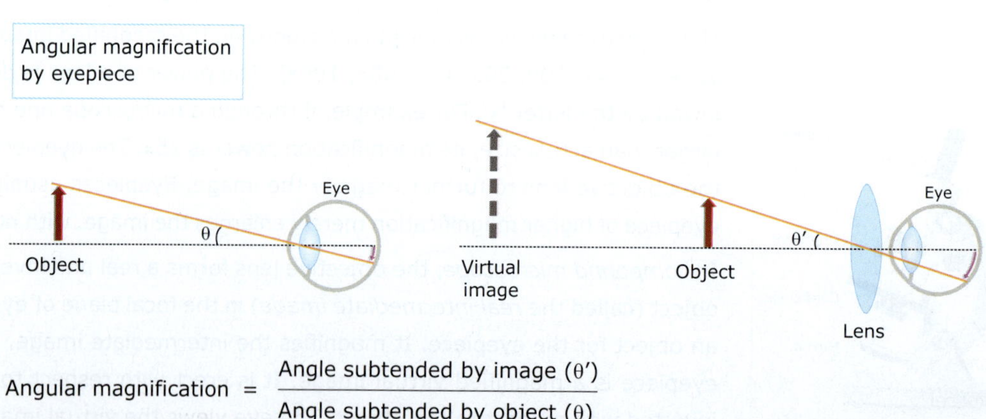

$$\text{Angular magnification} = \frac{\text{Angle subtended by image } (\theta')}{\text{Angle subtended by object } (\theta)}$$

The overall magnification is the product of the linear magnification of the objective lens and the angular magnification of the eyepiece with the first image at the focal length.

Light interacts with matter in a variety of ways. Interaction of light with an object may result in absorption, reflection, scattering, diffraction and refraction. The light microscope creates a magnified image of the specimen based on the absorption and diffraction of light waves. If light passes through an object but does not become absorbed or diffracted, it remains invisible.

Refractive index: When light enters from one medium to another, its speed changes. Light travels slowly in a matter as compared to a vacuum. The ratio of the speed of light in a vacuum to that in a matter is called *refractive index* (or *index of refraction*).

$$\text{Refractive index} = \frac{\text{Speed of light in vacuum}}{\text{Speed of light in matter}}$$

The refractive index of vacuum and water at room temperature is 1 and 1.33, respectively, which means that light travels 1.33 times slower in water than in a vacuum. As the speed of light reduces in the medium of higher refractive index, the wavelength is shortened proportionately, but the frequency remains unchanged.

Refraction: The bending of a light ray when it enters a medium where its speed is different is called *refraction*. The direction of bending depends on the refractive indices of the two media. When light enters the medium with the higher refractive index, it bends toward a line drawn normal to the surface. Conversely, when light exits the medium with the higher refractive index, it bends away from the normal line. The amount of bending depends on the differences in refractive indices of the two media. *Refraction is different from reflection.* **Reflection** occurs when a ray of light hits reflective surfaces and bounces back by changing the direction.

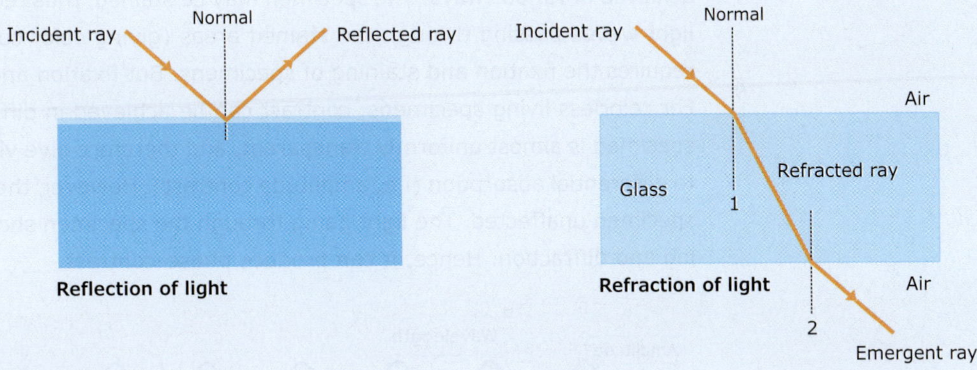

Scattering and diffraction: *Scattering* is the absorption of light by the particles followed by its re-radiation in different directions. It should be distinguished from the absorption of light as it passes through a medium. In absorption, the light energy is converted into internal energy of the medium, whereas, in scattering, the light energy is radiated in other directions. The strength of scattering depends on the wavelength of light and the size of the particles that cause scattering. The bending or spreading of light that occurs when light waves interact with objects is called **diffraction**. There are two primary sources of diffraction in the microscope: one at the specimen itself and another in the aperture of the objective lens.

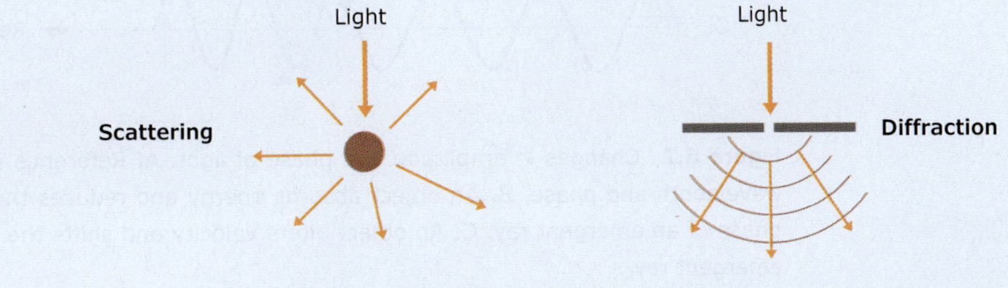

Refractive index

Vacuum = 1

Air = 1.0003

Water = 1.33

Glycerol = 1.47

Immersion oil = 1.51

Optical glass = 1.51

Diamond = 2.417

Amplitude contrast and phase-contrast

Both living and dead specimens are viewed with a light microscope. The visibility of the magnified specimen depends on *contrast* and *resolution*. In general, *contrast* is the differences in the light intensity within an image. It can be **amplitude contrast** (depend on the differences in amplitude of the transmitted light) or **phase-contrast** (depend on the phase difference of the transmitted light). As we know, the light intensity is related to the amplitude and it is proportional to the square of amplitude.

Specimens seen under transmission light microscope can be categorized into **amplitude objects**, (such as stained specimen) and **phase objects** (such as transparent colorless objects e.g. living cells). Amplitude objects directly produce amplitude differences in the image that are detected as differences in the intensity. This occurs due to differential absorbance of light. Amplitude object can produce *color contrast* or *brightness contrast*. **Color contrast** occurs when the degree of absorption depends on the wavelength and varies from point-to-point in the stained specimens or specimens that have natural pigmentation. Absorption of light may also occur if specimens are thick enough to absorb a significant amount of light despite being colorless. Such specimens give **brightness contrast** which arises due to different degrees of absorption at different points in the specimen. Although *phase objects* do not absorb light, they do diffract light and cause a phase shift in the rays of light passing through them. Some light microscopes transform differences in phase of light to amplitude differences in the image.

For colorless transparent specimens, as is the case for most biological material, contrast is achieved in various ways. The specimen may be stained, thus reducing the amplitude of certain light waves passing through the stained areas (giving color contrast). However, this usually requires the fixation and staining of specimens. But fixation and staining kills the specimens. For colorless living specimens, contrast can be achieved in different ways. A living biological specimen is almost uniformly transparent, and therefore give very low-intensity variation due to differential absorption (i.e. amplitude contrast). However, the light does not go through the specimen unaffected. The light going through the specimen shows phase shift due to scattering and diffraction. Hence, it can produce phase-contrast.

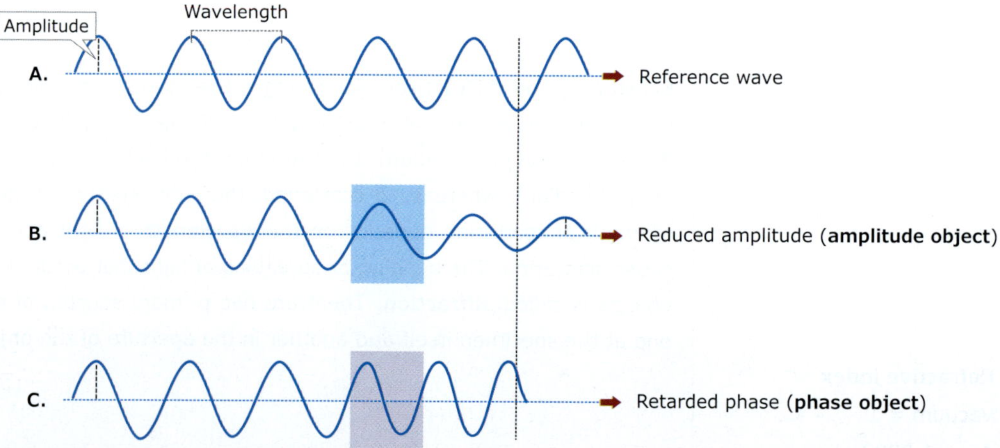

Figure 6.2 Changes in amplitude and phase of light. A. Reference wave with characteristic amplitude, wavelength and phase. B. An object absorbs energy and reduces the amplitude, but does not alter the phase of an emergent ray. C. An object alters velocity and shifts the phase, but not the amplitude of an emergent ray.

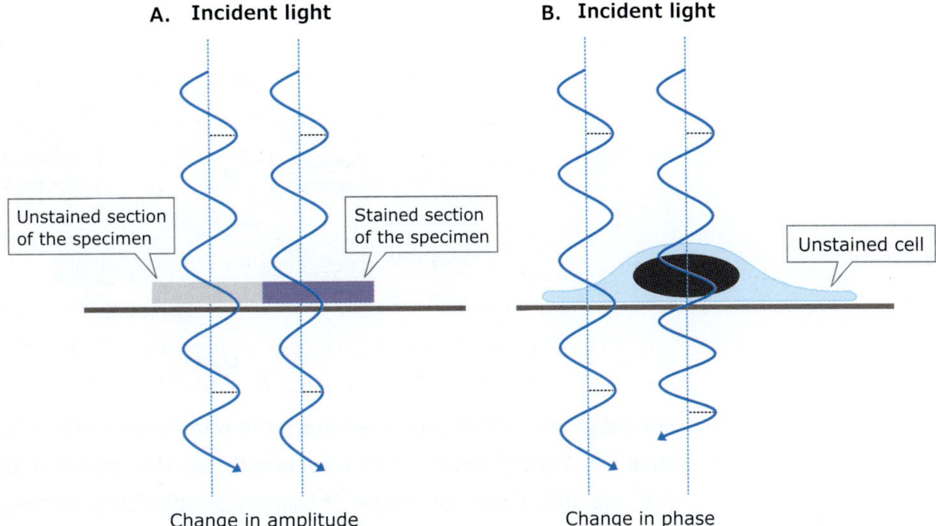

Figure 6.3 Two ways to obtain contrast in light microscopy. **A.** The stained portions of the specimen reduce the amplitude of light waves of particular wavelengths passing through them. A colored image of the specimen is, thereby, obtained that is visible in the ordinary way. **B.** Light passing through the unstained, living cell undergoes very little change in amplitude due to absorption and the structural details cannot be seen even if the image is highly magnified. The phase of the light, however, is altered by its passage through the cell, and small phase differences can be made visible by exploiting interference effects using a phase-contrast or a differential-interference-contrast microscope.

Resolving power

Resolving power is the ability of magnifying instrument to distinguish two objects that are close together. The resolving power is inversely related to the *limit of resolution*. The limit of resolution is defined as the minimum distance between two points that allows for their discrimination as two separate points. Thus, the higher the resolving power, the smaller the limit of resolution. The *limit of resolution* of the light microscope depends upon the three factors:

- *Wavelength* of the light used to illuminate the specimen
- *Angular aperture*
- *Refractive index* of the medium between the cover slip and the objective lens

The effect of these three variables on the limit of resolution is described quantitatively by the following equation known as the **Abbe equation**:

$$\text{Limit of resolution} = \frac{0.61\lambda}{n \times \sin\alpha} = \frac{0.61\lambda}{\text{Numerical aperture}}$$

The quantity $n \times \sin\alpha$ is called the **numerical aperture** of the objective lens, where *n* is the *refractive index* of the medium present between the cover glass and the objective lens and α is the *one-half angular aperture*. It is a measure of the ability of a lens to collect light from the specimen. Lenses with a low numerical aperture collect less light than those with a high numerical aperture. The human eye is itself a converging lens which forms the image of the object on the retina. The limit of resolution of the unaided human eye is 100 μm.

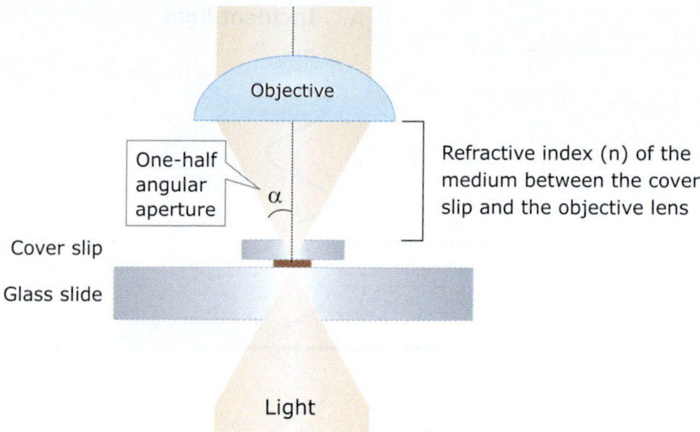

For small value of limit of resolution, the numerator of the equation should be as small as possible (i.e. light of the shortest wavelength) and the denominator should be as large as possible (i.e. maximum possible value of numerical aperture). Hence, resolution can be improved by:

- Increasing the *one-half angular aperture* (α)
- Increasing the *refractive index* (n) of medium between the objective lens and the specimen
- Shortening the *wavelength* (λ) of the illuminating light

Angular aperture

The angular aperture is the angle of the cone of light collected by the objective lens. It varies with the size and shape of the illumination cone entering the objective lens. As the light cones grow larger, the angular aperture increases. The light cone can be increased either by shortening the distance between the objective lens and the specimen or by increasing the diameter of the lens.

Theoretically, the highest angular aperture possible with a standard microscope objective lens would be 180°, resulting in a value of 90° for the one-half angular aperture. Thus, maximum value of $\sin \alpha$ will be 1. The working value of one-half angular aperture for the best objective lenses is about 70°. Hence, the maximum value for $\sin \alpha$ is about 0.94. Since the refractive index of air is about 1.0, no lens working in the air can have a numerical aperture greater than 1.0.

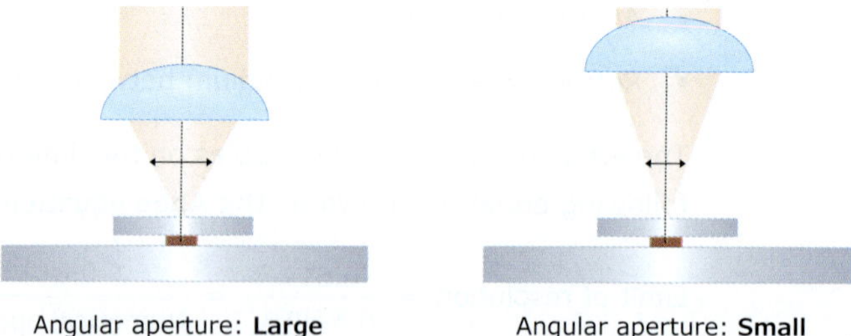

Figure 6.4 Angular aperture and working distance. The higher value of the one-half angular aperture (α) can be obtained by shortening the distance between the objective lens and the specimen.

Refractive index

The refractive index of air is about 1.0, so for a lens designed for use in air, the maximum numerical aperture is about 1.0. To increase the numerical aperture some microscope lenses are designed to be used with a layer of **immersion oil** between the objective lens and the specimen. Immersion oil has a higher refractive index (about 1.5) than air and, therefore, allows the lens to receive more of the light transmitted through the specimen. Thus, the limit of resolution for a microscope that uses visible light is about 300 nm in air and about 200 nm with immersion oil.

Wavelength of the illuminating light

For the minimum value of the numerator, the wavelength should be small. Thus, for the best resolution, the specimen is illuminated with the blue light of 450 nm.

Oil immersion lens

In most microscopes, air is present as a medium through which light rays pass between the coverslip protecting the sample and front lens of the objective. Objectives of this type are referred to as **dry objectives**. Air has a refractive index of 1.0003, very close to that of a vacuum and considerably lower than most liquids, including water (n = 1.33). In light microscopy, **immersion oil** is used to increase the resolving power. An objective lens specially designed to be used in this way is termed as an **oil immersion objective**.

In this system, air is replaced by transparent oil (termed *immersion oil*) of high refractive index (very similar to refractive index of glass). Immersion oil such as paraffin oil, cedarwood oil has been placed at the interfaces between the objective lens and the cover slip protecting the specimen (also between the condenser lens and the underside of the specimen slide).

If the air is present between the cover slip and the objective lens, light is refracted, scattered and effectively lost. This happens because the refractive index of air is very different from that of glass and light passing through a glass-air interface is refracted (bent) to a large degree. By reducing the amount of refraction at this point, more of the light can be directed to the narrow diameter lens of the high-power objective. The more the light, the clearer will be the image. Placing a material with a refractive index equal to that of glass in the airspace between cover slip and objective, more light can be directed through the objective which improves resolution. Immersion oils improve resolution by performing same function.

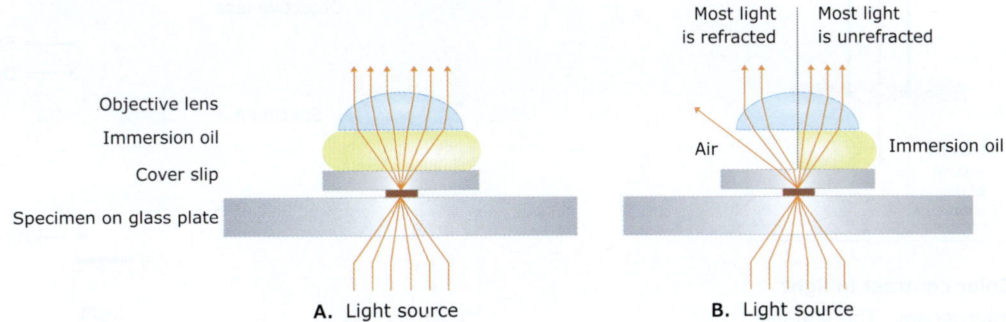

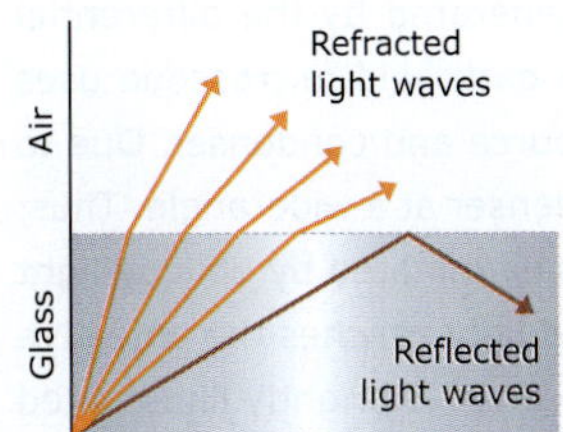

Figure 6.5 **A.** An oil immersion lens where immersion oil has been placed at the interfaces between the objective front lens and the specimen covered with a cover slip and also between the condenser lens and the underside of the specimen slide. **B.** In case of a dry objective, light rays pass through a specimen that is covered with a cover slip. These rays are refracted at the coverslip-air interface and only the light rays closest to the optical axis of the microscope have the appropriate angle to enter the objective lens. When air is replaced by oil of the same refractive index as glass, the light rays now pass straight through the glass-oil interface without any deviation due to refraction. In this case, the numerical aperture is, thus, increased by the factor of n, the refractive index of oil.

6.2 Types of light microscope

Bright-field microscopy

It is the original and most commonly used form of microscopy. When a specimen is illuminated with light, the transmitted light may undergo change in amplitude, phase and polarization. In bright-field microscopy, illuminating light is transmitted through the specimen and the contrast is generated due to differential absorption of light. In this microscopy, stained specimens or specimens that have natural pigmentation, give color contrast due to differential absorption of illuminating light. Absorption of light may also occur if specimens are thick enough to absorb a significant amount of light despite being colorless (*brightness contrast*). Absorption of light causes decrease in the amplitude of the transmitted light, resulting in **amplitude contrast**.

Dark-field microscopy

A dark-field microscope is a type of microscope in which objects are illuminated at a very low angle from the side so that the background appears dark and the objects show up against this dark background. It is a technique for improving the contrast of unstained, transparent specimens. Central circular disk stop that prevents direct condenser rays from entering the objective lens. To view a specimen in a dark field, a carefully aligned light source is used to minimize the quantity of directly transmitted light and collecting only the light scattered by the specimen. To achieve this an opaque disc is placed underneath the condenser lens, that blocks rays coming from condenser to enter the objective lens directly. Only those rays that have been scattered by the specimen enter the objective lens to generate the final image.

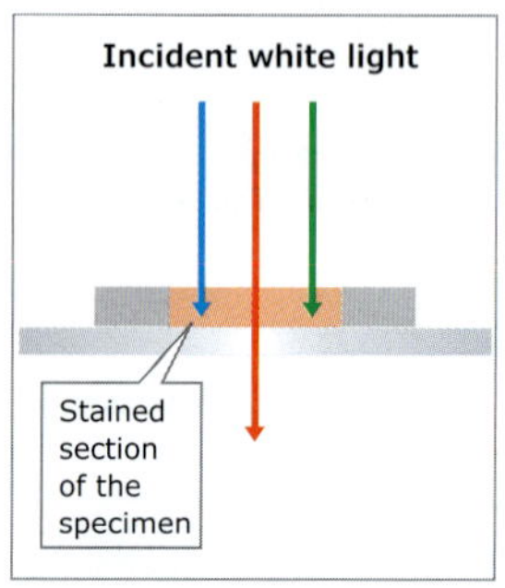

Color contrast in light microscopy. The stained portion of the cell absorbs light of specific wavelengths, depending on the stain types and allow other wavelengths to pass through it. As a result, a colored image of the cell is obtained that is visible in the normal bright-field light microscope.

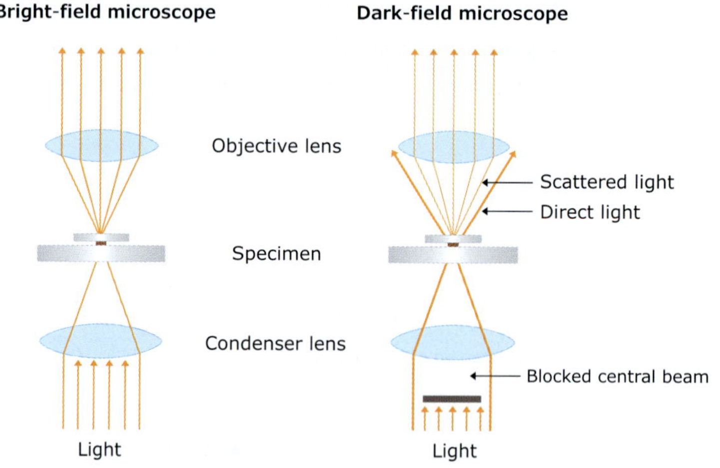

Figure 6.6 In bright-field microscope, the specimen is evenly illuminated by light directly coming from condenser. Light is transmitted through the specimen and the contrast is generated by the differential absorption of light. It forms a dark image against a brighter background. A dark-field microscope uses a dark-field condenser that contains an opaque disk, placed between light source and condenser. Due to presence of opaque disk, light can only pass through the outer edge of the condenser at a wide-angle. Thus, direct light is prevented from passing through the specimen. The specimen is illuminated by oblique light rays. Light passing through the specimen is scattered and only these scattered light reaches the objective lens for visualization. Direct light rays do not enter the objective thus the specimen is brightly illuminated on a dark background. Because there is no direct background light, the specimen appears light against a black background-the dark field.

Phase-contrast microscopy

Phase-contrast microscopy is used to study transparent and colourless specimens. This microscopic technique does not require cells to be killed, fixed or stained. When light passes through a living cell, the phase of the light wave changes due to variations in thickness and density within the cell. Light passing through a relatively thick or dense part of the cell, such as the nucleus, is retarded; its phase, consequently, is shifted relative to light that has passed through an adjacent thinner region of the cytoplasm.

Let us understand this by considering two beams of light from same source, one passing through an optically transparent material denser than air and other simply through air. When one beam of light passes through optically transparent material, its speed decreases and magnitude of change depends on both the density (as determined by the refractive index) and thickness of the material. The speed of second beam of light passing through air remain unchanged. Hence, these two beams, which was in phase are now out of phase; this difference is referred to as the phase difference.

When transparent unstained specimens are used for phase contrast imaging, they do not absorb light and are called **phase objects**. However, when light passes through the specimens, diffraction or scattering of light occurs. The diffracted and the non-diffracted beam of light have phase difference. Non-diffracted light is referred to as direct light as it continues unchanged through the sample. In phase-contrast microscopy additional phase differences is introduced between the non-diffracted and diffracted light. Interference of the two sets of beams leads to overall differences in the amplitude, which can be detected by the eye as differences in brightness. The specimen shows different degrees of brightness and contrast. Thus, phase contrast microscopy transforms differences in the relative phase of light waves to amplitude differences in the image.

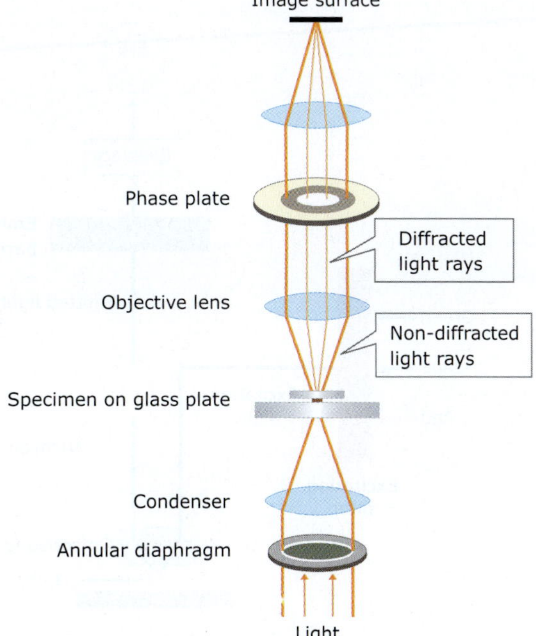

Figure 6.7 Phase contrast microscopy (ray diagram). When light passes through a transparent and colourless specimen, the phase of the light wave changes due to variations in thickness and density within the specimen. The thick lines represent the non-diffracted beams, while the diffracted beams are shown by thin lines. The diffracted light represents only a small part of the total light. This diffracted light beam arrives at the detector out of phase with the non-diffracted light. Phase plate introduces additional phase shift between the non-diffracted and diffracted light. These phase shifts are converted into changes in amplitude and can be observed as differences in image contrasts.

Fluorescence microscopy

In fluorescence microscopy, the specimen itself acts as a light source. The specimens used to study are either fluorescent materials or stained with fluorescent dyes. A chemical is fluorescent if it absorbs light at one wavelength and emits light (fluoresces) of a longer wavelength. Most fluorescent dyes (or fluorophores) emit visible light, but some emit infrared light. Fluorophores exhibit distinct excitation and emission spectra that depend on their atomic structure and electron resonance properties. Two fluorescent dyes commonly used are **fluorescein**, which emits an intense green fluorescence when excited with blue light, and **rhodamine** emits a deep red fluorescence when excited with green-yellow light.

A fluorescence microscope has three basic components – an *excitation filter*, a *dichroic mirror*, and an *emission* or *barrier filter*. To produce excitatory light of specific wavelength for a particular fluorophore, an **excitation filter** is inserted in the light path between the light source and the sample. The excitation filter is usually a bandpass filter, meaning that it transmits light of a narrow range of wavelengths while blocking others. Once filtered, the excitatory light is reflected towards the sample by a **dichroic mirror**. This is another filter that sits at a 45° angle relative to both the light from the light source and the sample. It reflects the excitatory light toward the sample. In this microscopy technique, only fluorescent light emitted by the fluorescent sample (or fluorescently stained sample) is allowed to form an image. The dichroic mirror allows fluorescent light of a longer wavelength (i.e. emitted light) emitted from the sample to reach the eyepiece. Although the dichroic mirror prevents excitatory light from reaching eyepiece, an **emission** or **barrier filter** is often inserted to block extraneous excitatory light or background fluorescence. Thus, an emission filter allows only fluorescent light emitted by the fluorescent sample to reach the eyepiece.

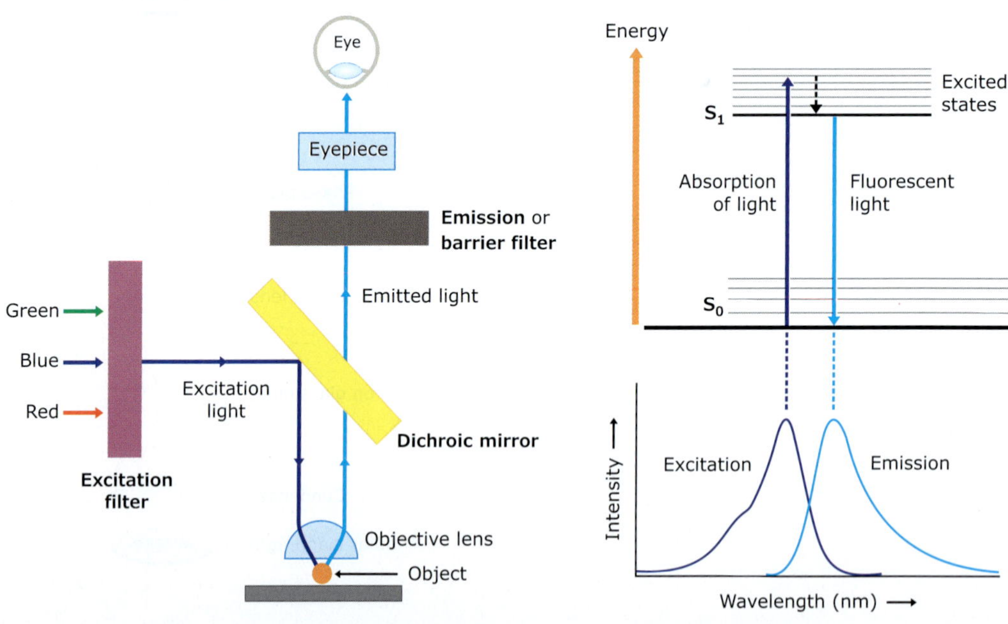

Figure 6.8 The optical system of a fluorescence microscope. Fluorescence microscopes contain special filters and employ a unique method of illumination to produce images of fluorescent light emitted from excited molecules in a specimen. It contains – excitation filter, dichroic mirror and barrier or emission filter. The excitation beam (blue line) passes through the excitation filter and is reflected by the dichroic mirror and directed towards the specimen. The return beam of emitted fluorescence wavelengths (cyan line) passes through the dichroic mirror and the emission filter to the eye.

Confocal microscopy

A confocal microscope creates sharp images of a specimen that would otherwise appear blurred when viewed with a conventional microscope. This is achieved by excluding most of the light from the specimen that is not from the microscope's focal plane. The image has less haze and better contrast than that of a conventional microscope and represents a thin cross-section of the specimen. Thus, apart from allowing better observation of fine details it is possible to build three-dimensional (3D) reconstructions of a volume of the specimen by assembling a series of thin slices taken along the vertical axis.

Confocal is defined as having the same focus. What this means in the microscope is that the final image has the same focus as or the focus corresponds to the point of focus in the object. The object and its image are confocal. The microscope is able to filter out the out-of-focus light from above and below the point of focus in the object. Normally when an object is imaged in the fluorescence microscope, the signal produced is from the full thickness of the specimen which does not allow most of it to be in focus to the observer. The confocal microscope eliminates this out-of-focus information by means of a confocal pinhole situated in front of the image plane which acts as a spatial filter and allows only the in-focus portion of the light to be imaged. Light from above and below the plane of focus of the object is eliminated from the final image. The confocal microscope uses a laser beam to illuminate a specimen, usually one that has been fluorescently stained. A diagram of the confocal principle is shown below.

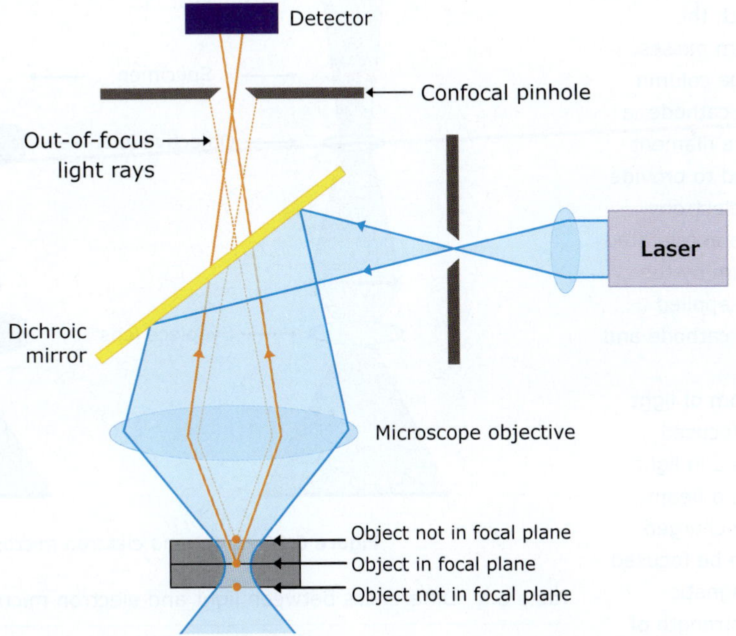

Figure 6.9 Ray path in a confocal microscope. A microscope objective is used to focus a laser beam onto the specimen, where it excites fluorescence. The fluorescent radiation is collected by the objective and efficiently directed onto the detector via a dichroic beam splitter. The wavelength range of the fluorescence spectrum is selected by an emission filter, which also acts as a barrier blocking the excitation laser line. The pinhole is arranged in front of the detector, on a plane conjugate to the focal plane of the objective. Light coming from planes above or below the focal plane is out-of-focus when it hits the pinhole, so most of it cannot pass the pinhole and, therefore, does not contribute to forming the image.

6.3 Electron microscope

The fundamental principles of electron microscopy are similar to those of light microscopy, except one major difference of using electromagnetic lenses, rather than optical lenses to focus a high-velocity electron beam instead of visible light. The relationship between the limit of resolution and the wavelength of the illuminating radiation hold true for both - a beam of light or a beam of electrons. Due to the short wavelength of electrons, the resolving power of the electron microscope is very high.

Electron microscopes are of two basic types: *transmission electron microscope* and *scanning electron microscope*. The most commonly used type of electron microscope is called the transmission electron microscope (**TEM**) because it forms an image from electrons that are transmitted through the specimen being examined. Scanning electron microscope (**SEM**) is fundamentally different from TEM because it produces image from electrons deflected from a specimen's outer surface (rather than electrons transmitted through the specimen).

Electron microscopes consist largely of a tall, hollow cylindrical column through which the electron beam passes. The top of the column contains the cathode, a tungsten wire filament that is heated to provide a source of electrons. Electrons are accelerated as a fine beam by the high voltage applied between the cathode and anode.

Just as a beam of light rays can be focused by a glass lens in light microscopes, a beam of negatively charged electrons can be focused by electromagnetic lenses. The strength of the magnets is controlled by the current provided them. Air is pumped out of the column, producing a vacuum through which the electrons travel. If the air were not removed, electrons would be prematurely scattered by collision with air molecules.

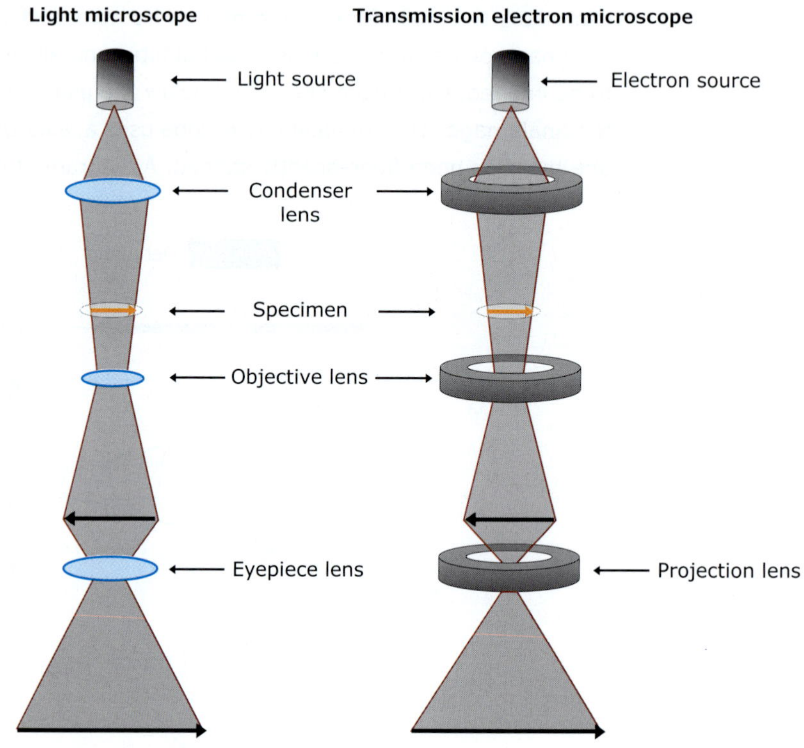

Figure 6.10 Light and electron microscopy.

Table 6.1 Differences between light and electron microscopes

Feature	Light microscope	Electron microscope (TEM)
Highest practical magnification	About 1,000-1,500	Over 100,000
Best resolution	0.2 μm	0.5 nm
Radiation source	Visible light	Electron beam
Medium of travel	Air	High vacuum
Type of lens	Glass	Electromagnet
Source of contrast	Differential light absorption	Scattering of electrons
Specimen mount	Glass slide	Metal grid (usually copper)

Transmission electron microscope

In light microscopy, *differential absorption* of light, which depends mainly on staining the specimen, results in the visible differences in various parts of the image. In the TEM, a condenser lens focuses the electron beam onto the specimen and electrons are transmitted through the specimen. The portion of the beam absorbed by specimen is minimal. To be absorbed, an electron must lose *all* its energy to the specimen. Although not absorbed, these electrons are scattered by the atoms of the specimen. Image formation in the electron microscope depends on *differential scattering* of electrons by parts of the specimen. Consider a beam of electrons focused on the screen. If no specimen were present in the column, the screen would be evenly illuminated by the beam of electrons, producing an image that is uniformly bright. By contrast, if a specimen is placed in the path of the beam, some of the electrons strike the specimen and are scattered away. When the electrons come in contact with the sample, it can either be scattered *elastically*, that is, without any loss of energy, or *inelastically*, i.e. transferring some of that energy to the atom. Interaction between the incoming fast electron and an atomic nucleus gives rise to *elastic scattering*. Interaction between the fast electron and atomic electrons results in *inelastic scattering*.

Some of the electrons passing through the specimen are scattered and the unscattered one is focused to form an image in a manner analogous to the way an image is formed in a light microscope. The image can be observed on a phosphorescent screen or recorded, either on a photographic plate or with a high-resolution digital camera. Because the scattered electrons are lost from the beam, the dense regions of the specimen show up in the image as areas of reduced electron flux, which look dark. *Scattering of electrons contributes to the contrast*. Because the electrons are absorbed by atoms in the air, the entire tube between the electron source and the detector is maintained under an ultrahigh vacuum.

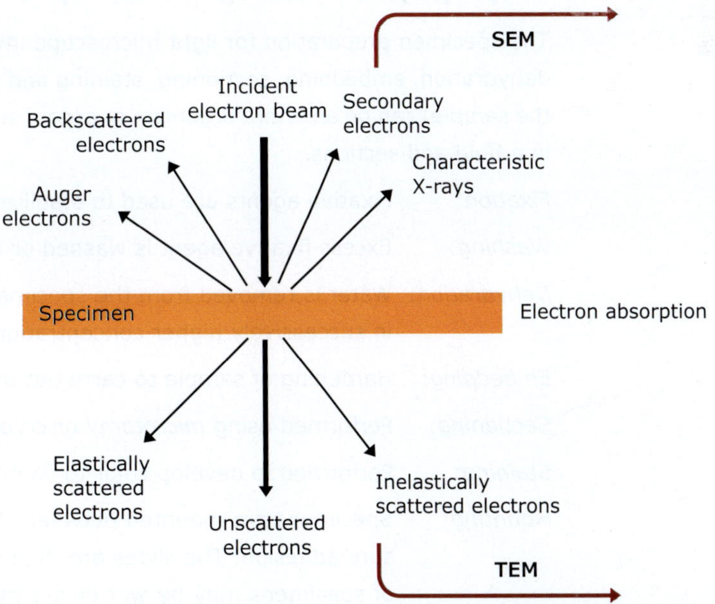

Figure 6.11 Interactions of the electron beam with the specimen.

Elastic and **inelastic scattering**

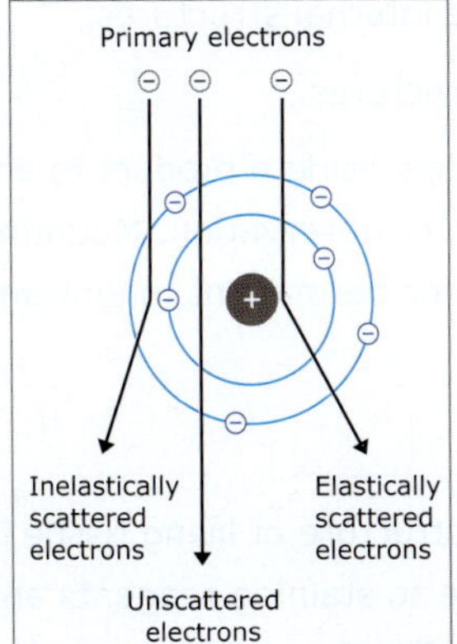

The amount of scattering which occurs at any particular specimen point is dependent on its density and overall thickness, and is relatively independent of other specimen properties. Specimen prepared for TEM has fairly uniform thickness, therefore, almost no contrast arises

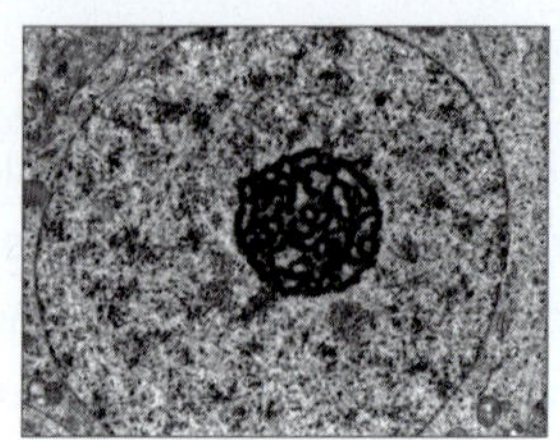

Credit: Jose Luis Calvo / Shutterstock

Transmission electron micrograph of section through a mast cell in connective tissue.

from the thickness. Their atomic number also remains approximately constant, so the overall contrast is very low and the specimen appears featureless in the TEM. To produce scattering contrast, the sample is generally stained with *electron dense materials*. So, before or after slicing, the specimen is immersed in a solution that contains salts of heavy metal such as uranium and lead. The degree of impregnation or staining, with these salts reveals different cellular constituents with various degrees of contrast. Darker areas in the image are where few electrons have been transmitted through the sample due to thickness or staining.

Scanning electron microscope

The scanning electron microscope (**SEM**) views the surfaces of specimens. The sample is fixed, dried, and coated with a thin layer of a heavy metal, such as gold or a mixture of gold and palladium. The specimen is, then, scanned with a very narrow beam of electrons. Molecules in the specimen are excited and they release secondary electrons that are captured by a detector, and generating an image of the specimen's surface. Contrast arises when different parts of specimen generate differing amounts of secondary electrons as the electrons beam strikes them. Areas which generate large numbers of secondary electrons will appear brighter than areas that generate fewer secondary electrons. The resolving power of scanning electron microscopes, which is limited by the thickness of the metal coating, is only about 10 nm, much less than that of transmission electron microscope. The image produced appear three-dimensional in SEM whereas transmission electron microscope produces two-dimensional images.

6.4 Sample preparation for microscope

Sample preparation for light microscope

The specimen preparation for light microscope involves sample collection, fixation, washing, dehydration, embedding, sectioning, staining and mounting. The specimen (sometimes called the *sample*) can be an entire organism or organ, squashed or crushed tissues, cells suspended in a fluid and sections.

Fixation: Fixative agents are used to stabilize the structure of the sample.

Washing: Excess fixative agent is washed or rinsed in clean water.

Dehydration: Water is removed from the specimen. It is performed by placing the specimen in successively higher concentrations of ethanol or acetone.

Embedding: Hardening of sample to carry out the sectioning.

Sectioning: Performed using *microtomy* or *cryotomy* to visualize internal structures.

Staining: Performed to develop color to contrast to various structures.

Mounting: Specimens are mounted between slides and cover slips using a product to ensure adhesion. The slides are then ready for storage or observation. Mounting of specimens may be wet or dry mount, temporary or permanent mount and hot or cold mount.

Fixation

It is the first step in sample preparation and aims to preserve the structure of living tissue in its original state (a 'life-like' state). Fixation makes cells permeable to staining reagents and cross-links their macromolecules to stabilize and lock in position. Fixation methods can be divided into two basic types—physical and chemical. **Physical fixation** includes *heating* and *freezing* (cryofixation). **Chemical fixation** is more commonly used fixation method. Under chemical fixation, specimens for light microscopes (also electron microscopes) are commonly fixed with a solution containing chemicals that cross-link most proteins and nucleic acids. Fixatives kill the cells while preserving their structural appearance.

The most widely employed fixatives are aldehydes such as **formaldehyde** (often referred to as *formalin*) and **glutaraldehyde**. Formaldehyde's mechanism of action is based on the reaction of the aldehyde group with primary amines in proteins. *Glutaraldehyde* is the most commonly used **primary fixative**. It contains two aldehyde groups separated by three methylene bridges. These two aldehyde groups and the flexible methylene bridge significantly increases the cross-linking potential of glutaraldehyde over formaldehyde. It penetrates rapidly and stabilizes proteins by forming cross-links but does not fix lipids. *Osmium tetroxide* is used as a **secondary fixative**, reacting with lipids and acting as a stain. It is especially useful for fixing cell membranes since it reacts with the C=C double bonds present in many fatty acids. Following each fixation step, excess fixative must be washed out of the tissue.

Staining

Staining is a biochemical technique of coloring specimens. **Dyes** are used to stain specimens. Generally, dyes are organic compounds consisting of two functional chemical groups - one group gives dyes their characteristic color, while other group contains an ionizable chemical structure, which helps to solubilize the dye and facilitates its binding to different structures. Dyes can be referred to as *anionic* or *cationic*. **Cationic** or **basic dyes** (such as crystal violet, methylene blue, safranin, malachite green) will react with negative charge groups. **Anionic** or **acidic dyes** (such as eosin, acid fuchsin) will react with groups that have a positive charge. Most biological materials show little contrast with their surroundings unless they are stained. In the case of light microscopy, contrast can be enhanced by using colored stains which selectively absorb certain wavelengths. The light that illuminates the specimen usually consists of a spectrum of colors. These colors become readily visible when the light strikes a specimen that has been stained with various dyes (stains). These dyes bind to specific molecules present in the specimens. For example, *hematoxylin* binds to basic amino acids (lysine and arginine) of different proteins, whereas *eosin* binds to acidic molecules such as DNA and side chains of aspartate and glutamate. These dyes absorb specific colors in the spectrum but transmit or reflect others to the eye. For example, in a specimen that has been stained with two commonly used biological stains, safranin and fast green, certain parts of the cell will appear red (safranin will stain acidic components of the cell such as DNA in the nucleus), while others will stain green (fast green stains basic components of the cytoplasm).

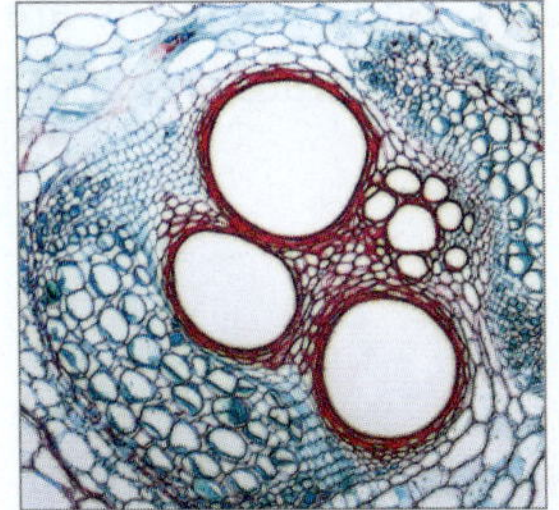

Credit: Digital Photo / Shutterstock

A section of stained young plant root - stained with two dyes, safranin and fast green. The fast green stains the cellulosic cell walls, while the safranin stains the lignified xylem cell walls.

Sample preparation for TEM

For TEM, samples must be cut into very thin cross-sections. This is to allow electrons to pass right through the sample. After fixation and dehydration, cutting of a sample is performed by an instrument called an **ultramicrotome**.

Contrast in the TEM depends on the atomic number of the atoms in the specimen: the higher the atomic number, the more electrons are scattered and the greater the contrast. Biological tissues are composed of atoms of very low atomic number (mainly carbon, oxygen, nitrogen, and hydrogen). Hence, heavy metal salts are used as stain in electron microscope. Colors do not exist in any electron microscope.

Light microscopy yields photomicrographs in color as well as in black and white, whereas electron microscopy produces electron micrographs in black and white only.

An electron stain is said to exhibit **positive contrast** when it increases the density of a particular biological structure as opposed to the adjoining areas. In positive staining, the heavy metal salts attached to various organelles or macromolecules within the specimen increase their density and, thereby, increase contrast differentially. The two most commonly used positive stains are uranyl acetate and lead citrate. The exact mode of action of these stains is not completely understood.

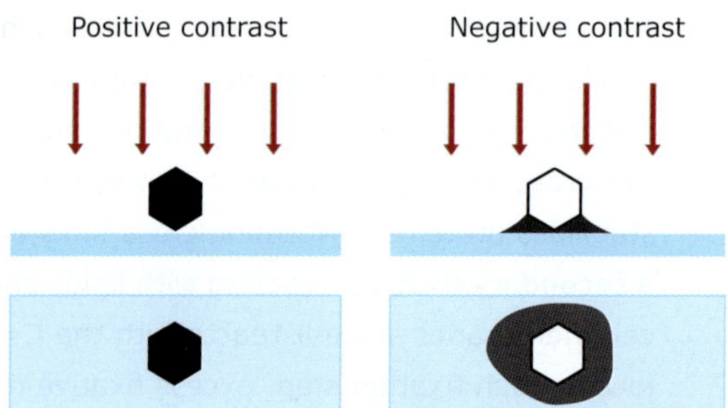

Figure 6.12 Comparison of virus appearance after positive and negative staining. To heighten the contrast between viruses and the background, electron-dense stains are used. These are usually compounds of heavy metals of high atomic number, that serve to scatter the electrons from regions covered with the stain. If virus particles are coated with stain (positive staining), fine detail may be obscured. Negative staining overcomes this problem by staining the background and leaving the virus relatively untouched.

It is known that uranyl ions react strongly with phosphate and amino groups so that nucleic acids and certain proteins are highly stained. With lead stains, it is thought that lead ions bind to negatively charged components. Positive staining differs from the *negative* stain situation, where the background area surrounding the specimen is made dense by a heavy metal salt so that the specimen appears lighter in contrast to the darkly stained background. If specimens are coated with stain (**positive staining**), fine detail may be obscured. **Negative staining** overcomes this problem by staining the background and leaving the specimens relatively untouched. The negative stain is moulded round the specimen. In the electron micrograph, the sample appears as a light region surrounded by a dark background originating from the stains. Negative stains are not used in sectioned materials, but are used to contrast whole, intact biological structures (viruses, bacteria, cellular organelles, etc.). Negative stains are salts of heavy metals such as uranium, tungsten and molybdenum. Uranyl acetate and phosphotungstic acid are the two most commonly used negative stains.

6.5 Freeze–fracture electron microscopy

Freeze-fracture electron microscopy technique is used to visualize the features of cell membranes. Cells are frozen and immobilize at the temperature of liquid nitrogen (–196°C) in the presence of a cryoprotectant (antifreeze) to prevent distortion from ice crystal formation. The most commonly used cryoprotectant is glycerol. The frozen specimen is usually fractured with a liquid-nitrogen-cooled microtome blade, which often splits the bilayer into monolayer and exposes the interior of the lipid bilayer and its embedded proteins. The fracture plane is not observed directly rather a replica or cast is made of the fractured surface. Making the replica involves two steps, shadowing and backing. An oblique, unidirectional shadowing is carried out by evaporating a fine layer of heavy metal (such as platinum) onto the specimen. A carbon layer is then deposited on the top of the metal layer to impart sufficient strength to the replica (backing). The topographical features of the frozen, fractured surface are thus converted into variations in thickness of the deposited platinum layer of the replica.

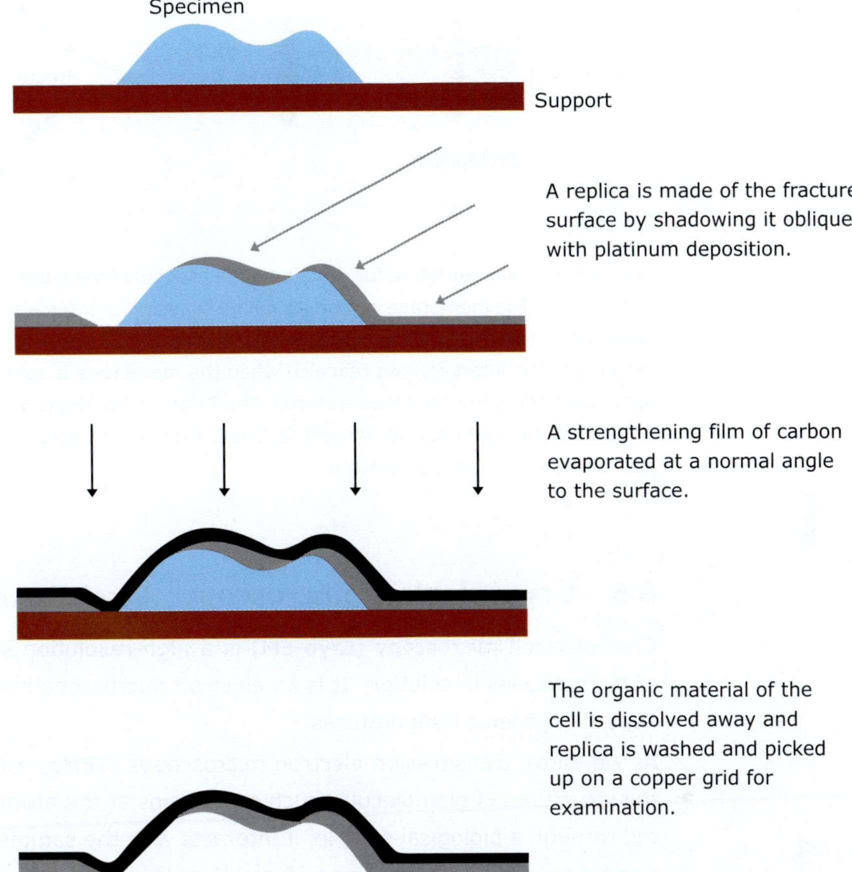

Figure 6.13 Replication of the fractured surface is accomplished by shadow-casting the surface with platinum at an oblique angle (45°). This produces a film 2-5 nm thick on the surface normal to the angle of the shadow. Following shadowing, a strengthening film of carbon evaporates at a normal (90°) angle to the surface. This imparts sufficient strength to the replica so that when it is removed from the surface it will not break. The thickness of the film is directly related to the slope of the shadowed surface. The topology of the surface is converted into variations in the thickness of the shadow and finally into variations in electron density recorded on photographic film.

After the replica has been made, the sample is brought to atmospheric pressure and allowed to warm to room temperature. The biological material is removed from the replica using sodium hypochlorite solution, chromic acid or other cleaning agents. After washing in distilled water, pieces of replica are mounted on grids for examination in the transmission electron microscope.

Hence, there are four essential steps in making a freeze-fracture replica: rapid freezing of the specimen, fracturing the specimen, making the replica of the frozen fractured surface by vacuum-deposition of platinum and carbon and cleaning the replica to remove all the biological material.

In addition, an optional etching step may be interposed between fracturing and making the replica. An optional etching step, involving vacuum sublimation of ice, may be carried out after fracturing. The term '**freeze etching**' is sometimes used synonymously with freeze fracturing. 'Etching' is defined as removal of ice from the surface of the fractured specimen by vacuum sublimation (freeze drying), before making the replica.

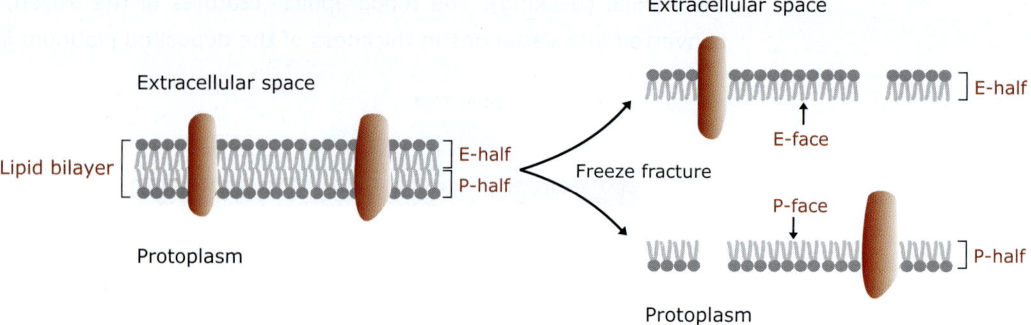

Figure 6.14 Nomenclature for describing the aspects of the plasma membrane revealed by freeze fracture and etching. The membrane comprises a lipid bilayer with intercalated proteins. The half-membrane leaflet adjacent to the extracellular space is termed the E-half and that adjacent to the protoplasm is termed the P-half. The interior views revealed when the membrane is split by freeze fracture are termed faces to distinguish them from the true surfaces. The P-face is the fracture face of the P-half of the membrane and will always have protoplasm beneath it. The E-face is the fracture face of the E-half of the membrane and has extracellular space beneath it.

6.6 Cryo–electron microscopy: Resolution revolution

Cryo-electron microscopy (**Cryo-EM**) is a high-resolution structure determination technique of biomolecules in solution. It is an electron microscope imaging of frozen biological samples kept at cryogenic temperatures.

As we know, transmission electron microscopes (**TEMs**) use a beam of electrons to examine the structures of biomolecules such as proteins at the atomic scale. As the beam is transmitted through a biological sample, it interacts with the samples, which projects an image of the sample onto the detector. Image formation depends on the differential scattering of electrons by different parts of the sample. Due to the short wavelength of the electron beam, the resolving power of the electron microscope is very high. But biological samples are not compatible with the high-vacuum conditions and high-energy electron beams used in TEMs. There are two reasons for this. *First*, the high-energy electron beam burns and destroys the biological sample. *Second*, the water that surrounds the biological sample evaporates and the vapour from the sample destroys the vacuum required to run the microscope. To overcome this, the biological sample could be dried out before being examined, but this can alter their structure.

Cryo-EM uses frozen samples and low-energy electron beams to overcome these problems. But low-energy electron beam would result in fuzzier images. This problem has been overcome by the use of a powerful image-analysis algorithm for merging multiple fuzzy 2D images from an electron microscope to create a 3D image of a molecule. This involves taking thousands of images of randomly oriented molecules and sorting them into groups of similar images. Each group is then processed to create a set of much sharper images. The spatial relationships between the groups are then calculated, leading to the assembly of a high-resolution 3D image.

Since liquid water evaporates in the case of TEM, thus to obtain an image using cryo-EM, biological samples are rapidly frozen at –190°C. Liquid nitrogen-cooled ethane freezes the thin film of water on the sample so quickly that the water molecules don't have time to arrange into a crystalline lattice. Due to rapid freezing, the water forms a disordered glass rather than crystalline ice. This **vitrification** of water is crucial because it does not cause electron diffraction and also allows the samples to retain their natural shape and organization. If the sample is frozen too slowly, then ice crystals form and the ordered ice crystals would strongly diffract the electron beam. Once the cryo-EM sample has been frozen, a focused beam of electrons reveals the shape of these very small, nanometer-sized 'vitrified' biological samples. Each image contains all the information required to determine the 3D structure. Specialized computer analysis then combines hundreds of thousands of individual, 2D snapshots from different angles into a composite that can be viewed in 3D.

Steps involved in structure determination by single particle cryo-EM

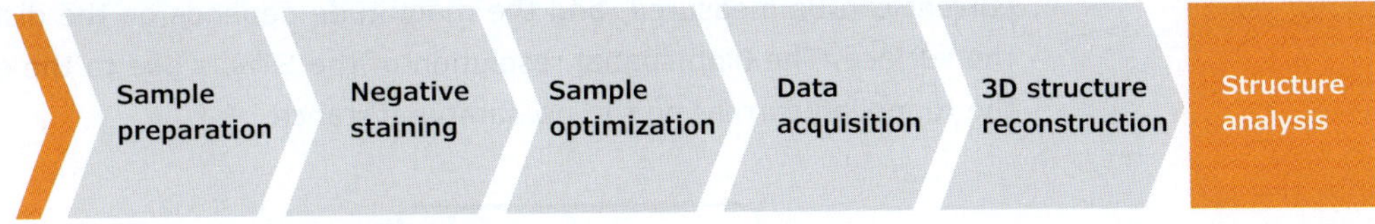

Cryo–EM Vs X–ray crystallography

X-ray crystallography gives very high-resolution structures of biomolecules. X-ray crystallography uses X-ray to determine the position and arrangement of atoms in a crystal. It is based on **X-ray diffraction**. When a beam of X-ray strikes a crystal, the beam may be diffracted. From the angles and intensities of these diffracted beams, a three-dimensional picture of the density of electrons within the crystal can be derived. But to get an x-ray diffraction structure, the biomolecule such as protein should be crystallised. Many proteins won't crystallise at all. And in some cases, the process of crystallisation alters the structure, so it's not representative of what the molecule looks like in real life. Cryo-EM doesn't require crystals, and it also enables scientists to see how biomolecules move and interact as they perform their functions.

Cryo–EM Vs NMR

NMR can also give very detailed structures, and investigate conformational changes or binding of small biological molecules. But it's limited to relatively small molecular mass (<30 kDa) and usually soluble intracellular molecules, rather than those membrane bound. If someone want to study high molecular mass proteins, membrane-bound receptors, or complexes of several biomolecules together, cryo-EM is where it's at.

6.7 Scanning probe microscopes

Scanning probe microscopes (**SPMs**) are a family of microscopes used to map topographical features of the sample surface at atomic level (resolution in nanometer scale). In all SPMs, a physical probe is used to scan the surface of a sample. During the scanning process, interactions of the probe tip with the surface of a sample are measured and data are used to generate an image of the surface. Various types of interaction between the probe tip and the surface are exploited in different types of SPMs. Hence, there are several types of SPMs. **Scanning tunneling microscope** (**STM**) and **Atomic force microscope** (**AFM**) are two major types of SPMs. AFMs measure the interactive forces between the probe tip and the sample whereas STMs measure the tunneling current flowing between the probe tip and the conducting sample. Application of the STM for surface imaging requires conductive sample. Hence, most biological samples due to poor electron conductivity are not suitable for STM imaging. Unlike STMs, the AFM does not need a conducting sample.

Scanning tunneling microscope

Historically, the first microscope in the family of SPMs is the scanning tunneling microscope (STM). It was invented by Binning and Rohrer in 1981. STM is based on the concept of electrons tunneling through a narrow potential barrier between a metal tip and a conducting sample in an external electric field. If the probe tip and sample are connected to a voltage source and the tip is close enough to the sample surface and the sample is electrically conductive, electrons will begin to leak or tunnel across the gap between the probe and the sample. This current can be measured, and the magnitude depends on the distance between the tip and the surface. The high spatial resolution of the STM is due to the exponential dependence of the tunneling current on the tip-sample distance.

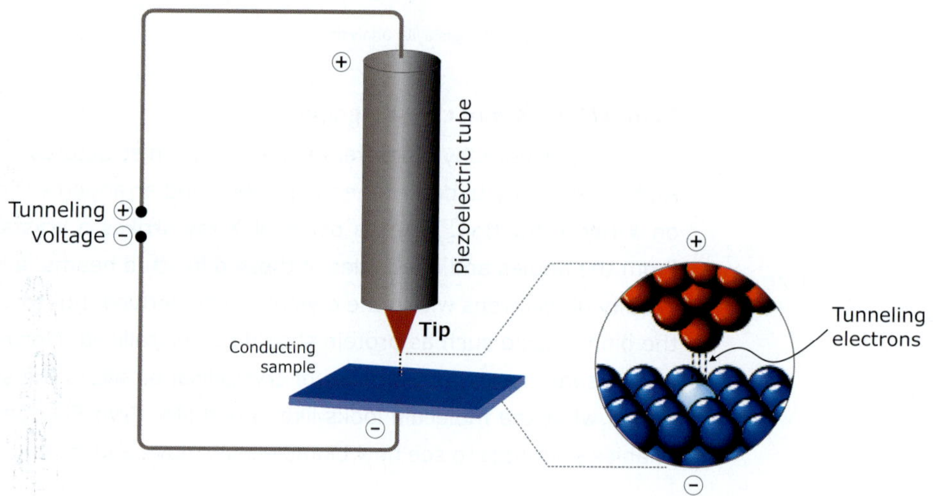

Figure 6.15 The metal probe tip and the sample are both connected to a voltage supply. The sample is scanned by a very fine metallic probe tip mounted on a piezoelectric tube. A tunneling current starts to flow when the tip is close to the sample. This feeble tunneling current is amplified and measured.

Atomic–force microscope

Atomic-force microscope (**AFM**) is based on the principle of the measurement of the interactive forces between a probe tip and a sample surface. The interactive forces (like capillary forces, electrostatic forces, van der Walls forces, etc.) have been classified as short- or long-range forces, attractive or repulsive forces. As the tip scans the surface, the interactions between the tip and the sample experience attractive or repulsive forces. This causes the tip to move up and down, which in turn causes the entire **cantilever** to deflect. The attractive force between the surface and the tip causes the cantilever to deflect towards the surface whereas repulsive force causes the cantilever to deflect away from the surface. By measuring the deflection of the cantilever, the topographical features of the surface can be mapped out.

Atomic force microscopy does not use lenses to form an image, but instead uses a sharp probe tip at the end of a flexible cantilever to scan and sense the topography of a sample.

The methods used in AFM to acquire images can be categorized in three groups: *contact mode, tapping* or *intermittent mode* and *non-contact mode*. In **contact mode**, the probe tip directly touches the sample surface during scanning and the force acting between the atoms of probe tip and sample is counterbalanced by the elastic force. The probe tip predominately experiences repulsive forces. The contact mode may be carried out either at *constant force* (fixing the tip–sample interaction force through a feedback loop) or at *constant distance* between probe tip and sample (fixing the height of the scanner). A drawback of contact modes is the direct mechanical interaction of the probe tip with the sample. The **tapping mode** is similar to contact mode. However, in this mode intermittent contact occurs between the probe tip and the sample surface during scanning. The probe tip experiences both attractive and repulsive forces. The cantilever is oscillated at its resonant frequency. *Resonant frequency* is a natural frequency of vibration determined by the physical parameters of the vibrating object. By maintaining a constant oscillation amplitude, a constant tip-sample interaction is maintained and an image of the surface is obtained. In **non-contact mode**, probe tip does not contact the sample surface, but oscillates above the surface during scanning. In this mode, the probe tip predominately experiences attractive forces.

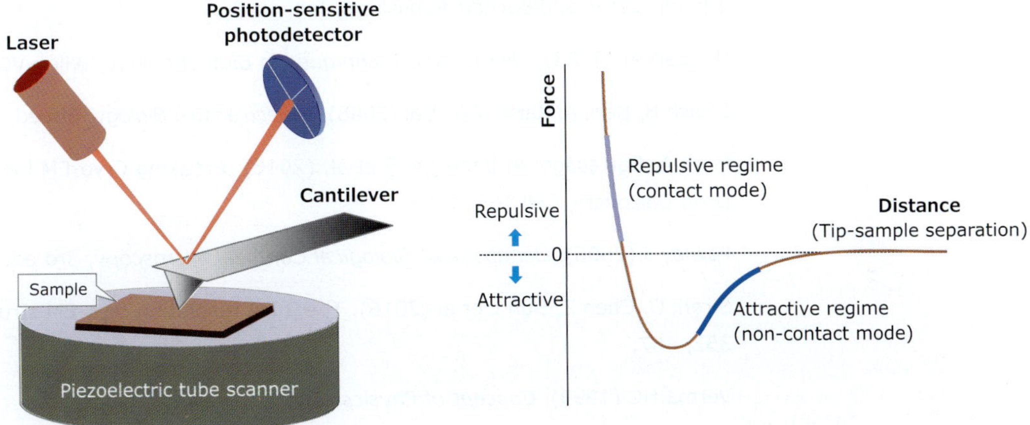

Figure 6.16 **A.** Schematic diagram of an AFM. The essential elements of an AFM are: a probe tip attached to a cantilever; a laser deflection system to measure deflections of the cantilever; a piezoelectric tube scanner for moving the sample and probe relative to each other. The deflection of the cantilever in response to interactions between probe tip and surface is detected by a laser deflection system. It measures the position of a laser on the photodetector which is reflected from the back of a cantilever. **B.** Tip-sample interaction force curve. The Graph shows different force regimes which correspond to inter-atomic forces. As seen in this figure, the atoms are separated by a large distance on the right side of the curve. As the atoms are brought close, they attract each other. This attraction increases until they come so close together that their electrons begin to repel each other electrostatically. This repulsive electrostatic force exceeds the attractive force as the inter-atomic distance decreases.

References

Alberts B et al (2008), *Molecular Biology of the Cell*, 5th ed. Garland Science Publishing.

Bai XC, Yan C, Yang G et al (2015), An atomic structure of human γ-secretase, Nature, 525, 212.

Becker WM, Kleinsmith LJ and Hardin J (2006), *The world of the cell*, 6th ed. Pearson Education Inc.

Behrmann E, Loerke J, Budkevich TV et al (2015), Structural snapshots of actively translating human ribosomes, Cell, 161, 845.

Bozzola JJ and Russell LD (1998), *Electron Microscopy: Principles and Techniques for Biologists*, 2nd ed. Jones and Bartlett Publishers, Inc.

Hoppert M (2003), *Microscopic Techniques in Biotechnology*, Wiley-VCH Verlag.

Lodish H, Berk A, Kaiser CA et al (2008), *Molecular Cell Biology*, 6th ed. W.H. Freeman and Company.

Merk A, Bartesaghi A, Banerjee S et al, (2016), Breaking Cryo-EM Resolution Barriers to Facilitate Drug Discovery, Cell 165, 1698.

Pawley J (2006), *Handbook of Biological Confocal Microscopy*, 3rd ed. Springer.

Sirohi D, Chen Z, Sun L et al (2016), The 3.8Å resolution cryo-EM structure of Zika Virus, Science, 352, 467.

Verma HC (1999), *Concept of Physics*, Bharati Bhawan, India.

7

Flow cytometry

Flow cytometry is a technique used for sorting, counting, and examining cells (or microparticles) suspended in a fluid by their optical properties (light scattering and fluorescence). The term **cytometry** refers to the measurement of cell number and cell characteristics such as cell size, shape, cell cycle stage, DNA content, and more. It simultaneously measures and analyzes multiple physical characteristics of a single cell as they flow through a fluid stream and pass through a beam of light.

Components of a flow cytometer

A *flow cytometer* consists of three main systems: *fluidics*, *optics*, and *electronics*. The **fluidic system** transports the cells in a stream to the light beam for analysis. The **optical system** consists of excitation optics and collection optics. The *excitation optics* include light sources generating scattered and fluorescent light signals, and *collection optics* include lenses that collect light scattered or emitted from the interaction between cells and the light beam. They also include a system of optical mirrors and filters to separate and then direct specific light signals to the appropriate optical detectors. The **electronics system** converts light signals into electronic signals that can be processed by the computer. Essentially, the electronics serve as the brain of the flow cytometer.

Fluidic system

The flow cytometer measures both the physical and chemical characteristics of a cell or microscopic object. This is achieved by introducing cells suspended in a fluid to pass through an *interrogation point* one by one. The location where cells interact with light beam is termed the **interrogation point**. At this point, we interrogate each cell using a specific light beam, and observe the light responses electronically.

The primary function of the fluidics system is to present cells in suspension to the interrogation point for examination one cell at a time. This is achieved by a process known as '*hydrodynamic focusing*'. For this, the sample stream is injected into a faster-moving stream of **sheath fluid** (usually phosphate-buffered saline) in the flow cell (flow chamber). Differences in the pressure, velocity, and density of the two fluids prevent them from mixing. The flow of sheath fluid propels and confines the cells to the center of the sample core, creating a single-file stream of cells.

Both sample pressure and sheath fluid pressure are maintained at different levels, with the sample pressure always greater than the sheath pressure. Altering the rate at which the cell suspension is injected into the center of the sheath fluid will have a direct effect on the width of the sample stream and the number of cells passing through the interrogation point.

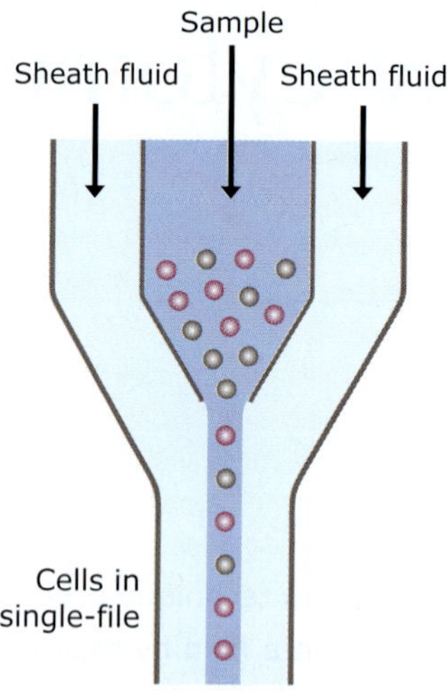

Figure 7.1
The fluidics system is to present cells (or particles) in suspension to the interrogation point (the point at which cells pass through the light beam) one cell at a time. This is achieved by a process known as hydrodynamic focusing.

Optical system

After hydrodynamic focusing, each cell passes through a beam of light. Lasers are the most common light sources used in flow cytometers. **Laser light** is coherent (has a synchronized, identical wave frequency), monochromatic (has a single wavelength), and energetic. These properties ensure that the cells are illuminated with uniform light of a specific wavelength.

For analysis, individual cells are interrogated by a beam of light, and the resulting *fluorescence* and *scattered* light are collected. Light scattering or fluorescence emission (when the particle is labeled with a *fluorochrome*) provides information about the cell's properties. The amount of light scattered is directly related to the structural and morphological properties of the object, while fluorescence emission, which originates from a fluorescent probe, is proportional to the amount of the fluorescent probe bound to the object.

The optical system is responsible for excitation and light collection within the instrument. This system comprises *excitation optics* and *collection optics*. The excitation optics include light sources for excitation and lenses that shape and focus the light beam. When cells pass through the interrogation point, they scatter light. If any fluorescent molecules are present on the cell, they exhibit fluorescence. The scattered and fluorescent light is collected by appropriately positioned lenses. A system of mirrors and filters is used to direct this light to the detection system, which converts the light signal to electronic signals.

Optical detectors capture the fluorescent and scattered light, converting them into **photocurrent**. This photocurrent is then passed to the electronics system for processing. Several types of detectors can be used on a flow cytometer, the most common being the **photodiode** (**PD**) and the **photomultiplier tube** (**PMT**). *Photodiodes* are less sensitive to light signals than *photomultiplier tubes* and are thus used for detecting stronger signals. Conversely, photomultiplier tubes are employed for detecting weaker signals.

The specificity of detection is controlled by **optical filters**, which block certain wavelengths while transmitting (passing) others. These filters, designed for specific wavelength ranges, are positioned in front of the detector. They enable only a particular range of wavelengths to reach the detector. Generally, there are three main types of optical filters:

Longpass: Transmits wavelengths of light equal to or greater than the spectral band of the filter.

Shortpass: Transmits wavelengths of light equal to or shorter than the spectral band of the filter.

Bandpass: Transmits wavelengths of light within a specific range of wavelengths.

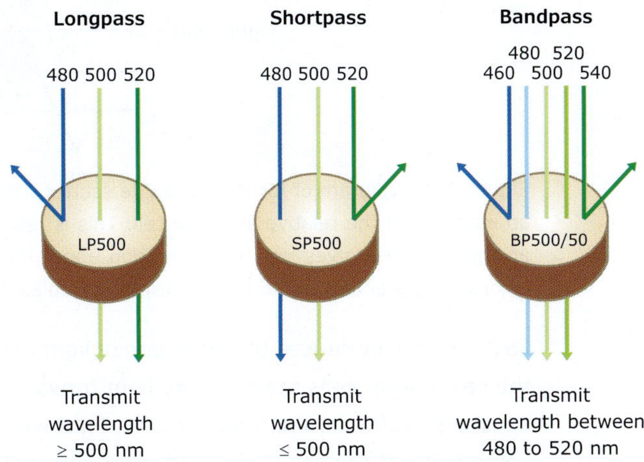

Figure 7.2 Different types of optical filters. 'Longpass' filters allow through light above a cut-off wavelength, 'shortpass' permit light below a cut-off wavelength and 'bandpass' transmit light within a specified narrow range of wavelengths (termed a band width).

Dichroic filters/mirrors, also known as **dichroic beamsplitters**, are specialized optical filters that combine the features of longpass or shortpass filters with reflective capabilities. These filters serve dual functions: they permit specific wavelengths to pass in the forward direction and can also reflect light at a 90° angle. This characteristic enables the light path to traverse a series of filters.

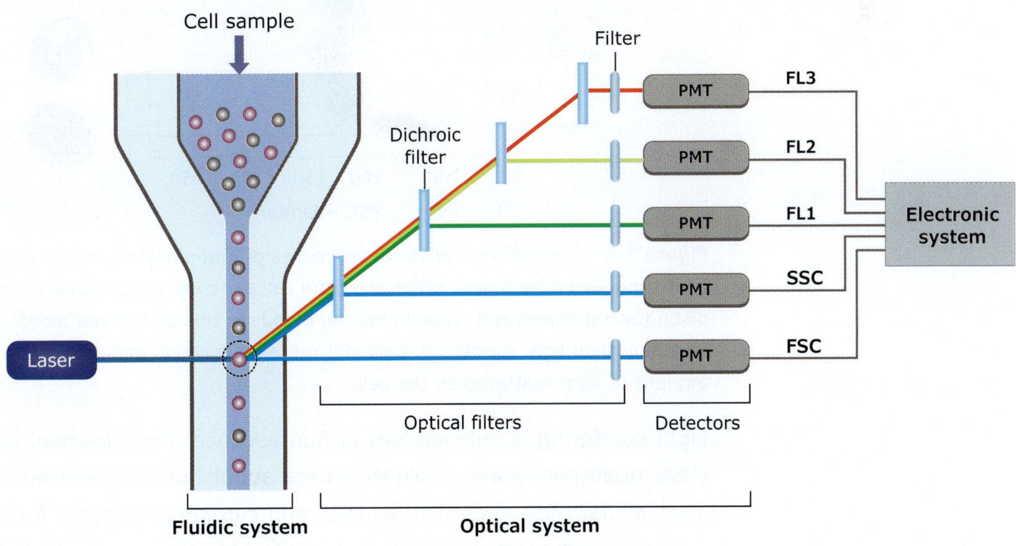

Figure 7.3 Optical system of a typical flow cytometer: The optical system consists of excitation optics and collection optics. The excitation optics consist of the laser and lenses that are used to shape and focus the laser beam. The collection optics consist of a collection lens to collect light emitted from the particle–laser beam interaction and a system of optical mirrors and filters to route specified wavelengths of the collected light to designated optical detectors.

Scattering and fluorescence

Flow cytometers measure light scattering in two distinct directions: forward direction (**forward scatter** or **FSC**) and side direction (**side scatter** or **SSC**). FSC is measured in the same direction as the incident light beam, while SSC is measured at approximately 90° to the incident light beam. Both FSC and SSC provide unique information about an object, and a combination of the two can be used to differentiate different objects in a heterogeneous sample.

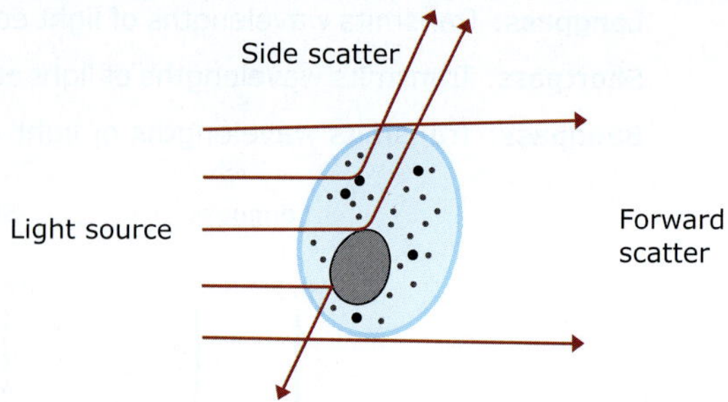

Figure 7.4 Diagram of forward and side scatter. The amount of light scattered in a forward direction is an indicator of the relative size of a cell. This is referred to as *forward scatter.* Likewise, light scattered at higher angles gives rise to information about internal complexity and granularity of a cell, such as cytoplasmic granules, multi-lobed nucleus, cytoplasmic vacuoles, etc. This is referred to as *side scatter*.

FSC is primarily caused by *diffraction* of light around the cell, which is dependent on the size of the cell. Larger cells scatter more light forward, resulting in higher FSC signals. SSC is mainly caused by *refraction* and *reflection* of light within the cell, which is influenced by the internal complexity of the cell. Cells with more internal structures, such as granules and organelles, tend to generate higher SSC signals. Light that is scattered by an object is detected by different detectors. One detector is placed in line with the light beam to measure the FSC from the objects. Detectors perpendicular to the beam measure SSC.

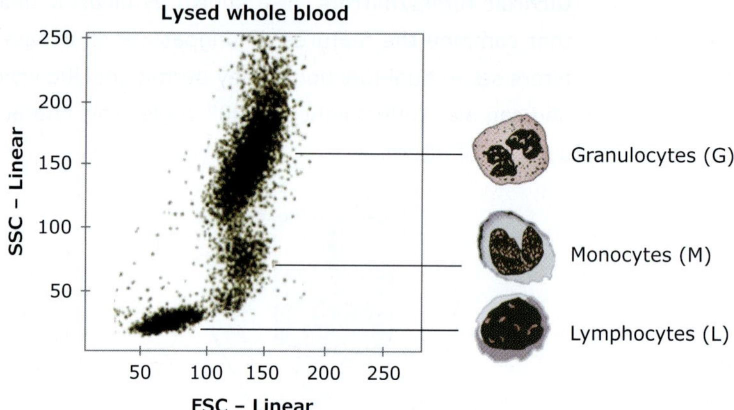

Figure 7.5 The different physical properties of granulocytes, monocytes and lymphocytes allow them to be distinguished from each other and from cellular contaminants. Forward and side scatter can be used to distinguish different cell types in human blood. In this plot, three blood cell populations can be identified based on their light scattering. Each dot represents a cell, and its location in the graph corresponds to the amount of light scattered by the cell.

Light scattering is independent of fluorescence. Measurement of fluorescence emission provides quantitative and qualitative data about fluorochrome-labeled cell surface receptors or intracellular molecules such as DNA and cytokines. When a fluorescent dye is conjugated to a monoclonal antibody, it can be used to identify a particular cell type based on the specific antigenic surface markers of the cell. In a mixed population of cells, different fluorochromes can be used to distinguish between different subpopulations. Detectors perpendicular to the beam measure fluorescence.

The staining pattern of each subpopulation, along with FSC and SSC data, can be analyzed to identify the specific types of cells present in a sample and determine their relative proportions. This information allows researchers to differentiate between various cell types and study their biological functions.

Fluorescence-activated cell sorting (FACS)

FACS is a technique used for the physical separation of specific sub-populations of cells from a heterogeneous mixture based on their fluorescence characteristics. However, there's a common misunderstanding where FACS is wrongly used interchangeably with the term 'flow cytometer.' While a flow cytometer is used for cell analysis, it does not conduct cell sorting. FACS employs fluidics and fluorescence components similar to those in flow cytometers, but they are able to divert a specific population from within a heterogeneous sample into a separate tube, typically based on specified fluorescence characteristics.

The most commonly used cell sorting method is the electrostatic deflection of droplets. In this method, the cell suspension stream is focused in a vibrating nozzle. As the cell continues to travel down the stream, the stream eventually breaks into droplets and the particle of interest is captured in a drop. To prevent the break-off point happening at random distances from the nozzle and to maintain consistent droplet sizes, the nozzle is vibrated at high frequency. The droplets containing a target cell are charged electrically (positively or negatively) at a specific moment called the '**break-off point**'. This break-off point describes the instant the droplet containing the target cell separates from the stream. When a charged droplet passes through a high voltage electrostatic field between deflection plates, it is deflected and collected into the corresponding collection tube. The deflection of the charged droplet is towards the oppositely charged plate, so that it is separated from uncharged and oppositely charged droplets. The speed of cell sorting depends on several factors, including particle size and the rate of droplet formation.

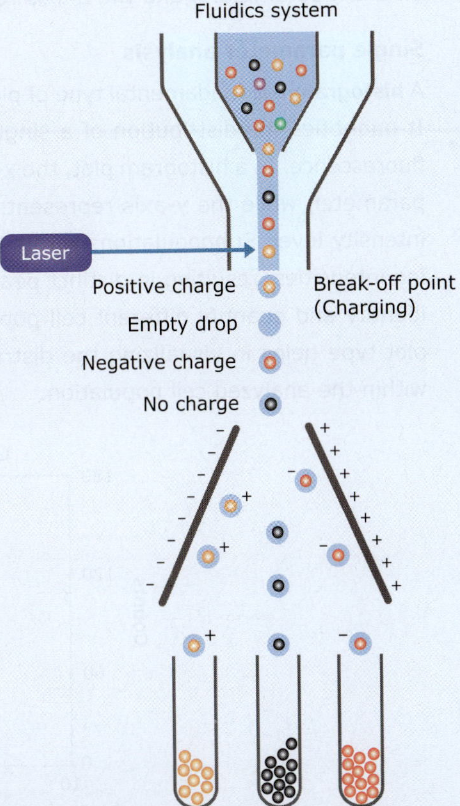

Figure 7.6 Fluorescence-activated cell sorter (FACS) separates cells that are labeled differentially with a fluorescent reagent. A mixture of fluorescently stained cells is passed through a small aperture so that each drop that emerges contains just one cell. The fluorescence detector identifies the signal from drops containing the correct cell and applies an electric charge to these drops. When the drops reach the electric plates, the charged ones are deflected into a separate beaker.

Data analysis

Data analysis in flow cytometry involves processing the information collected from the cells passing through the cytometer. This analysis includes interpreting various parameters, such as fluorescence intensity, scatter signals, and other characteristics measured by the instrument. Researchers use specialized software to analyze and interpret the data generated by flow cytometers. One of the key processes involved in this analysis is **gating**, a process of selecting specific subsets of cells within a heterogeneous sample, allowing researchers to focus on cells of interest while excluding unwanted cells or debris. For example, in a mixed cell population containing lymphocytes, monocytes, and granulocytes, if the researcher is interested in examining only the fluorescence characteristics of lymphocytes, they can create an analysis gate around the lymphocyte population. This gate essentially filters the data, displaying fluorescence properties solely from the lymphocyte subset, facilitating a more targeted analysis of the desired cell type.

Additionally, statistical analysis and visualization tools are commonly employed to extract meaningful information and draw conclusions from the analyzed data. Flow cytometry data can be visualized using various types of plots. Analysis of a single parameter is typically displayed as a single parameter histogram. Two parameters can be plotted simultaneously in a 2-D plot, with one parameter on the x-axis and the other parameter on the y-axis. In a 3-D plot, three parameters can be visualized simultaneously, where the x and y-axes represent different parameters, and the z-axis represents the number of events per channel.

Single parameter analysis

A **histogram** is a fundamental type of plot used in flow cytometry for single parameter analysis. It quantifies the distribution of a single parameter, such as forward scatter, side scatter, or fluorescence. In a histogram plot, the x-axis represents the range of intensities for the selected parameter, while the y-axis represents the number of events (cell count) at each particular intensity level. Subpopulations within the sample will have different fluorescence or scattering intensities, resulting in distinct peaks in the histogram. These peaks allow researchers to identify and quantify different cell populations based on their specific properties. Thus, this plot type helps in visualizing the distribution and intensity variations of a single parameter within the analyzed cell population.

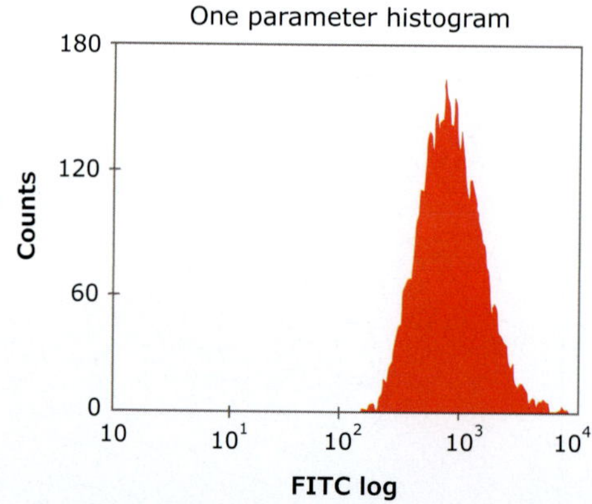

Figure 7.7 Histogram plot displaying a single parameter (such as forward scatter, side scatter, or fluorescence) on the x-axis and the number of events (cell count) on the y-axis. The histogram shows the total number of cells in a sample that possess certain selected physical properties. Cells with the desired properties are known as the positive dataset.

Two parameter analysis

Two parameter plots of flow cytometry data display two measurement parameters, one on the x-axis and another on the y-axis. These plots can be presented as either dot plots, density plots, or contour plots. In a **dot plot**, each dot represents a cell/particle, and the position of the dot on the plot reflects the values of the two measured parameters. The measurement from each detector is referred to as a 'parameter' e.g. forward scatter, side scatter or fluorescence. The parameters can be any combination of scatter and fluorescence. Three common modes of use for the dot plot are: 1. Forward scatter versus side scatter; 2. Single color versus side scatter; and 3. Two-color fluorescence plot.

In the dot plot of FSC versus SSC, one is able to look at the distribution of cells based upon size. The distribution of the dots in the plot can distinguish one type of cell from another. Larger cells are represented as higher FSC values, while the cells with high internal complexity are represented as higher SSC values.

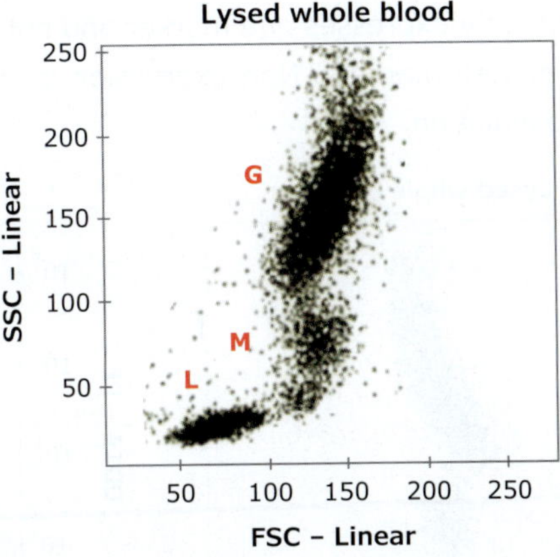

Figure 7.8
A dot plot of lysed whole blood showing the distribution of blood cells based on their size and granularity. The plot is drawn between FSC and SSC data. Each dot or point represents an individual cell that has passed through the instrument. The clusters labeled G, M, and L correspond to granulocytes, monocytes, and lymphocytes, respectively.

In a **density plot**, the frequency distribution of the two parameters is visualized using color to encode the event density. A **contour plot** is similar to a density plot, but instead of color, it uses contour lines to represent the event density. Density and contour plots are not generally used for routine analysis.

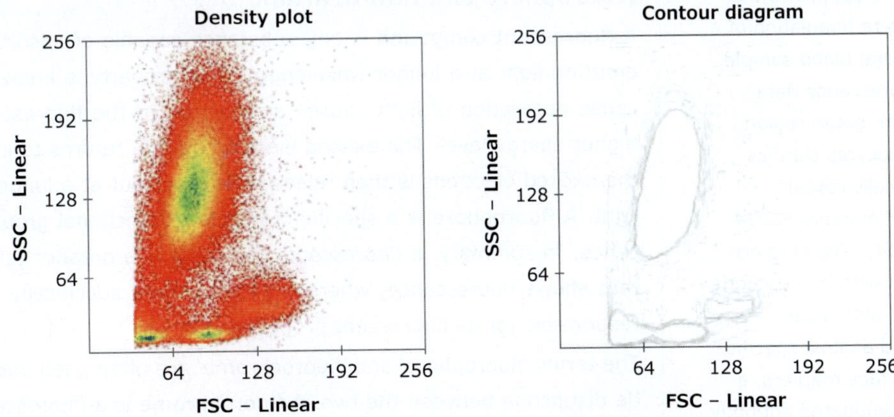

Figure 7.9 Analysis of lysed whole blood using FSC and SSC.

A quadrant marker can be used to divide two-parameter dot plots into four quadrants to differentiate subpopulations as negative, single positive, or double positive. The upper-right quadrant of the plot indicates cells positive for both fluorescent markers, also known as double positive. The lower-left quadrant shows events that are negative for both markers. The upper-left quadrant indicates cells only positive for the parameter represented on the y-axis. The lower-right quadrant contains events that are positive for the parameter represented on the x-axis but negative for the parameter represented on the y-axis. The specific function of quadrant markers is to identify and quantify subpopulations within a sample based on their expression of different markers.

For example, if a researcher analyzing a sample of peripheral blood that has been stained with fluorochromes to identify the CD4 and CD14 surface markers but were only interested in knowing the percentage of monocytes that contained those markers, they might place a gate around the monocyte population of the FSC versus SSC scatter plot thereby limiting the data they visualize to monocytes as compared to the entire population of cells.

Next, using a dot plot to graph the gated region they would be able to gain a clearer understanding of what percentage of monocytes express only CD4, only CD14, neither or both. In this case, two different markers are paired and assessed together. Mutually exclusive marking is where a population only expresses one marker and not the other. Co-expression is where a population expresses both markers. Non-expression is when the population of interest does not express either population.

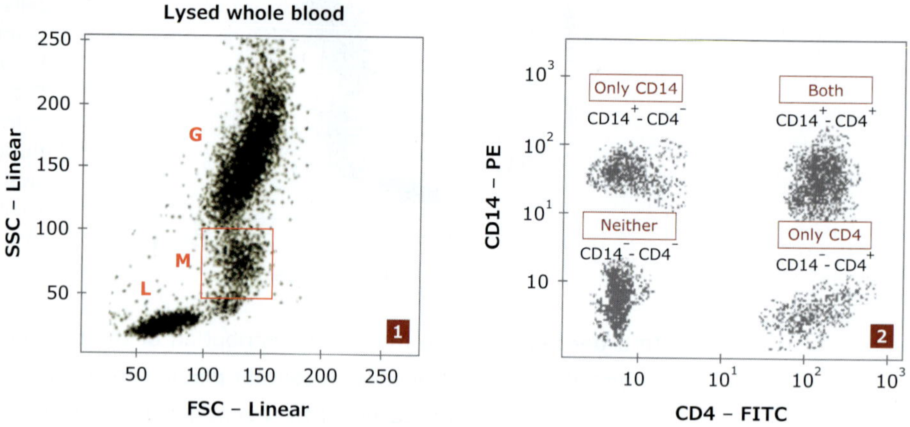

Lysed whole blood

SSC - Linear

250

200

150

100

50

G

M

L

1

50 100 150 200 250

FSC - Linear

CD14 - PE

10^3

10^2

10^1

10

Only CD14

$CD14^+ - CD4^-$

Both

$CD14^+ - CD4^+$

Neither

$CD14^- - CD4^-$

Only CD4

$CD14^- - CD4^+$

2

10 10^1 10^2 10^3

CD4 - FITC

Figure 7.10

1. Light scatter plot – the clusters labeled G, M and L arise from granulocytes, monocytes and lymphocytes respectively. A gate is applied to the monocyte population of a peripheral blood sample.
2. Fluorescence data from the gated region of monocytes clarifies which cells contain surface markers (CD14 and CD4). The cells are stained with fluorescently labeled antibodies directed against specific cell-surface markers, a FITC-conjugated antibody against CD4 and PE conjugated antibody against CD14.

FITC–Fluorescein isothiocyanate dye, PE–Phycoerythrin dye.

Fluorophore and fluorochrome

A **fluorescent compound** is any substance capable of absorbing light at one wavelength and emitting light at a longer wavelength. This property is known as **fluorescence**. It occurs because absorption of light causes an electron in the fluorescent compound to be raised to a higher energy level. The excited electron quickly returns to its ground state. The energy from the excited electrons is then released as light, but at a longer wavelength than the absorbed light. A **fluorophore** is a specific molecule or functional group that is responsible for fluorescence. In summary, a *fluorescent compound* is a broader term encompassing any substance that shows fluorescence, whereas a *fluorophore* specifically refers to the part of a molecule responsible for its fluorescent properties.

The terms 'fluorophore' and 'fluorochrome' are often used interchangeably, but there is a subtle distinction between the two. A **fluorochrome** is a fluorescent dye that is used to stain biological materials, such as cells or tissues. Fluorochromes are typically conjugated to antibodies or other targeting molecules, which allow them to bind to specific structures within the cells or tissues. Hence, all fluorochromes are fluorophores, but not all fluorophores are fluorochromes. A fluorophore can be any molecule or functional group that is capable of fluorescence, regardless of whether or not it is used for staining biological materials.

Application of flow cytometry

Flow cytometry is a powerful tool that has applications in multiple disciplines such as immunology, virology and cancer biology. For example, it is very effective for the study of the immune system and its response to infectious diseases and cancer.

Immunophenotyping

It is the most used application of flow cytometry, involving the identification and differentiation of cell types present in a heterogeneous mixture of cells based on specific markers or antigens.

Cell health and growth

Flow cytometry plays a crucial role in determining cell cycle phases, proliferation rates, and apoptosis. It is instrumental in studying cellular health, growth patterns, and responses to various stimuli.

Apoptosis analysis: Flow cytometry helps in understanding the mechanisms behind cell death. It allows researchers to distinguish between necrosis and early apoptosis by identifying morphological, molecular, and biochemical changes in dying cells.

Cell proliferation assays: These assays measure the rapid expansion of cell populations due to cell division. They are employed in studying cellular metabolic activity and assessing the activation of cells, particularly T cells undergoing rapid clonal expansion.

Cell cycle analysis: Flow cytometry enables the quantification of DNA content in cells. By assessing DNA levels, it's possible to determine stages of the cell cycle.

Transfection efficiency assays

Transfection generally refers to the introduction of foreign DNA into animal cells. Flow cytometry assesses the efficiency of transfection by determining the percentage of cells that have successfully taken up and expressed foreign genes.

Cell sorting

An additional capability of specialized flow cytometers is the ability to sort cells and recover the subsets for post experimental use. This specialized flow cytometer is called a fluorescence activated cell sorter (FACS), a term that is sometimes erroneously used interchangeably with 'flow cytometer'. This usage is incorrect. A flow cytometer is an analytical machine that does not perform cell sorting.

Apoptosis analysis

Apoptosis is a finely tuned mechanism for the control of cell number in eukaryotes. The process is operative during embryogenesis, in tumor regression and in the control of immune response. In most cases, it consists of an ordered sequence of cellular events that start with the transcription of specific genomic sequences, synthesis of specific proteins, and activation of classes of nucleases that cut DNA. These events are paralleled by specific morphological changes in both the cell nucleus and cytoplasm. Cells undergoing apoptosis display typical changes in their morphological, chemical and physical properties, which are well measurable by flow cytometry. The following features of the apoptotic cascade can be observed using flow cytometry:

- Altered phospholipid composition in the plasma membrane
- Activation of caspases
- Cell shrinkage
- Chromatin condensation

- DNA fragmentation
- Expression of proteins involved in apoptosis
- Changes in the mitochondrial membrane potential
- Decrease cytosolic pH
- Altered membrane permeability

Detection of apoptotic cells based on changes in forward scattering

Interaction of a cell with the laser beam produces FSC that correlates with cell size and SSC that correlates with granularity and/or cell density. While necrotic death is characterized by a reduction in both FSC and SSC (probably due to a rupture of plasma membrane and leakage of the cell's content), during apoptosis there is an initial increase in SSC (probably due to the chromatin condensation) with a reduction in FSC (due to the cell shrinkage). In many cases, however, the forward light scattering histograms of apoptotic and live cells overlap and make it difficult to discriminate apoptotic cells based solely on this parameter. Thus, scatter changes alone are not specific to apoptosis and should be accompanied by an additional characteristic associated with cell death.

Detection of apoptotic cells based on Annexin V binding

Live cells have phospholipids asymmetrically distributed in the inner and outer leaflet of plasma membrane, with most of phosphatidylserine on the inner leaflet. Early during apoptosis, cells lose asymmetry, exposing phosphatidylserine on the outer leaflet of plasma membrane.

Annexin V is a calcium-dependent phospholipid-binding protein that binds preferentially to negatively-charged phosphatidylserine. Fluorophore (or fluorochrome) conjugated Annexin V is a useful reagent for the detection of apoptotic cells. The assay involves incubating cells briefly in a solution containing FITC-conjugated Annexin V in a buffer that facilitates its binding. The concentration of $CaCl_2$ in the buffer can be varied to obtain optimal binding for individual cell lines. Apoptotic cells can be detected on the basis of increased binding of FITC-conjugated Annexin V.

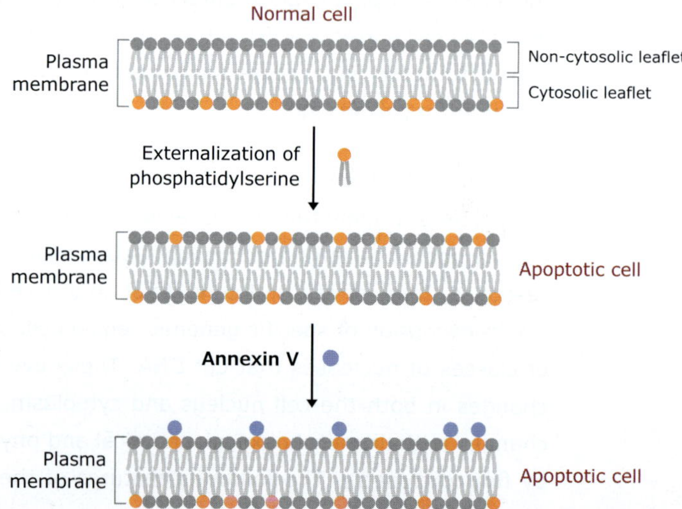

Figurem 7.11 Changes in the plasma membrane are one of the first characteristics of the apoptotic process detected in living cells. Apoptosis can be detected by the presence of phosphatidylserine (PS), which is normally located on the cytoplasmic face of the plasma membrane. During apoptosis PS translocates to the outer leaflet of the plasma membrane and can be detected by flow cytometry through binding to fluorochrome-labeled Annexin V when calcium is present.

The Annexin V binding assay has many of the advantages of other flow cytometric assays in that it is simple, rapid, sensitive, and objective. Unfortunately, it is not specific only for apoptosis in that whenever cell membrane integrity is disrupted (even nonionic detergents), cells may stain with Annexin V, probably because sites on the inner membrane become accessible. Thus, even though loss of membrane asymmetry is an early event characteristic of apoptosis, Annexin V alone cannot distinguish cells dying by apoptosis from those dying by necrosis.

Detection of apoptotic cells based on PI binding

Another characteristic of plasma membranes associated with live cells is that they exclude charged cationic dyes such as propidium iodide (PI) and 7-amino-actinomycin D (7-AMD). PI staining is a good method to distinguish apoptotic, necrotic and normal live cells. The intact membrane of living cells excludes cationic dyes such as PI. In contrast, due to their extensive membrane damage, necrotic cells are quickly stained by short incubation with PI. Apoptotic cells show an uptake of PI that is much lower than that of necrotic cells. It is, therefore, possible to distinguish live (PI-negative), apoptotic (PI-dim) and necrotic (PI-bright) cells from each other. However, apparent PI staining only appears in the late phase of apoptosis due to more damaged cell membranes. Thus, the combined use of cationic dyes (e.g. PI) with annexin V allows the discrimination between live cells (annexin V negative/PI negative), early apoptotic cells (annexin V positive/PI negative), late apoptotic cells (annexin V positive/PI positive) and necrotic cells (annexin V negative/PI positive).

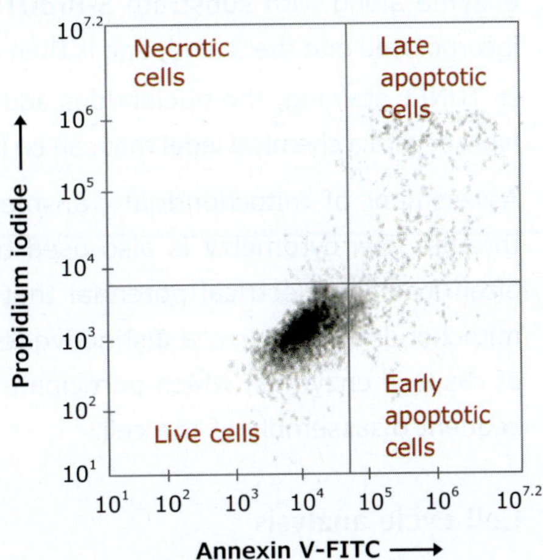

Figure 7.12
Measurement of apoptosis by Annexin V/PI dual stain. Early apoptosis is defined as cells positive for Annexin V only. Late apoptosis is defined as cells positive for Annexin V and PI. Necrotic is defined as cells positive for PI only.

Detection of apoptotic cells based on DNA fragmentation

The late stages of apoptosis are characterized by changes in nuclear morphology, including DNA fragmentation, chromatin condensation, degradation of nuclear envelope, nuclear blebbing and DNA strand breaks. Hence, using flow cytometry, a variety of nuclear changes during apoptosis can be analyzed. Cells undergoing apoptosis display an increase in nuclear chromatin condensation. As the chromatin condenses, cell-permeable nucleic acid stains become hyperfluorescent, thus enabling the identification of apoptotic cells when combined with a traditional dead-cell stain.

The late stage of apoptosis is marked with the activation of endonucleases, which translocate into the nucleus and cause DNA fragmentation and internucleosomal cleavage. These nucleases degrade the higher order chromatin structures into fragments of 50-300 kbp and subsequently into smaller DNA pieces of about 200 bp in length, forming a characteristic 'ladder' of DNA fragments. In order to quantify the percentage of cells in the culture that displays such fragmentation pattern, we can perform *terminal dUTP nick end labeling assay* (**TUNEL assay**).

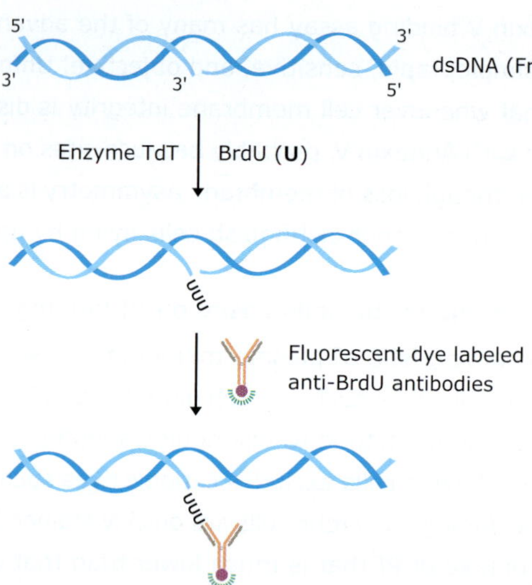

Figure 7.13
TUNEL assay. Apoptotic cells undergo DNA fragmentation. The TUNEL assay uses enzyme terminal deoxynucleotidyl transferase (TdT) to add BrdU onto the 3' ends exposed on DNA fragments. BrdU is added at the 3' end of the DNA, and is then detected with fluorescent dye (FITC) labeled anti-BrdU antibodies.

The TUNEL assay is based on the following principles: the genomic DNA fragments in apoptotic cells contain a 3′-hydroxyl group (and 5′-phosphate) that arise from the cleavage of the phosphodiester bond of the DNA helix. These breaks can be detected on the basis of the ability of the enzyme **terminal deoxynucleotidyl transferase** (TdT) to add nucleotides to free 3′-hydroxyl group in fragmented genomic DNA. Thus, in the TUNEL assay, cells are incubated with this enzyme along with substrate **5-BrdUTP** (5-Bromo-2′-deoxyuridine 5′-triphosphate). BrdU is incorporated into the 3′ end, and is then detected with fluorescently labeled anti-BrdU antibodies. In TUNEL staining, the nucleotides added by TdT are tagged either directly with a fluorescent label or with a chemical label that can be indirectly linked to either a fluorescent label or an enzyme.

Assessment of mitochondrial transmembrane potential and caspases level within the cell through flow cytometry is also used to analyze apoptotic cells. Cells undergoing apoptosis often lose the electrical potential that normally exists across the inner membrane of their mitochondria. Similarly, a distinctive feature of the early stages of apoptosis is the activation of caspase enzymes, which participate in the cleavage of protein substrates and in the subsequent disassembly of the cell.

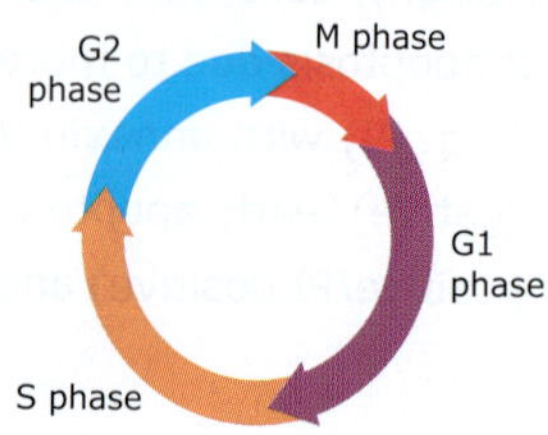

Cell cycle is the period between successive divisions of a cell. Standard cell cycle of eukaryotic cells is divided into four phases:

G1 phase: Cells prepare for DNA replication and the DNA content is 2C.

S-phase: Replication of DNA occurs at this phase. During this phase, there is a DNA content between G1 and G2 phase. The DNA content is doubled by the end of this phase (2C to 4C).

G2 phase: Replication of the DNA completes in this phase. The cells contain twice the amount of DNA (i.e. 4C) that are found in G1 (i.e. 2C).

M-phase: It is the period of actual division, corresponding to the visible mitosis. At this phase, the cells contain 4C amount of DNA.

Cell cycle analysis

In addition to surface immunophenotyping and cytoplasmic characterization, flow cytometry is also used in cell cycle analysis. One of the important applications of flow cytometry is the measurement of DNA content in cells. Cells spend most of time in the G0/G1 phase. In diploid cell, the amount of DNA in this phase is 2C. The duplication of the DNA at the time of cell division occurs during the S-phase. Once the cell has finished S-phase, it has twice the amount of DNA in each cell (4C). Next the cell goes through G2/M phase.

To measure the DNA content, the cells have to be stained with a fluorescent dye that binds to DNA in a stoichiometric manner (the amount of stain is directly proportional to the amount of DNA within the cell). The cells are treated with a fluorescent dye that stains DNA quantitatively. The fluorescence intensity of the stained cells at certain wavelengths will therefore correlate with the amount of DNA they contain. Some dyes possess an intercalative binding mode, such as propidium iodide, whereas others present an affinity for DNA A.T rich regions such as DAPI or G.C rich regions such as chromomycin A_3.

An accurate method for detection of cell cycle progression also uses the incorporation of thymidine analog bromodeoxyuridine (BrdU) during new DNA synthesis. This thymidine analog is used to directly determine the percentage of S-phase cells and also, when used in kinetic studies it permits a determination of the individual times for the components of the cell cycle. In this technique, BrdU is incorporated into newly synthesized DNA in cells entering and progressing through the S-phase of the cell cycle. The incorporated BrdU is then stained with specific fluorescently labeled anti-BrdU antibodies, and the levels of cell-associated BrdU measured using flow cytometry. A dye that binds to total DNA, such as 7-aminoactinomycin D (7-AAD), is often used in conjunction with immunofluorescent BrdU staining.

Typically, in a population of cells that are all proliferating rapidly but asynchronously, about 30–40% will be in S-phase at any instant and become labeled by a brief pulse of BrdU. From the proportion of cells in such a population that are labeled, we can measure the **labeling index**. Once the labeling index is determined, the duration of S-phase can be calculated by multiplying the labeling index by the mean generation time. Similarly, we can estimate the duration of M-phase, by measuring the **mitotic index** (the proportion of cells in mitosis). The duration of M-phase can be calculated by multiplying the mitotic index by the mean generation time.

If the mean generation time for a given population is known and cells are reasonably homogenous with respect to their cell cycle kinetics, the length of G1, S and G2 can be estimated by flow cytometry. The mean generation time of cells in a population can be calculated from the change in cell number with incubation time, using the formula:

$$N = N_0 \times 2^{T/mgt}$$

where, N is the final cell number, N_0 is the initial cell number, T is the elapsed time and mgt is the mean generation time.

Fluorescence-based detection

Live cell measurement

Hoechst 33342

Binds to the minor groove of double-stranded DNA.

Fixed cell measurement

Bromodeoxyuridine (BrdU)

A thymidine analog is incorporated into the genome during the S-phase of the cell cycle. Detected using anti-BrdU antibodies.

DAPI

Binds to the A.T rich regions of the DNA.

Propidium Iodide (PI)

An intercalating agent that stains the cellular genome upon cell fixation.

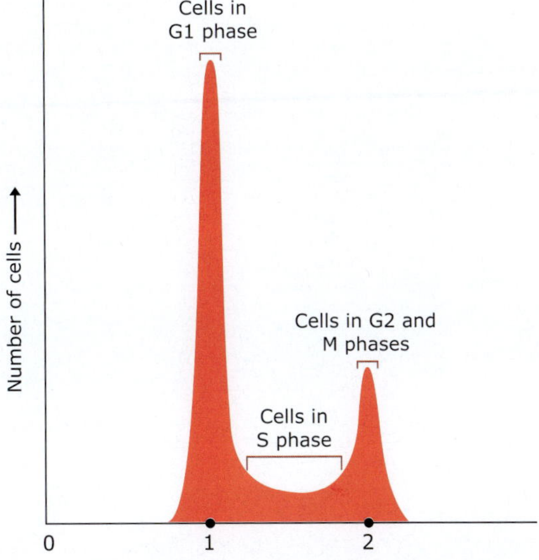

Figure 7.14 Relationship between the cell cycle and the DNA histogram. The cells analyzed here were stained with a fluorescent dye (PI) that binds to DNA, so that the amount of fluorescence is directly proportional to the amount of DNA in each cell. The cells fall into three categories: those that have an unreplicated complement of DNA and are therefore in G1, those that have a fully replicated DNA (twice the G1 DNA content) are in G2 or M-phase, and those that have an intermediate amount of DNA and are in S-phase. The distribution of cells in the case illustrated also indicates that there are greater numbers of cells in G1 than in G2 + M-phase, showing that G1 is longer than G2 and M-phase of this population. Labeling DNA with propidium iodide (PI) allows for fluorescence based analysis of cell cycle. PI stains the nucleus light to dark red in color. Amount of PI fluorescent intensity is correlated to the amount of DNA within each cell. Since the amount of DNA doubles (2C → 4C) between G1 and G2 phases, the amount of fluorescent intensity of the cell population also doubles.

References

Alberts B et al (2008), *Molecular Biology of the Cell*, 5th ed. Garland Science Publishing.

Baumgarth N and Roederer M (2000), A practical approach to multicolor flow cytometry for immunophenotyping. *J Immunol Methods* 243:77-97.

Nunez R (2001), DNA measurement and cell cycle analysis by flow cytometry. *Curr. Issues Mol. Biol* 3(3): 67-70.

Ormerod MG (2000), *Flow Cytometry. A practical approach*. 3rd ed. Oxford University Press.

Riccardi C and Nicoletti I (2006), Analysis of apoptosis by propidium iodide staining and flow cytometry, *Nature Protocols* 1458-1461.

Shapiro HM (2003), *Practical Flow Cytometry*, John Wiley and Sons, Inc.

Watson JV (2005), *Flow Cytometry Data Analysis: Basic Concepts and Statistics,* Cambridge University Press.

8

X-ray crystallography

X-rays are electromagnetic radiation of short wavelength. Like all electromagnetic radiation, X-rays are absorbed, scattered, and diffracted by matter. The *scattering* and *diffraction* of X-rays in gases, liquids, or disordered solids is caused by interaction with electrons. Electrons in ordered arrays of atoms in crystals scatter X-rays only in particular directions; in other directions, the scattering is negligible. Diffraction is the scattering of X-rays in a few specific directions by crystals. **X-ray crystallography** is a method of determining the arrangement of atoms within a crystal. This technique is based on **X-ray diffraction**, a nondestructive technique. When a beam of X-ray strikes a crystal, the beam may be diffracted. From the angles and intensities of these diffracted beams, a three-dimensional picture of the density of electrons within the crystal can be derived. From this electron density, the mean positions of the atoms in the crystal can be determined. This method acts as an atomic microscope, using X-rays instead of visible light to determine the three-dimensional structure of crystals. X-rays cannot be focused by lenses to form an image of a molecule. Instead, the X-rays are diffracted from a single crystal. This technique requires three distinct steps:

1. growing a crystal,
2. collecting the X-ray diffraction pattern from the crystal, and
3. constructing and refining a structural model to fit the X-ray diffraction pattern.

When X-rays interact with a single particle, it scatters the incident beam uniformly in all directions. However, when X-rays interact with a solid material, the scattered beams can add together in a few directions and reinforce each other to yield *diffraction*. The regularity of the material is responsible for the diffraction of the beams. Hence, for X-ray crystallography, molecule must be crystallized. A **crystal** is built up of many billions of small identical units called **unit cells**. The *unit cell* is the smallest and simplest volume element that is completely representative of the whole crystal. Each unit cell may contain one or more molecules. A *crystal* is a stacking of unit cells repeated in three dimensions to build a *lattice*, leaving no space between the unit cells.

The terms **diffraction** and **scattering** are often used interchangeably and are considered to be almost synonymous. Scattering simply refers to the ability of objects to change the direction of a wave. Diffraction describes a specialized case of light scattering in which an object with regularly repeating features produces an orderly pattern.

Theory of X-ray diffraction

A molecular structure resolved at atomic level means that the positions of each atom can be distinguished from those of all other atoms in three-dimensional space. The closest distance between two atoms in space is the length of a covalent bond, and a typical length of a covalent

Refraction refers to the bending of light rays due to differences in refractive indices when it moves from one medium to another, while **diffraction** is a phenomenon that is often described as 'waves bending around corners'.

bond is approximately 0.12 nm. If we require to resolve the atoms of a macromolecule, the wavelength of light required for our atomic microscope would necessarily be <0.24 nm. This falls into the X-ray range of the electromagnetic spectrum. However, X-rays cannot be focused and thus cannot form an image of an object in the same manner as a light microscope. We rely on the constructive and destructive interference caused by scattering radiation from the regular and repeating lattice of a single crystal to determine the structure of macromolecules.

A good way to understand X-ray diffraction is to draw an analogy with visible light. Light has certain properties that is best described by considering wave nature. Whenever wave phenomena occur in nature, interaction between waves can occur. If waves from two sources are in same phase with one another, their total amplitude is additive (**constructive interference**); and if they are out of phase, their amplitude is reduced (**destructive interference**). This effect can be seen in figure 8.1 when light passes through two pinholes in a piece of opaque material and then falls onto a white surface. Interference patterns result, with dark regions where light waves are out of phase and bright regions where they are in phase.

If the wavelength of the light (λ) is known, one can measure the angle α between the original beam and the first diffraction peak and then calculate the distance d between the two holes with the formula

$$d = \frac{\lambda}{\sin \alpha}.$$

The same approach can be used to calculate the distance between atoms in crystals. Instead of visible light, which has longer wavelength to interact with atoms, we can use a beam of X-rays. X-rays, like light, are a form of electromagnetic radiation, but they have a much smaller wavelength. The wavelengths of X-rays (typically around 0.1 nm) are of the same order of magnitude as the distances between atoms or ions in a molecule or crystal. If a narrow beam of X-rays is directed at a crystalline solid, most of the X-rays will pass straight through it. A small fraction, however, scatters by the atoms in the crystal. The electrons of an atom are primarily responsible for the scattering of X-rays. The number of electrons in a given volume of space (the *electron density*) determines how strongly an atom scatters X-rays. The interference of the scattered X-rays leads to the general phenomenon of *diffraction*.

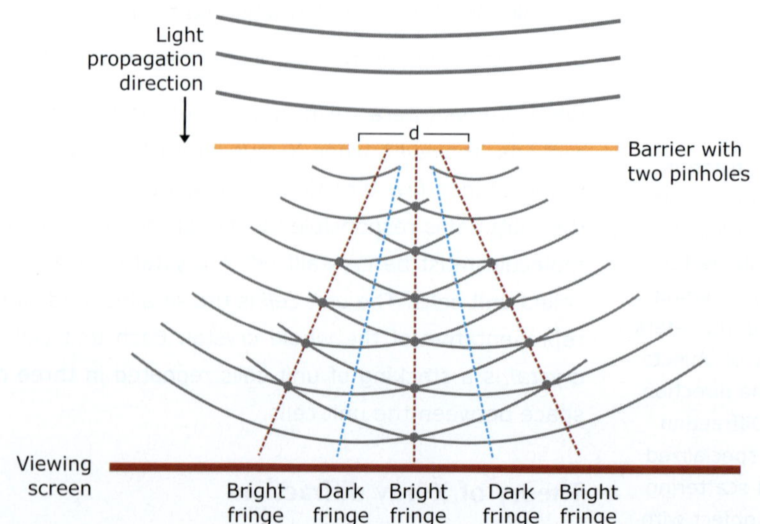

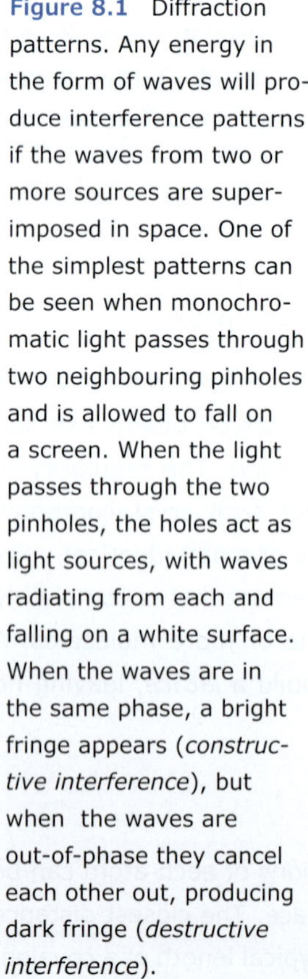

Figure 8.1 Diffraction patterns. Any energy in the form of waves will produce interference patterns if the waves from two or more sources are superimposed in space. One of the simplest patterns can be seen when monochromatic light passes through two neighbouring pinholes and is allowed to fall on a screen. When the light passes through the two pinholes, the holes act as light sources, with waves radiating from each and falling on a white surface. When the waves are in the same phase, a bright fringe appears (*constructive interference*), but when the waves are out-of-phase they cancel each other out, producing dark fringe (*destructive interference*).

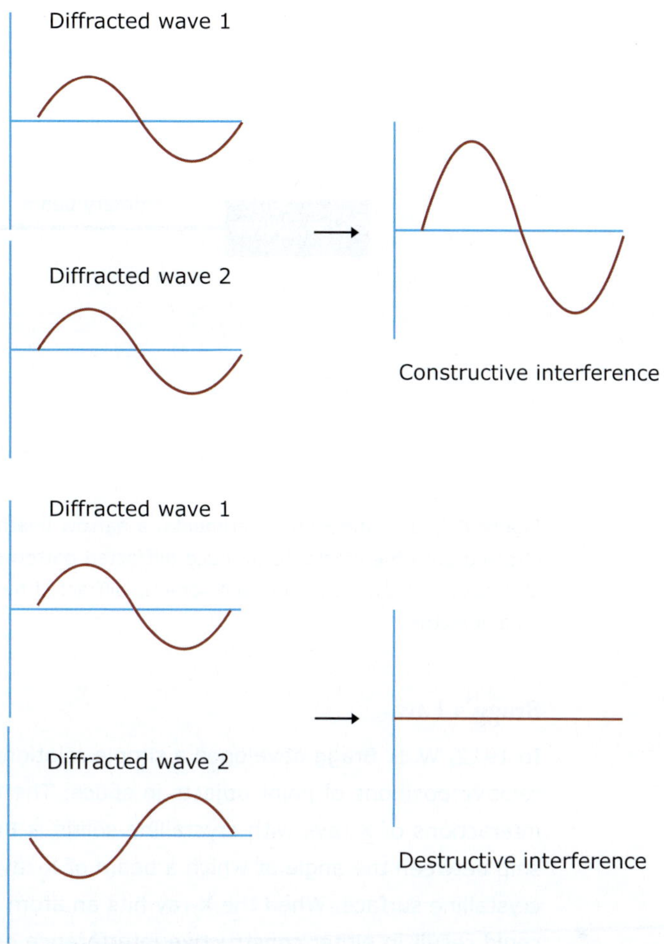

Figure 8.2 Interference occurs among the waves scattered by the atoms when crystalline solids are exposed to X-rays. There are two types of interferences depending on how the waves overlap one another. Constructive interference occurs when the waves are moving in phase with each other; and destructive interference occurs when the waves are out-of-phase.

X-rays are electromagnetic radiation of wavelengths 0.1–100 Å. X-rays can be produced by bombarding a metal target (most commonly copper, molybdenum or chromium) with electrons produced by a heated filament and accelerated by an electric field. A high-energy electron collides with and displaces an electron from a lower orbital in a target metal atom. Then an electron from a higher orbital drops into the resulting vacancy, emitting its excess energy as an X-ray photon.

A crystal is placed in an X-ray beam between the X-ray source and a detector, and a regular array of spots called reflections is generated. The spots are created by the diffracted X-ray beam; and each atom in a molecule makes a contribution to each spot. The electrons that surround the atoms are the entities which physically interact with the incoming X-ray to diffract them, not the atomic nuclei. The diffraction pattern is recorded on a photographic plate and then analyzed to reveal the nature of that lattice. The position and intensity of each spot in the X-ray diffraction pattern contain information about the positions of the atoms in the crystal that gave rise to it. An optical scanner precisely measures the position and the intensity of each reflection and transmits this information in digital form to a computer for analysis. The position of a reflection can be used to obtain the direction in which that particular beam was diffracted by the crystal. The intensity of a reflection is obtained by measuring the optical absorbance of the spot on the film, giving a measure of the strength of the diffracted beam that produced the spot.

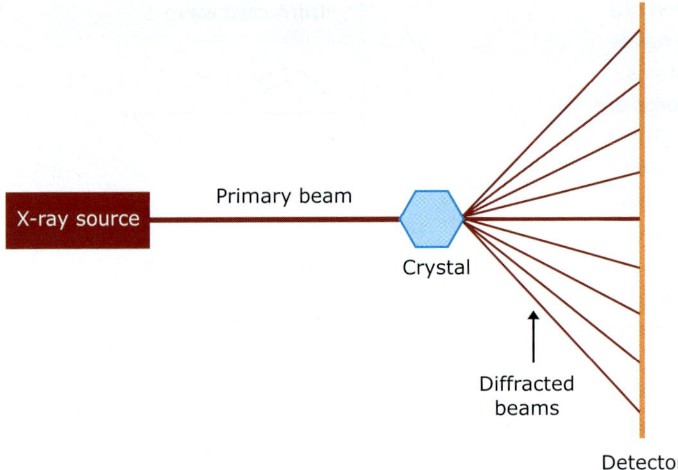

Figure 8.3 In diffraction experiments, a narrow beam of X-ray is taken out from the X-ray source and directed onto the crystal to produce diffracted patterns. When the primary beam hits the crystal, most of it passes straight through, but some is diffracted by the crystal. These diffracted beams are recorded on a detector.

Bragg's Law

In 1912, W. L. Bragg developed a simple relationship to explain how diffraction relates to the relative positions of point objects in space. The mathematical expression that describes the interactions of X-rays with crystalline solids is **Bragg's Law**. This law describes the relationship between the angle at which a beam of X-rays of a particular wavelength diffracts from a crystalline surface. When the X-ray hits an atom it can be diffracted. However, the diffraction could result in either constructive interference or destructive interference. Diffracted beams which have constructive interference will be observed and from that observation, the spacing between planes of atoms, d, can be determined. Knowing the plane spacings can lead us to different pieces of information about the material.

To obtain constructive interference, the path difference between the two incidents and the scattered waves, which is $2d \sin\theta$, has to be a multiple of the wavelength λ. For this case, the Bragg's equation gives the relation between interplanar distance, d and reflection angle, θ.

$$n\lambda = 2d \sin\theta$$

where, n is an integer,

λ is the wavelength of X-ray,

d is the spacing between the planes and

θ is the reflection angle between the incident ray and the scattering planes.

Consider, a plane lattice crystal with interplanar distance, d. Suppose a beam of X-rays of wavelength λ is incident on the crystal at an angle θ, the beam will be diffracted in all possible atomic planes. The path difference between any two diffracted waves is equal to the integral multiple of wavelength. The ray P gets diffracted from the surface, while the ray Q has to undergo some path difference. The extra distance traveled by the ray Q from the figure is (BC + CD). From the diagram, either BC or CD is equal to $d \sin\theta$. So the path difference is:

$$d \sin\theta + d \sin\theta = n\lambda$$

$$2d \sin\theta = n\lambda$$

Here, n is the order = 1, 2, 3,.... This is Bragg's law.

Fiber diffraction

Many important biological substances do not form crystals. Among these are most membrane proteins and fibrous materials like collagen, DNA and muscle fibers. Like crystals, fibers are composed of molecules in an ordered form. When irradiated by an X-ray beam perpendicular to the fiber axis, fibers produce distinctive diffraction patterns that reveal their dimensions at the molecular level. The order in a fiber is one-dimensional (along the fiber) rather than three dimensional, as in a crystal.

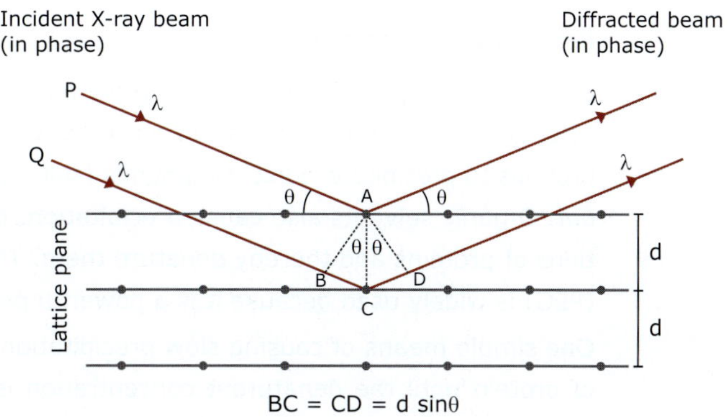

When **n** is an integer (1, 2, 3 etc.), the diffracted waves from different layers are perfectly in phase with each other and produce constructive interference. Otherwise, the waves are not in phase. Diffracted beams which have constructive interference will be observed and from that observation, the spacing between planes of atoms, d, can be determined.

The relationship between the reflection angle, θ, the distance between the planes, d, and the wavelength, λ, given by Bragg's law can be used to determine the size of the unit cell.

To determine the size of the unit cell, the crystal is oriented in the beam so that diffraction is obtained from the specific set of planes in which any two adjacent planes are separated by the length of one of the unit cell axes. This distance, d, is then equal to $\lambda/(2 \sin\theta)$. The wavelength (λ), of the beam which is known (since we use monochromatic radiation) as the reflection angle, θ, can be calculated from the position of the diffracted spot on the film.

The reflection angle for a diffracted beam can be calculated from the distance (r) between the diffracted spot on a film and the position where the primary beam hits the film. From the geometry shown in the figure 8.4, the tangent of the angle $2\theta = r/A$. A is the distance between crystal and film while r can be measured on the film. r is the distance from the original axis to that point where diffracted spots were observed. Hence reflection angle, θ, can be calculated. The angle between the primary beam and the diffracted beam is 2θ. This angle is equal to the angle between the primary beam and the reflecting plane plus the reflection angle, both of which are equal to θ.

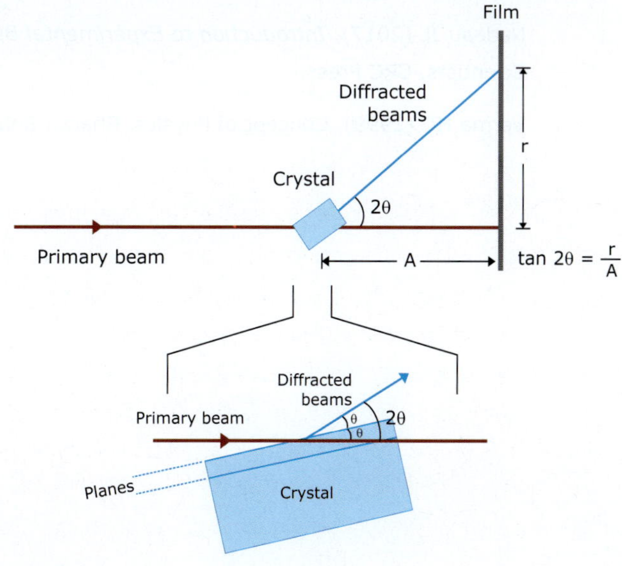

Figure 8.4 Measurement of reflection angle, θ, for a diffracted beam.

Protein crystallization

Crystallographers grow crystals of proteins by slow, controlled precipitation from aqueous solution under conditions that do not denature the protein. A number of substances cause proteins to precipitate. Ionic compounds (salts) precipitate proteins by a process called *salting out*. Organic solvents also cause precipitation, but they often interact with hydrophobic portions of proteins and thereby denature them. The water-soluble polymer polyethylene glycol (PEG) is widely used because it is a powerful precipitant and a weak denaturant.

One simple means of causing slow precipitation is to add denaturant to an aqueous solution of protein until the denaturant concentration is just below that required to precipitate the protein. Then water is allowed to evaporate slowly, which gently raises the concentration of both protein and denaturant until precipitation occurs. Whether the protein forms crystals or instead forms a useless amorphous solid depends on many properties of the solution, including protein concentration, temperature, pH and ionic strength. Finding the exact conditions to produce good crystals of a specific protein often requires many careful trials and is perhaps more art than science.

References

Berg JM, Tymoczko JL and Stryer L (2006), *Biochemistry*, 6th ed. W.H. Freeman and Company.

Branden C and Tooze J (1999), *Introduction to protein structure*, 2nd ed. Garland Publishing, Inc.

Drenth J (2007), *Principles of protein X-ray crystallography*, 3rd ed. Springer.

Glusker JP, Lewis M and Rossi M (1994), Crystal structure analysis for chemists and biologists. New York, VCH Publishers.

Ladd FC and Palmer RA (2003), Structure Determination by X-ray Crystallography, Kluwer 4th ed. Academic/plenum publishers.

Lodish H, Berk A, Kaiser C et al (2007), *Molecular Biology of the Cell*, 4th ed. W.H. Freeman and Company.

Nadeau JL (2017), *Introduction to Experimental Biophysics*: Biological Methods for Physical Scientists, CRC Press.

Verma HC (1999), Concept of Physics, Bharati Bhawan, India.

9 Patch clamp techniques

The introduction of the patch clamp technique has revolutionized the study of cellular physiology by providing a method of observing the function of individual ion channels in a variety of cell types. It permits high resolution recording of the ionic currents flowing through a cell's plasma membrane.

The patch clamp technique has been invented by Sakmann and Neher in the 1976, for which they received the Nobel Prize in Physiology and Medicine in 1991. This technique is based on a very simple idea. A micropipette with a very small opening is used to make tight contact with a small area, or patch, of cell membrane. After the application of a small amount of suction to the back of the pipette, the contact between pipette and membrane becomes so tight that no ions can flow between the pipette and the membrane. Thus, all the ions that flow when a single ion channel opens must flow into the pipette. The resulting electrical current, though small, can be measured with an ultra-sensitive electronic amplifier connected to the pipette. Based on the geometry involved, this arrangement usually is called the **cell-attached patch clamp recording**.

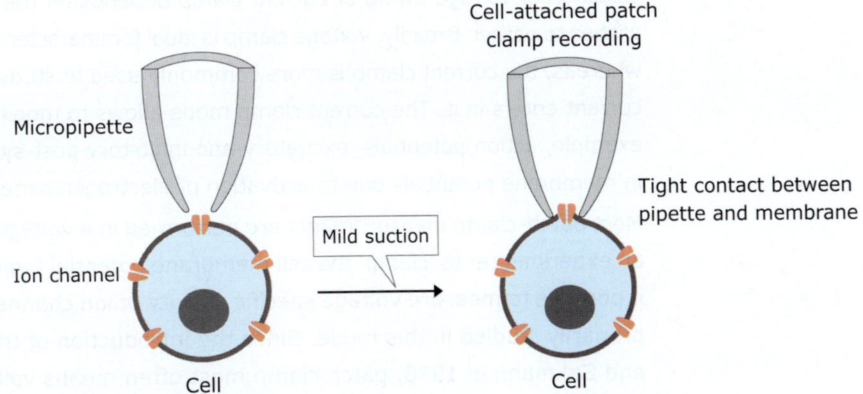

Figure 9.1 Cell-attached patch clamp recording. When the pipette is in closest proximity to the cell membrane, mild suction is applied to gain a tight seal between the pipette and the membrane.

In order to form the cell-attached mode, a pipette tip is placed on the surface of the cell, forming a low resistance contact (seal) with its membrane. Slight suction applied to the upper end of the pipette results in formation of a tight seal. Such a seal with a resistance in the range of gigaohms is called '**giga-seal**'. In the cell-attached mode, recordings are made from the membrane area under the pipette, while the structure of the cell remains intact. A metal electrode inside the glass pipette, containing a salt solution resembling the fluid normally found within the cell, accomplishes transduction of the ionic current into the electrical current while another one in the bath solution serves as ground. Pt and Ag/AgCl electrodes are especially used for their low junction potentials and weak polarization. The tight seal between pipette and cell membrane isolates the membrane patch electrically, which means that all ions fluxing the membrane patch flow into the pipette and are recorded by an electrode connected to a highly sensitive electronic amplifier. A bath electrode is used to set the zero level.

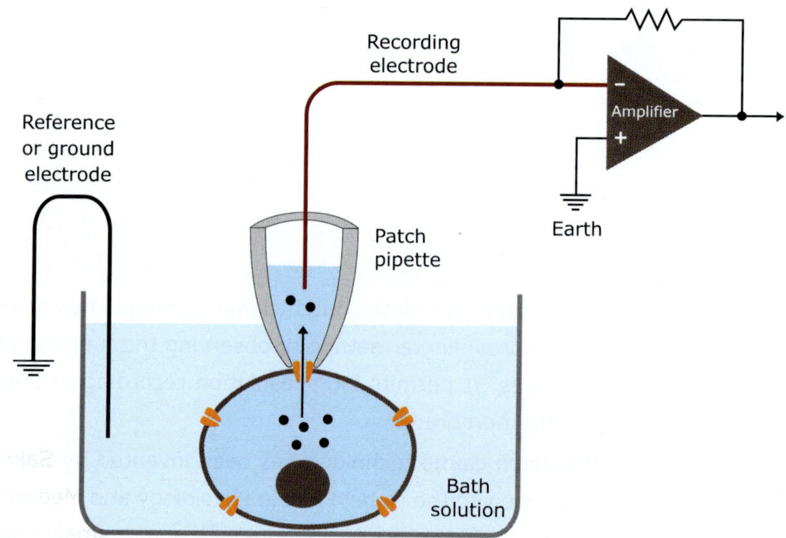

Figure 9.2 General principle of patch-clamp recordings. A micropipette containing electrolyte solution is tightly sealed onto the cell membrane and, thus, isolates a membrane patch electrically. Currents fluxing through the channels in this patch, hence, flow into the pipette and can be recorded by an electrode that is connected to a highly sensitive differential amplifier. In the voltage-clamp configuration, a current is injected into the cell via a negative feedback loop to compensate changes in membrane potential. Recording of this current allows conclusions about the membrane conductance.

Recording mode

Patch clamp technique can be operated in one of two modes – *current clamp* and *voltage clamp* modes. **Current clamp** mode applies a constant current at the tip of the pipette, and measures the membrane potential over time. In contrast, **voltage clamp** mode holds a cell (or patch of membrane) at a specified potential, and measures the current required to maintain that set voltage. The selection of voltage clamp or current clamp depends on the type of information the investigator wishes to gather. Broadly, voltage clamp is ideal for characterizing individual ion channel properties whereas, the current clamp is more commonly used to study how a cell responds when electrical current enters in it. The current clamp mode allows to monitor different forms of cell activity, for example, action potentials, excitatory and inhibitory post-synaptic potentials as well as changes in membrane potentials due to activation of electrogenic membrane transporters.

Most patch-clamp measurements are performed in a *voltage clamp mode*. Voltage clamp allows an experimenter to 'clamp' the cell membrane potential (voltage) at a chosen value. This makes it possible to measure voltage specific activity of ion channels. The voltage-gated channels are primarily studied in this mode. Since the introduction of the patch clamp technique by Neher and Sakmann in 1976, patch clamp most often means voltage clamp of a membrane patch.

The voltage-clamp mode is used to control the voltage of the membrane. It takes advantage of a patch-clamp amplifier which allows maintaining (clamping), through a feedback circuit, a specified membrane voltage and measuring, at the same time, the current across the membrane. In practice, during a voltage-clamp experiment, the electronic feedback system of the amplifier measures the membrane voltage and compares it to a pre-set voltage defined by the experimenter. When a current is activated, the voltage of the membrane changes. To compensate for this change and bring the voltage to the pre-set value, a current of equivalent magnitude (but opposite direction) is injected through the pipette.

Patch clamp recording configurations

A variety of recording configurations can be used depending on the type of activity the investigator is interested in recording. The different recording configurations are:

- Cell-attached configuration
- Whole-cell configuration
- Perforated configuration
- Inside-out configuration
- Outside-out configuration

Cell-attached configuration

In a cell-attached configuration, the micropipette is sealed onto the cell membrane to obtain a gigaseal. The cell remains intact and single channel current can be recorded within the membrane patch. The current flowing through ion channel enclosed by the pipette tip within that patch is measured by means of a connected patch clamp amplifier. This configuration is ideal for recording the activity of single ion channel located in a small patch of cell membrane of an intact cell. Cell-attached configuration is the precursor to all other patch clamp recording configurations.

Whole-cell configuration

If the membrane patch within the pipette is disrupted by briefly applying strong suction, the interior of the pipette becomes continuous with the cytoplasm of the cell. This arrangement allows measurements of electrical potentials and currents from the entire cell and is, therefore, called the *whole-cell configuration.* The whole-cell configuration also allows diffusional exchange between the pipette and the cytoplasm, producing a convenient way to inject substances into the interior of a 'patched' cell. There are various ways to break the patch membrane to obtain access to the cell interior. In the 'conventional' whole-cell technique, the membrane is disrupted with extra suction or with a brief, high-voltage pulse. In either case, a diffusional pathway is created that allows quite effective dialysis of the cell.

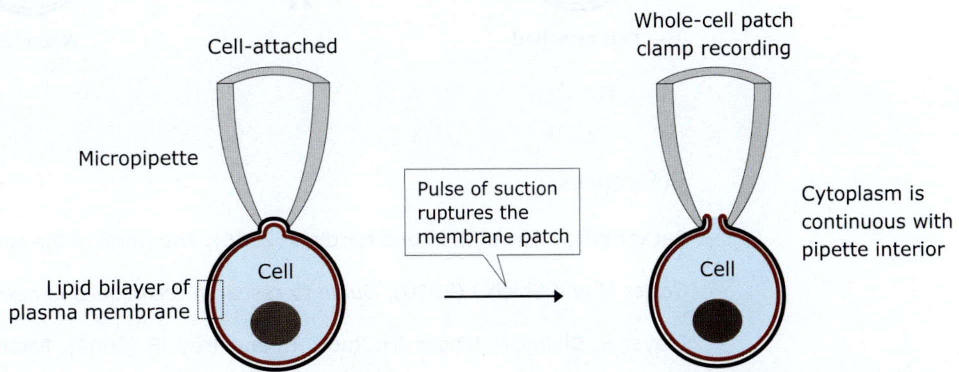

Figure 9.3 Whole-cell patch clamp recording configuration. By applying brief but strong suction, the cell membrane is ruptured and the pipette gains access to the cytoplasm.

This method is used to record the electrical potentials and currents from the entire cell. The researcher can choose voltage clamp or current clamp recording mode in whole-cell configuration. In *voltage clamp mode*, the voltage is kept constant and current is recorded, whereas, in the *current clamp mode,* the current is kept constant and changes in the membrane potential is observed.

Perforated configuration

An alternative variant of *whole-cell* configuration is the perforated patch clamp recording which was developed to overcome the dialysis of cytoplasmic constituents that occurs with conventional whole-cell recording. In perforated-patch clamp recordings, *perforants*, such as the antibiotics nystatin, amphotericin and gramicidin, are included in the pipette solution. These perforants form channels in the membrane attached to the patch pipette. These pores allow certain monovalent ions to permeate, enabling electrical access to the cell interior but prevent the dialysis of larger molecules and other ions.

Inside-out and outside-out configuration

The *cell-attached* configuration may be used (as such) to record single-channel activity, or one may proceed to isolate the patch of the membrane by withdrawing the pipette from the cell. If a small piece of membrane is pulled away from the cell without disrupting the seal, this yields a condition where a small patch of the membrane with its intracellular surface is exposed. This arrangement is called the **inside-out** configuration. The cytosolic side of the patch now faces the outside bath solution. This is often used to investigate single channel activity with the advantage that the medium that is exposed to the intracellular surface can be modified. From the *whole-cell* configuration, if the pipette is retracted, a membrane patch is detached, which forms a convex loop structure around the tip of the pipette with the extracellular face exposed to the outside solution. This geometric orientation is termed as **outside-out** configuration. This configuration allows one easily to change the extracellular side of the patch. It is, therefore, often used to study receptor-operated ion channels. As for the inside-out configuration, the cytosolic environment of the channels is lost on patch excision.

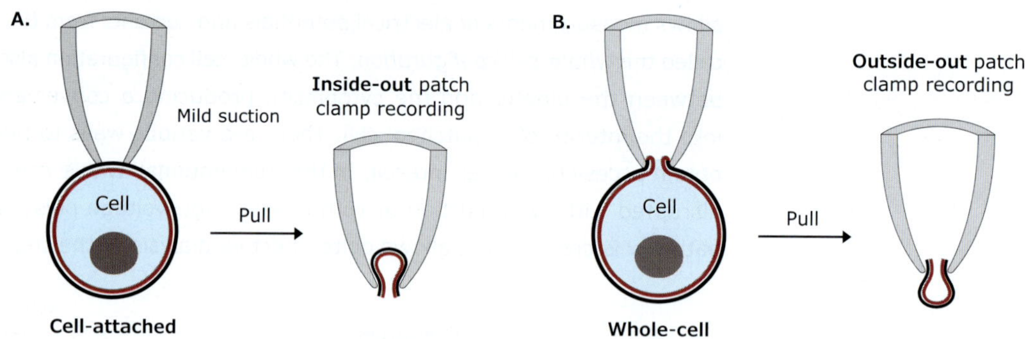

Figure 9.4 A. *Inside-out configuration.* In the cell-attached configuration, the pipette is retracted and the patch is separated from the rest of the membrane and exposed to air. The cytosolic face of the membrane is exposed. B. *Outside-out configuration.* If the pipette is pulled slowly away after whole-cell mode, the outside-out configuration will be accomplished. It allows a vesicle to form from the patch of the cell membrane due to resealing, so the cytosolic side faces the pipette solution.

References

Becker WM, Kleinsmith LJ and Hardin J (2006), *The world of the cell*, 6th ed. Pearson Education Inc.

Carter M and Shieh J (2010), *Guide to research techniques in neuroscience*, Academic Press.

Inayat S, Dikin DA, Singer JH, Ruoff RS and Troy JB (2009), Patch clamp technique: review of the current state of the art and potential contributions from nanoengineering. *J Nanoengineering and Nanosystems* vol. 222 no. 3-4 121-131.

Kandel ER, Schwartz JH and Jessell TM (2000), *Principles of neural science*, 4th ed. McGraw-Hill.

Molleman A (2003), *Patch Clamping: An introductory guide to patch clamp electrophysiology*, John Wiley and Sons, Inc.

Sakmann B and Neher E (1984), Patch clamp techniques for studying ionic channels in excitable membranes. *Annual Review of Physiology* Vol. 46: 455-472.

10

Immunotechniques

10.1 Immunoprecipitation reaction

Antigen-antibody interaction is highly specific and occurs in a similar way as a bimolecular association of an enzyme and a substrate. The binding between antigen (Ag) and antibody (Ab) involves weak and reversible non-covalent interactions consists mainly of van der Waals forces, electrostatic forces, H-bonding and hydrophobic forces. The smallest unit of antigen that is capable of binding with antibodies is called an *antigenic determinant* (or **epitope**). The corresponding area on the antibody molecule combining with the epitope is called **paratope**. The number of epitopes on the surface of an antigen is its **valence**. The valence determines the number of antibody molecules that can combine with the antigen at one time. If one epitope is present, the antigen is **monovalent**. Most antigens, however, have more than one copy of the same epitope and are termed **polyvalent**.

Immunoprecipitation reaction results from the interaction of a *soluble* antibody with a *soluble* antigen to form an *insoluble* complex. Antibodies that aggregate soluble antigens are called **precipitins**. Formations of an antigen-antibody lattice depend on the valency of both antibody and antigen. The antibody must be bivalent for precipitation reaction to occur. Monovalent Fab fragments cannot form precipitate with antigen. Similarly, the antigen must be either bivalent or polyvalent. If the antigen is bi- or polyvalent, it can bind with multiple antibodies. Eventually, the resulting cross-linked complex becomes so large that it falls out of solution as a precipitate. Immunoprecipitation reaction can be performed in *solution* or in *gel.*

Immunoprecipitation reaction in solution

Precipitation occurs maximally only when there are optimal proportions of the two reacting substances – antigen and antibody. Hence, an insoluble antigen-antibody complex formation occurs within a narrow concentration range known as the **zone of equivalence**. This represents the conditions under which antigen-antibody complexes are formed that are sufficiently large to be precipitated. Outside the equivalence concentration, conditions known as *antigen* or *antibody excess* occur, which result in the formation of small, soluble complexes. When increasing concentrations of antigen are added to a series of tubes that contain a constant concentration of antibodies, variable amounts of precipitate form. If the amount of the precipitate is plotted against the amount of antigen added, a **precipitin curve**, as shown in the figure 10.1, is obtained. This immunoprecipitation reaction can be used to remove particular antigens from a solution.

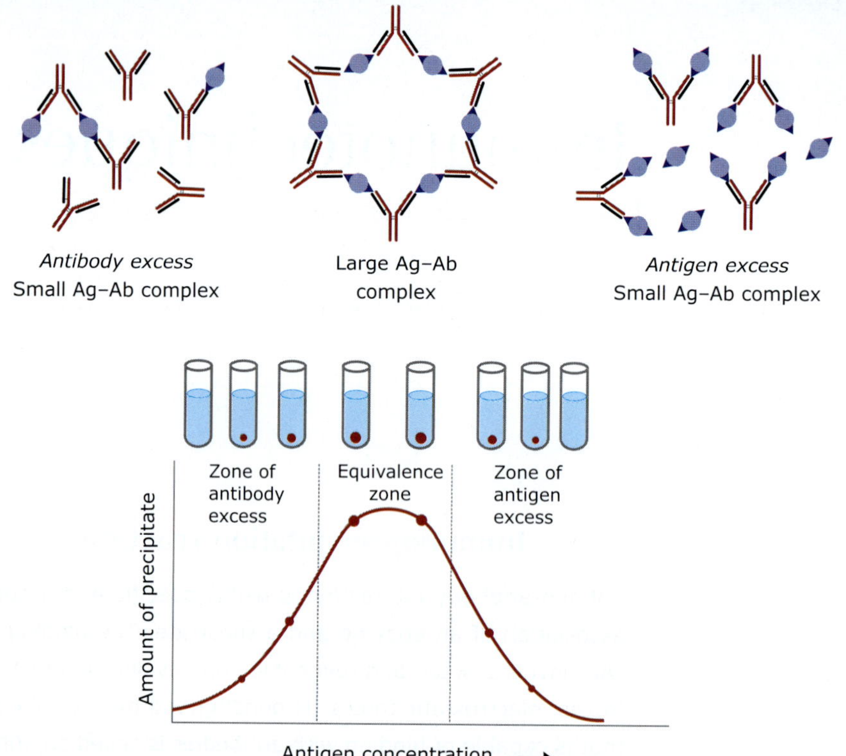

Figure 10.1 A precipitation reaction can be performed by placing a constant amount of antibody (Ab) in a series of tubes and adding increasing amounts of antigen (Ag) to the tubes. After the precipitate forms, each tube is centrifuged to pellet the precipitate, the supernatant is poured off, and the amount of precipitate is measured. Plotting the amount of precipitate against increasing antigen concentrations yields a precipitin curve. Under conditions of antibody excess or antigen excess, extensive lattices do not form and precipitation is inhibited. In the equivalence zone, the ratio of antibody to antigen is optimal. As a result, a large multimolecular lattice is formed at equivalence, the complex increases in size and precipitates out of solution.

Immunoprecipitation reactions in gels

Immunoprecipitation reactions carried out in agar gels are referred to as *immunodiffusion reactions*. When antigen and antibody diffuse toward one another in gel, or when an antibody is incorporated into the gel and antigen diffuses into the antibody-containing matrix, a visible line of precipitation (**precipitin line**) will form. Visible precipitation occurs in the region of equivalence. No visible precipitate forms in regions of antibody excess or antigen excess. Two types of immunodiffusion reactions can be used to determine the relative concentrations of antibodies or antigens as well as the identity of antigens. These immunodiffusion techniques are *radial immunodiffusion* and *double immunodiffusion*.

Radial immunodiffusion (Mancini method)

The relative concentration of an antigen can be determined by a simple quantitative assay in which an antigen sample is placed in a well and allowed to diffuse into agar gel containing antibody. In agar gel, antibody is uniformly distributed. At the region of equivalence, a precipitation ring forms around the well. The diameter of the ring is proportional to the log of the concentration of antigen since the amount of antibody is constant.

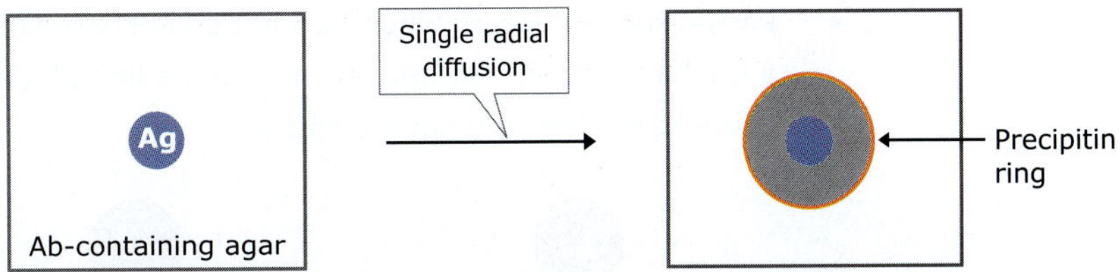

It is used for the quantitative estimation of antigen. By running different concentrations of a standard antigen on the gel and by measuring the diameters of their precipitin rings, a standard calibration graph is plotted. Antigen concentrations of unknown samples, run on the same gel, can be found by measuring the diameter of precipitin rings and extrapolating this value on the calibration graph.

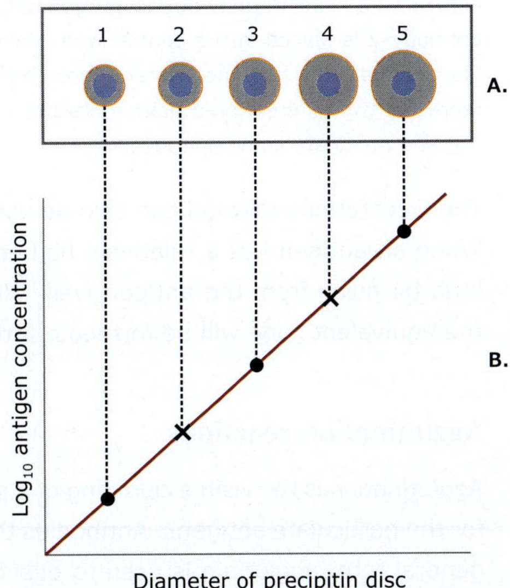

Figure 10.2 Radial immunodiffusion. **A.** Precipitin ring formed in gel containing monospecific antiserum. Wells 1, 3 and 5 contained standard antigen solutions of increasing concentration. Wells 2 and 4 contained samples of the antigen at unknown concentrations. **B.** Semilog plot of diameter of the precipitin discs of standard antigen solutions (•) against concentration. Measurement of the diameters of the precipitin discs of the unknown solutions (×) allows an estimation of the antigen concentration to be made by simple interpolation.

Ouchterlony double immunodiffusion

In the Ouchterlony double immunodiffusion (developed by Ouchterlony), both antigen and antibody diffuse radially from wells toward each other, thereby establishing a concentration gradient. Antigens and antibodies diffuse toward each other at rates that increase in proportion to their concentration in the well, but decrease in proportion to their sizes. They form a precipitin line where they meet at equivalence. It is an immunological technique used for the detection, identification and quantification of antibodies and antigens.

Using this technique, the antigenic relationship between two antigens can be analyzed. Distinct precipitation line patterns are formed against the same antibody depending on whether two antigens share all antigenic epitopes or partially share their antigenic epitopes or do not share their antigenic epitopes at all.

- **Identity** occurs when two antigens share identical epitopes.
- **Non-identity** occurs when two antigens are unrelated i.e. share no common epitopes. The antisera form an independent precipitin line with each antigen, and the two lines cross.

- **Partial identity** occurs when two antigens share some epitopes but one of the other has a unique epitope. The antiserum forms a line of identity with the common epitope and a curved spur with a unique epitope.

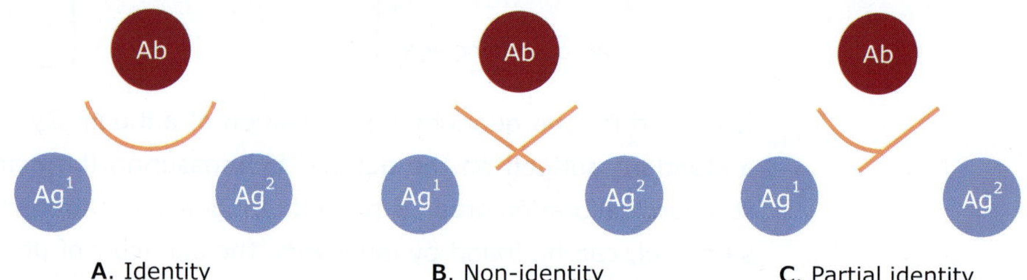

Figure 10.3 Ouchterlony double immunodiffusion patterns. Antibody that is a mixture of antibody-1 and antibody-2 is placed in the central well. Unknown antigens are placed in the outside wells. **A.** *Identity*: The arc indicates that the two antigens (Ag^1 and Ag^2) are identical. **B.** *Non-identity*: Two crossed lines represent two different precipitation reactions. The antigens share no common epitopes. **C.** *Partial identity*: Ag^1 and Ag^2 share some epitopes.

The Ouchterlony method can also be used to estimate the relative concentration of antigens. When an antigen has a relatively higher concentration, the equivalent zone will be formed a little bit away from the antigen well. When an antigen has a relatively lower concentration, the equivalent zone will be formed a little bit closer to the antigen well.

Agglutination reactions

Agglutination is the visible clumping of a particulate antigen when mixed with antibodies specific for the particulate antigens. Antibodies that produce such reactions are called **agglutinins**. The general term agglutinin is used to describe antibodies that agglutinate particulate antigens. When the antigen is an erythrocyte, the term *hemagglutination* is used. All antibodies can theoretically agglutinate particulate antigens; but IgM, due to its high valence, is a particularly good agglutinin.

Agglutination reactions are similar, in principle, to precipitation reactions. Just as an excess of antibody inhibits precipitation reactions, such excess can also inhibit agglutination reactions; this inhibition is called the **prozone effect**. In the case of antigen excess, the **postzone effect** occurs.

The agglutination reactions can be *direct* (active) or *indirect* (passive). When the antigen is an integral part of the surface of a cell or other insoluble particle, the agglutination reaction is referred to as **direct agglutination**. The agglutination test only works with particulate antigens. However, it is possible to coat cell or other insoluble particles with a soluble antigen (e.g. viral antigen, a polysaccharide or a hapten) and use the coated cells in an agglutination test for antibody to the soluble antigen. This is called **passive agglutination**.

Zeta potential

The surfaces of certain particulate antigens may possess an electrical charge, as, for example, the net negative charge on the surface of erythrocytes caused by the presence of sialic acid. When such charged particles are suspended in saline solution, an electrical potential, termed the zeta potential, is created between particles, preventing them from getting very close to each other. This introduces a difficulty in agglutination of charged particles by antibodies.

Coombs test

When antibodies bind to erythrocytes, they do not always result in agglutination. This can result from the antigen-antibody ratio (either antigen excess or antibody excess); or in some cases, **zeta potential** on the erythrocytes preventing the effective cross-linking of the cells. In order to detect the presence of non-agglutinating antibodies on erythrocytes, one simply adds a second antibody directed against the antibodies attached to their respective epitopes on erythrocytes. This anti-immunoglobulin can now cross-link the erythrocytes and result in agglutination. This test is known as the *Coombs test* (anti-immunoglobulin test).

The Coombs test is based on two important facts: 1. that antibodies of one species (e.g. human) are immunogenic when injected into another species (e.g. rabbit) and lead to the production of antibodies against the antibodies, and 2. that many of the anti-immunoglobulins (e.g. rabbit anti-human Ab) bind with antigenic determinants present on the Fc portion of the antibody and leave the Fab portions free to react with antigen. For example, if human IgG antibodies are attached to their respective epitopes on erythrocytes, then the addition of rabbit antibodies to human IgG will result in their binding with the Fc portions of the human antibodies bound to the erythrocytes by their Fab portions. These rabbit antibodies not only bind with the human antibodies that are bound to the erythrocyte but also, by doing so, form cross-links.

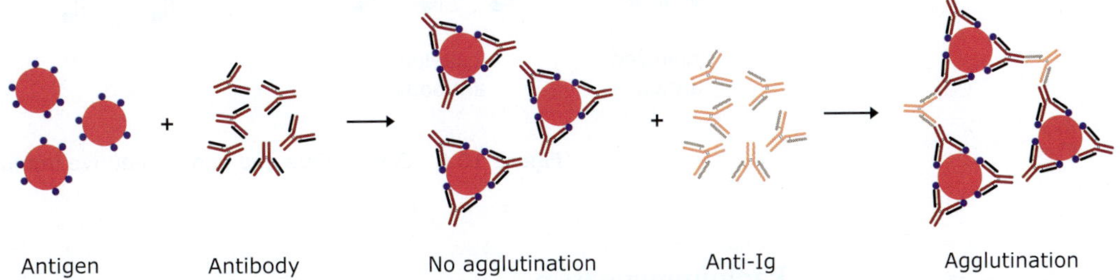

Figure 10.4 Indirect Coombs test.

There are two versions of the Coombs test: *direct* and *indirect Coombs test*. The two versions differ somewhat in the mechanics of the test, but both are based on the same principle. In the direct test, anti-immunoglobulins are added to the erythrocytes that are suspected of having antibodies bound to antigens on their surfaces. The indirect test is used to detect the presence of antibodies specific to antigens on the erythrocytes in the serum.

10.2 Immunoassays

Immunoassays are based on the specific antibody-antigen reactions. It is a biochemical method that measures the presence or concentration of an analyte (referred to as an *antigen*) in a solution through the use of an antibody. The assay relies on the ability of an antibody to recognize and bind epitopes of antigen in the solution. The analytes bind to the antibody and form an immune complex. Immunoassays are the methods of choice because of their high specificity, sensitivity and low limits of detection. The detection system in immunoassays depends on readily detectable labels (e.g. radioisotopes or enzymes) coupled to one of the immunoanalytical reagents (i.e. analyte or antibody).

Immunoassays can be *competitive* and *noncompetitive*. The **competitive immunoassay** relies on the competition between the labeled and unlabeled antigens for a limited amount of specific antibody. In this immunoassay, an unlabeled antigen of interest competes with a constant added amount of labeled similar antigen for a limited amount of specific antibody. Because the two antigens compete for the same antibody, the labeled antigen must react identically to the unlabeled one. Competitive immunoassays require only a small amount of antibody. One example of a competitive immunoassay is traditional radioimmunoassay (RIA). In **noncompetitive immunoassay**, the labeled antibody detects the antigen of interest. This immunoassay requires an excess of labeled antibody towards the antigen of interest. In competitive immunoassays, the antigen is labeled, while in noncompetitive immunoassays, the antibody is labeled. Maximal sensitivity will be reached by decreasing the amount of antibodies in competitive assays and increasing the antibody concentration in noncompetitive assays.

Competitive immunoassay

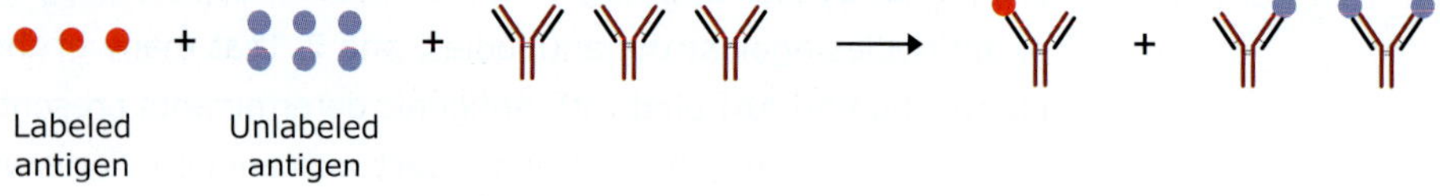

Noncompetitive immunoassay

Unlabeled
antigen

Labeled
antibody

Figure 10.5 Competitive and noncompetitive immunoassays.

Radioimmunoassay

Radioimmunoassay (RIA) uses antibodies to detect and quantitate the amount of antigen (analyte) in a sample. It is a very sensitive and specific assay for determining the concentration of a particular antigen in a sample, based on competitive binding between unlabeled and radioisotope labeled antigen for its specific antibody. The first RIA was developed by Yalow and Berson in 1959. They used radiolabeled insulin to assess the concentration of insulin in human plasma.

The basic principle of radioimmunoassay is competitive binding, where a radiolabeled antigen (usually with ^{125}I, which binds to exposed tyrosine residues of proteins) competes with an unlabeled antigen for a fixed number of antibody. Since the initial description of Yalow's assay, many technical variations have been developed to make the method more rapid and reliable.

An RIA requires a sample containing the antigen of interest, a complementary antibody, and a radiolabeled version of the antigen. The sample antigen and antibody are incubated together, allowing the sample antigens to bind with the antibody. The radiolabeled antigen is then added. The radiolabeled antigen competes with the sample antigen and displaces it from the antibody. The more sample antigen present, the less the radiolabeled antigen is able to bind to the antibody.

In order to calibrate the measurements, known amounts of unlabeled antigen are added to the tubes that contain the same fixed amount of antibody attached to the bottom, and in addition, the same fixed amount of radiolabeled antigen in the tube. Now in the tube, there is a competition between the unlabeled and radiolabeled antigen to bind to the antibody binding sites. The relative amount of unlabeled or radiolabeled antigen that binds to the antibody is strictly a function of their relative amounts in the tube. Because the radiolabeled and the unlabeled antigen bind to the antibody with the same affinity, the higher the concentration of the unlabeled antigen, the better it can compete for the binding sites. Therefore, less radiolabeled antigen will bind, which will result in a smaller CPM (counts per minute). In this fashion, a standard curve can be obtained by measuring counts from tubes, which have increasing known concentrations of the unlabeled antigen. The concentration of antigen in an unknown sample can be determined by reference to a standard curve. Each experiment must therefore have its own standard curve.

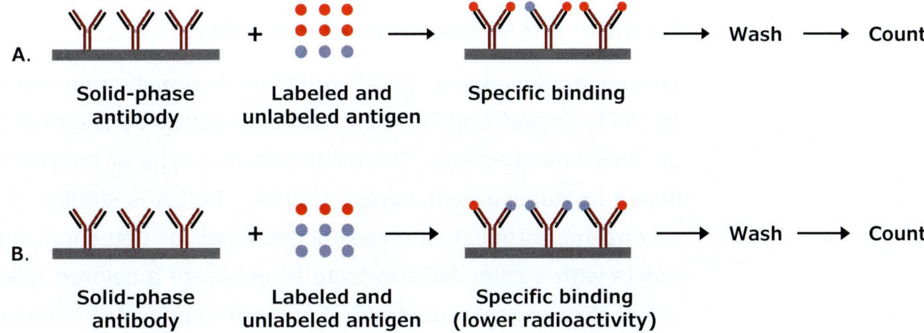

Figure 10.6 Principle of RIA. Radiolabeled antigen competes with unlabeled antigen for a limited number of binding sites on solid-phase antibody. **A.** Very little unlabeled antigen is present, making radioactivity of the solid phase high. **B.** More unlabeled antigen is present, and the radioactivity of the solid phase is reduced in proportion to the amount of unlabeled antigen bound.

Let's assume that we want to determine the concentration of a particular cytokine in the blood of a patient. First, the wells of a microtiter plate are coated with a constant amount of antibody specific for the cytokine. For purposes of quantitation, a standard curve based on unlabeled cytokines is generated by adding increasing, known concentrations of cytokine to the wells of one row of the antibody-coated plate. Then, a known, constant amount of radiolabeled cytokine is added to each well of that row. As more unlabeled antigen competes with the labeled antigen, less and less radiolabeled cytokine will bind. After a predetermined incubation period, the amount of plate-bound radiolabeled material is assessed by washing off the unbound material and measuring the remaining, antibody-bound radioactivity in individual wells.

A standard curve is generated in this way. Measurement of the amount of cytokine in experimental samples is accomplished by treating unknown samples in exactly the same way as the standard curve. The investigator then compares the amount of radioactivity bound to the plate in the experimental wells with the radioactive signal obtained in the standard curve wells containing known amounts of unlabeled cytokine.

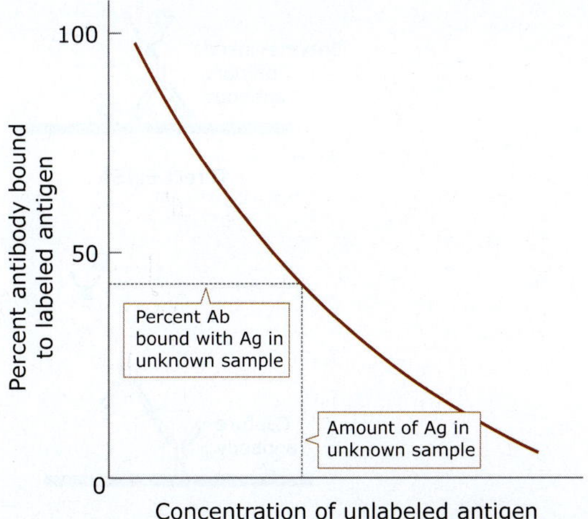

Figure 10.7 A standard curve used in radioimmunoassay for quantitative estimation of antigen in an unknown sample. The concentration of antigen in an unknown sample can be determined by reference to a standard curve constructed from data obtained by allowing varying amounts of unlabeled antigen to compete.

Enzyme–linked immunosorbent assay

Enzyme immunoassays (EIA) exploit an enzymatic reaction for detecting the immune reaction. In 1971, Engvall and Perlmann (independently by Weemen and Schuurs) described the use of enzyme-linked agents. The most common type of enzyme immunoassays in use is **enzyme-linked immunosorbent assays** (ELISA). ELISA is similar in principle to RIA; but depends on an enzyme rather than a radioactive label for detection. An enzyme-linked with an antibody reacts with a colorless substrate to generate a colored reaction product. Such a substrate is called **chromogenic substrate**. A number of enzymes have been employed for ELISA, including alkaline phosphatase, horseradish peroxidase, urease and beta galactosidase. Detection can also occur by fluorescently-labeled antibodies; here the assay is usually termed a *fluorescence-linked immunosorbent assay* (**FLISA**).

Table 10.1 Enzymes used for conjugation of antibodies

Enzyme	Source	Reaction catalyzed
Peroxidase	Horseradish	H_2O_2 + Oxidisable substrate $\longrightarrow$ Oxidized product + $2H_2O$
Alkaline phosphatase	Calf intestine	$R{-}O{-}P_i + H_2O \longrightarrow R{-}OH + P_i$
β-Galactosidase	*E. coli*	β-D-Galactoside + $H_2O \longrightarrow$ Galactose + Alcohol
Urease	Jack bean	$(NH_2)_2\,CO + 3H_2O \longrightarrow CO_2 + 2NH_4OH$

Different variants of ELISA have been developed, allowing qualitative detection or quantitative measurement. Hence, ELISA can be divided into two types, *qualitative* and *quantitative*. **Qualitative ELISA** provides a simple positive or negative result for a sample, while **quantitative ELISA** reflects the concentration of the target molecule in a sample via a standard curve. In general, ELISAs can be grouped into the four main categories: direct, indirect, sandwich, and competitive ELISAs.

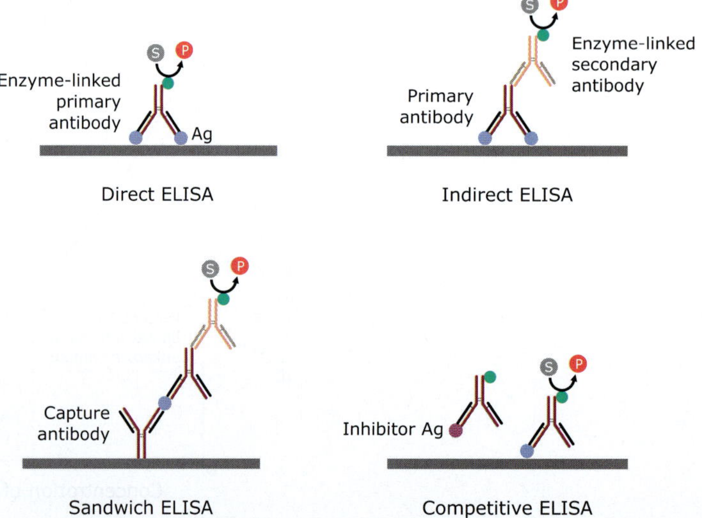

Figure 10.8 Common ELISA formats. In this assay, the antigen of interest is immobilized by direct adsorption to the assay plate or by first attaching a capture antibody to the plate surface. Detection of the antigen can then be performed using an enzyme-linked primary antibody (direct detection) or a matched set of unlabeled primary and enzyme-linked secondary antibodies (indirect detection).

Direct ELISA

This is the simplest ELISA technique. The antigen present in the sample is first immobilized to the wall of the wells of a microtiter plate. The wells are then washed thoroughly, leaving only the adsorbed antigen. An enzyme-linked antibody, complementary to the antigen of interest, is then added to the wells where it binds to the antigen. The well is again washed. This leaves a bound antigen-antibody complex on the surface of the well. A substrate is then added, which will be converted by the enzyme-linked with antibodies into a detectable product. Detection may be based on color, fluorescence, or luminescence. This method has the advantage of being quicker and simpler than the other ELISA methods, with fewer steps, and just one antibody. It does, however, have some limitations. In complex samples containing a range of different antigens, there will be a variety of antigens adsorbed onto the well that is not the antigen of interest. This proves problematic when the antigen of interest is in low abundance as the sensitivity of the test is reduced. Also, conjugating the antibody with an enzyme has the potential to reduce the affinity of the antibody to the antigen and thus reduces sensitivity.

Indirect ELISA

Sample containing the antigen of interest is adsorbed onto the microtiter well. A complementary antibody (**primary antibody**) is then added, which binds to the antigen forming a complex. After any free primary antibody is washed away, the presence of antibody bound to the antigen is detected by adding an enzyme-linked **secondary antibody**, which binds to the primary antibody. Any free secondary antibody then is washed away, and a substrate for the enzyme is added. The amount of colored reaction products that forms is measured by specialized spectrophotometric plate readers, which can measure the absorbance of all of the wells.

This method differs from the direct method in that the antibody binding to the antigen does not have attached to it an enzyme. Instead, the purpose of this antibody is to act as a bridge between the antigen and an enzyme-linked secondary antibody. The indirect ELISA is used to detect the presence of antibodies against HIV. In this test, the antigens are adsorbed to the bottom of a well. Then antibodies (act as primary antibodies) from a patient are added to the coated well and allowed to bind to the antigen. Finally, the enzyme-linked secondary antibodies to the human antibodies are allowed to react in the well, and unbound antibodies are removed by washing. A substrate is then applied. An enzyme reaction suggests that the enzyme-linked antibodies were bound to human antibodies, which in turn, implies that the patient had antibodies to the viral antigen.

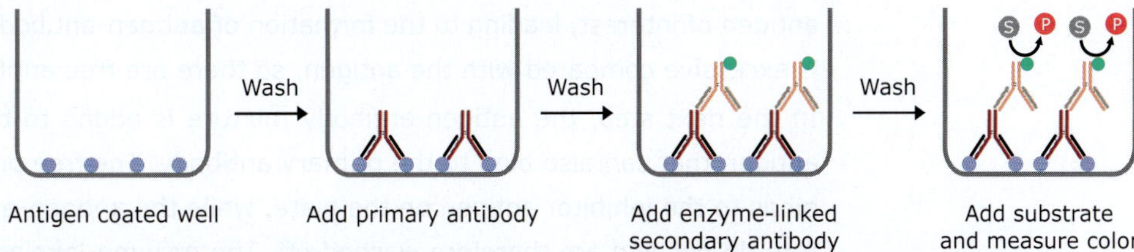

Figure 10.9 Indirect ELISA. The purified antigen of interest is first immobilized on the bottom of a well. Test antiserum is added and allowed to incubate. If any antibody in the test antiserum has bound to the immobilized antigen, their presence is detected by adding an enzyme-conjugated secondary antibody. Enzyme substrate (S) is then added and the amount of colored reaction product (P) that forms is measured.

Sandwich ELISA

An antigen can be detected or measured by a sandwich ELISA. In this ELISA, the antibody (rather than the antigen) is immobilized on a microtiter well. Immobilized antibodies (act as *capture antibodies*) are able to bind to specific antigen present in the sample with high affinity. A separate enzyme-linked antibody that recognizes a different epitope from that recognized by the immobilized first antibody is then used to detect the bound antigen. Thus, in this case, two antigen-specific antibodies are required, and they need to bind to different epitopes of the same antigen. To measure the presence and amount of specific antigen, a sample containing antigen is added and allowed to react with the immobilized antibody. After the well is washed, a second enzyme-linked antibody specific to a different epitope on the antigen is added and allowed to react with the bound antigen. Any free secondary antibodies are then washed away, and a substrate for the enzyme is added. Finally, the amount of colored reaction products that form is measured.

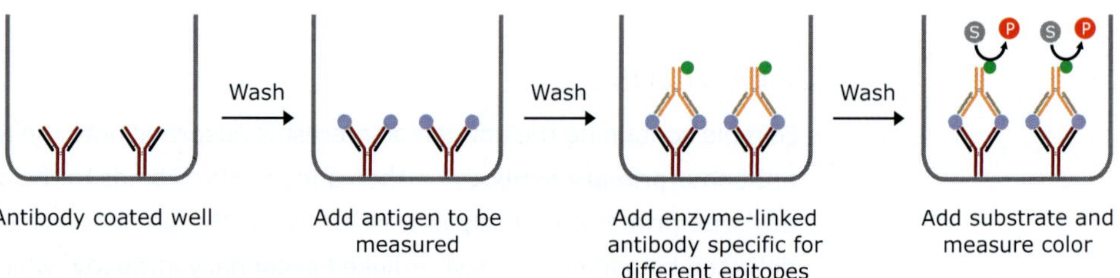

Figure 10.10 Sandwich ELISA. Antibody to a particular antigen is first immobilized on the bottom of a well. Next, the sample containing antigen is added to the well, which binds to the antibody. Finally, a second, different antibody to the antigen is added. This antibody is enzyme-linked. After any free second antibody is removed by washing, substrate (S) for enzyme-linked with antibody is added, and the colored reaction product (P) is measured. The extent of reaction is directly proportional to the amount of antigen present.

Competitive ELISA

The competitive ELISA is perhaps the most complex of all the ELISA techniques. It involves the use of **inhibitor antigen**, so competitive ELISA is also known as **inhibition ELISA**. In fact, each of the three formats, direct, indirect, and sandwich, can be adapted to the competitive format. In competitive ELISA, the inhibitor antigen and the antigen of interest compete for binding to the primary antibody i.e. two antigens to compete with each other for binding to antibodies.

In this ELISA, the unlabeled primary antibody is first incubated with the sample containing the antigen of interest, leading to the formation of antigen-antibody complex. Since the antibody is excessive compared with the antigen, so there are free antibodies left.

In the next step, the antigen-antibody mixture is added to the plate coated with inhibitor antigen that can also bind to the primary antibody. The free primary antibody in the mixture binds to the inhibitor antigen on the plate, while the antigen-antibody complexes in the mixture do not and are therefore washed off. The enzyme-labeled secondary antibody is added to the plate and binds to the primary antibody bound to the inhibitor antigen on the plate. Finally, a substrate is added to react with the enzyme and emit a visible signal for detection. Through this procedure, one can find that the final signal is inversely associated with the amount of the antigen of interest in the sample, meaning that the more antigen in the sample, the weaker the final signal. This is because primary antibodies bound to sample antigen will be washed off, while free primary antibodies left will be captured by inhibitor antigen immobilized to the plate and be measured by an enzymatic reaction. Competitive ELISA described here is based on antibody capture, in which the plate is coated with antigen. There is another type of competitive ELISA that is based on antigen capture, in which the plate is coated with unlabeled antibody.

Immunofluorescence

Immunofluorescence is a powerful technique that utilizes fluorescent-labeled antibodies to detect specific target antigens. Fluorescent molecules absorb light of one wavelength (excitation) and emit light of another wavelength (emission). If antibody molecules are tagged with a fluorescent dye or fluorophore, immune complexes containing these fluorescently labeled antibodies can be detected by the colored light emission when excited by the light of the appropriate wavelength. Antibody molecules bound to antigens in cells or tissue sections can similarly be visualized. This technique is known as *immunofluorescence*. The most commonly used fluorescent dyes are **fluorescein** and **rhodamine**. Other dyes such as phycoerythrin and phycobiliproteins are also used nowadays.

Immunofluorescence is of two types: One is the **direct fluorescent antibody method** (or direct immunofluorescence) and the other is the **indirect fluorescent antibody method** (or indirect immunofluorescence). In direct fluorescent antibody method, the specific antibody is directly conjugated with fluorophore, whereas in indirect fluorescent antibody method, the primary antibody is unlabeled and is detected with a fluorophore-labeled secondary antibody. It is a two-step process in which a primary, unlabeled antibody binds to the target, after which a fluorophore-labeled secondary antibody (directed against the Fc portion of the primary antibody) is used to detect the first antibody.

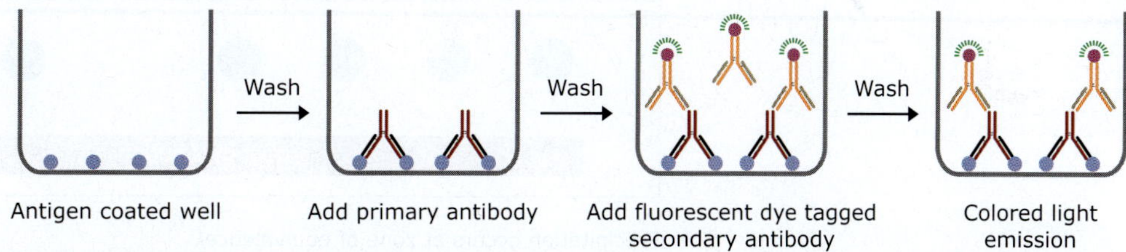

Figure 10.11 Indirect or secondary immunofluorescence. It uses two antibodies; the unlabeled first (primary) antibody specifically binds the target molecule, and the secondary antibody, which carries the fluorophore, recognizes the primary antibody and binds to it. Multiple secondary antibodies can bind a single primary antibody. This provides signal amplification by increasing the number of fluorophore molecules per antigen.

10.3 Immunoelectrophoresis

Immunoelectrophoresis is an immunotechnique that combined electrophoresis and immunoprecipitation for identifying and characterizing antigens within complex mixtures. *Electrophoresis* is conducted in the first stage and *immunoprecipitation* using antibodies against specific antigens in the second stage without removing the antigens from the gel.

There are several variants of immunoelectrophoresis like classical immunoelectrophoresis, crossed immunoelectrophoresis, rocket immunoelectrophoresis and immunofixation electrophoresis.

Classical immunoelectrophoresis

Classical immunoelectrophoresis combines the process of electrophoresis as well as double immunodiffusion. Electrophoresis is used for the separation of antigens on the basis of charge and double immunodiffusion for the purpose of their identification. It is also used in clinical laboratories to detect the presence or absence of antigen in the serum.

In this technique, antigen mixture is placed in a well cut in the center of an agar gel. Then the antigen mixture is subjected to electrophoresis, which separates the various components according to their charge in the electrical field. After electrophoresis, a trough is cut along the side of the gel, and antibodies are applied in the trough. The antibodies and the separated antigen mixture diffuse in the agar. At an optimal antigen-to-antibody ratio for each antigen and its corresponding antibodies, precipitin lines form.

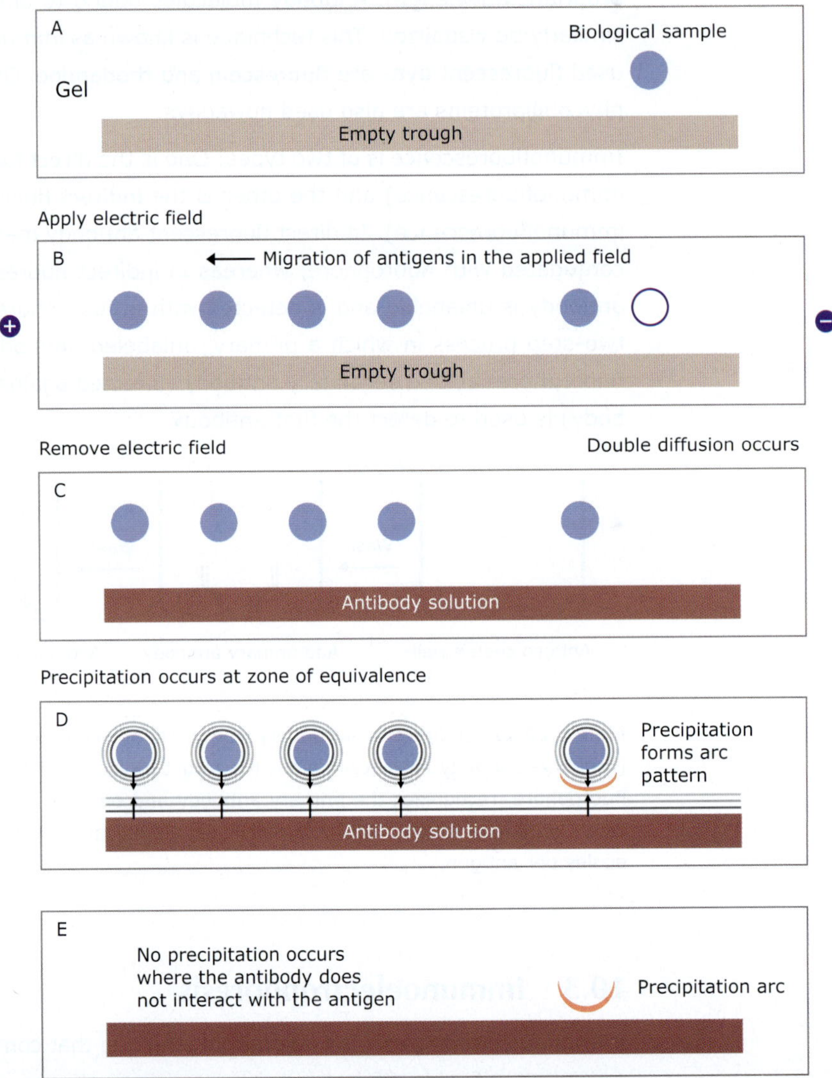

Figure 10.12 Classical immunoelectrophoresis. Sample (Ag) is introduced into a well cut in the gel and subjected to a voltage gradient which causes the various antigens to migrate different distances through the gel depending on their charge. After electrophoresis, antibody is introduced in the trough that is located parallel to the separated antigens. Both the antigens that have been electrophoretically separated and the antibodies present in trough diffuse in the gel. When they meet at the appropriate concentrations (equivalence), precipitation occurs.

Crossed immunoelectrophoresis

Crossed immunoelectrophoresis (also known as two-dimensional immunoelectrophoresis) is a two-stage technique in which antigens are electrophoretically separated in one dimension and then subjected to a second electrophoresis perpendicular to the first electrophoresis. The second stage differs from the first in that the antigens move through an agarose gel containing antibodies specific to the antigen. This leads to the formation of a precipitin arc when antigen and antibody meet at equivalence.

This technique is also used for the separation of proteins that have similar electrophoretic mobility. *This technique should not be confused with two-dimensional gel electrophoresis, a technique in which proteins are first separated according to their isoelectric points, and then separated according to their masses.*

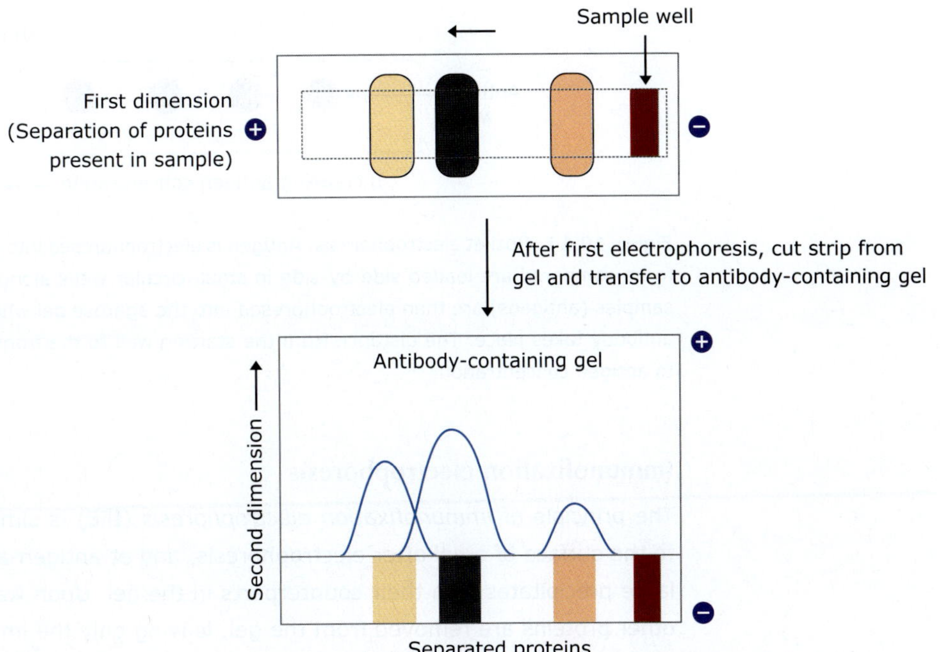

Figure 10.13 Crossed immunoelectrophoresis. The first stage of crossed immunoelectrophoresis consists of protein separation under the influence of an applied field. In preparation for the second stage, liquid agarose, containing the antibody solution of interest, is poured onto a large plate in a manner that allows the liquid agarose to fuse with the piece of gel that has been transferred to the plate. When the gel is hardened, an electric field is applied. The electric field is perpendicular to that applied to the protein sample in the first gel. Precipitation occurs when the proteins (antigens) and antibodies interact in appropriate concentration (equivalence).

Rocket immunoelectrophoresis

Rocket immunoelectrophoresis (also referred to as electroimmunoassay) is a simple, quick, and reproducible method for determining the concentration of a specific protein antigen in a mixture. In this method, a negatively charged antigen is electrophoresed in an antibody-containing gel. During electrophoresis, as the antigen starts to leave the well and move in the gel, antigen molecules will start to interact, and bind with antibody molecules. However, at this early stage, there is considerable antigen excess over antibody and no precipitation occurs. As the antigen samples electrophoreses further through the gel, more antibody molecules are encountered that interact with the antigen, until eventually there is sufficient antibody-antigen cross-linking such that 'equivalence' is reached and the antigen-antibody complex precipitates. The precipitin lines appear like the shape of a rocket. The majority of the antibody-antigen precipitate is, indeed, at the head of this rocket, but the fine precipitation lines up the side of the rockets are formed by a small amount of antigen diffusing sideways as the antigen passes through the gel. This small amount of antigen very quickly meets sufficient antibody to reach equivalence and precipitate.

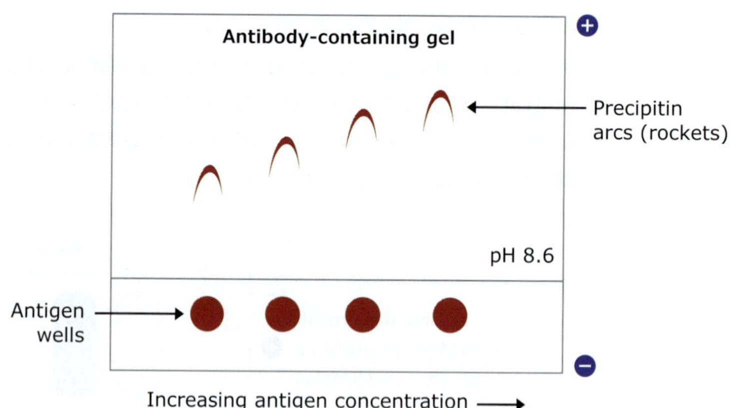

Figure 10.14 Rocket electrophoresis. Antigen is electrophoresed into gel containing antibody. The samples to be compared are loaded side-by-side in small circular wells along the edge of an agarose gel. These samples (antigens) are then electrophoresed into the agarose gel where interaction between antigen and antibody takes place. The distance from the starting well to the front of the rocket shaped arc is related to antigen concentration.

Immunofixation electrophoresis

The principle of *immunofixation electrophoresis* (IFE) is simple. The antibodies are applied to the surface of a gel after electrophoresis, and at antigen-antibody equivalence, they form large precipitates with their counterparts in the gel. Upon washing, unbound antibodies and other proteins are removed from the gel, leaving only the immunoprecipitates which are too large and insoluble to be washed out from the pores. The immunoprecipitates remaining in the gel can be then directly stained or identified by other techniques such as by means of fluorescein, enzyme, or radiolabeled primary or secondary antibodies.

10.4 Monoclonal antibodies and Hybridoma technology

Antibodies produced ordinarily by infection or immunization are **polyclonal** because natural antigens have multiple epitopes (or *antigenic determinants*), each of which generates clones of B-lymphocytes (B-cells). This results in antisera containing antibodies from different clones of B-cells with specificities against different epitopes of the antigens. On the other hand, **monoclonal antibodies** (mAb) are monospecific antibodies. These antibodies are produced from clone of single B-cell directed against a single epitope.

Hybridoma technology is a method of forming hybrid cell lines (called *hybridomas*) by fusing a specific antibody-producing normal B-cell with a *myeloma cell* (*cancerous B-cell*). The antibodies produced by the hybridoma are all of a single specificity and are therefore monoclonal antibodies. The production of monoclonal antibodies was invented by Cesar Milstein and Georges J. F. Köhler in 1975. They shared the Nobel Prize in 1984 for Medicine and Physiology.

Hybridomas are somatic cell hybrids produced by fusing antibodies forming spleen cells with myeloma cells. Antibody-producing B-cells normally die after several weeks in cell culture *in vitro*. Therefore, antibody-producing B-cells are fused with cancerous B-cells called *myelomas*. These myelomas are capable of dividing indefinitely and are, therefore, often called *immortal cell* lines. The immortal cell lines that result from the B cell-myeloma fusion are hybrid cell lines called *hybridomas*. The hybridoma cell lines share the properties of both fusion partners. They grow indefinitely *in vitro* and produce antibodies.

To produce a monoclonal antibody, a mouse is immunized with the antigen of interest. During the next several weeks, antigen-specific B cells proliferate and begin producing antibodies in the mouse. Spleen tissue, rich in B cells, is then removed from the mouse, and the B cells are fused with myeloma cells.

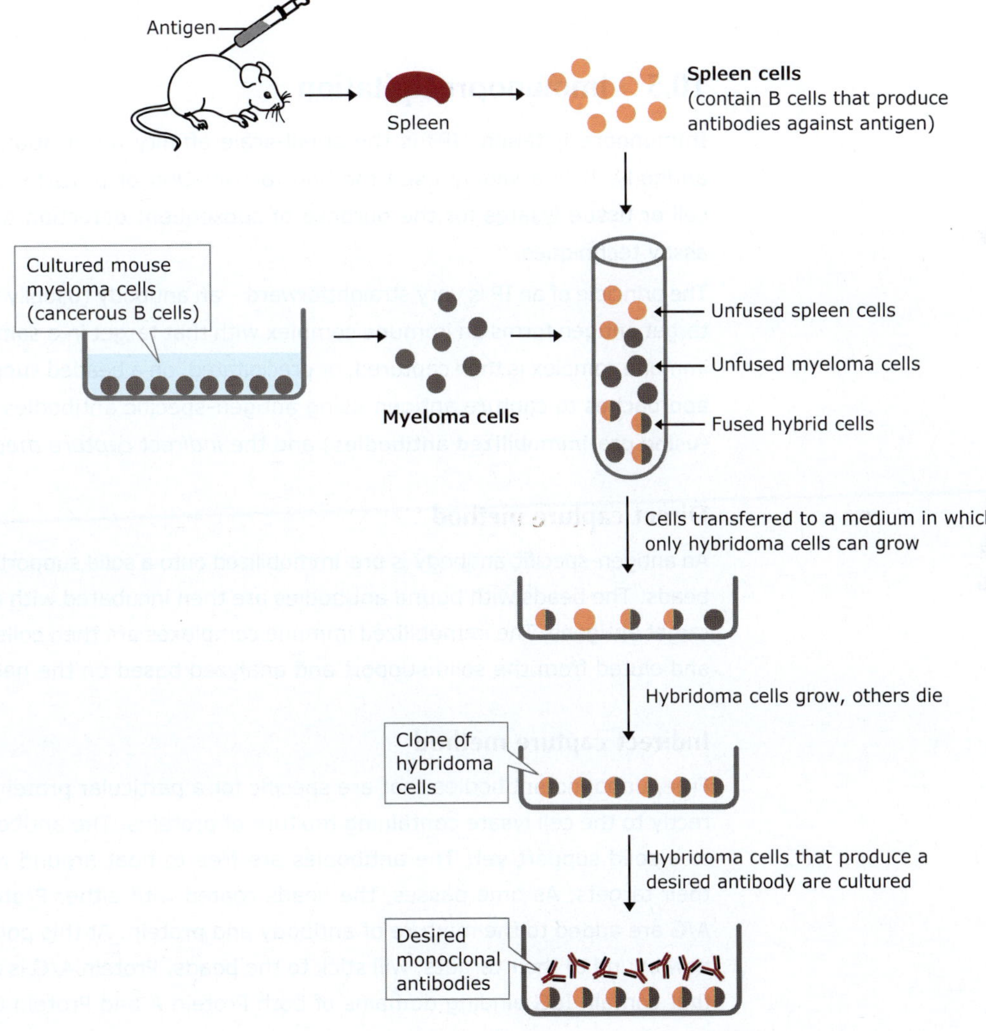

Figure 10.15 Production of monoclonal antibodies.

Procedure for monoclonal antibody production

Immunization
(antigen injection to host)
↓
B cells
↓ ← Myeloma cells
Fusion
↓
Screening of hybrid cells
↓
Clonal propagation
↓
Characterization
(ELISA, FACS, etc.)
↓
Large-scale culture
↓
Purification

Hybridomas are selected by the use of a selective medium in which the myeloma cells die, but hybridomas survive. The most widely used selective systems involve the inclusion of the antibiotic aminopterin in the growth medium. **Aminopterin** is a synthetic derivative of pterin. This folate analog acts as a competitive inhibitor for enzyme *dihydrofolate reductase* which catalyzes the reduction of dihydrofolate into tetrahydrofolate. Normal animal cells synthesize purine nucleotides and thymidylate for DNA synthesis by a *de novo* pathway requiring tetrahydrofolate. Thus, the addition of aminopterin inhibits the *de novo* nucleotide synthesis pathway. However, normal cells survive in this medium as they are able to use the salvage pathway for nucleic acid synthesis. But if the cells are unable to produce the enzyme *hypoxanthine-guanine phosphoribosyltransferase* (**HGPRT**), they are unable to utilize the salvage pathway and, therefore, die in the aminopterin-containing medium.

In the procedure, myeloma cells are engineered to be deficient in enzyme HGPRT. After fusion of lymphocytes with HGPRT-negative myeloma cells, aminopterin-containing medium, supplemented with hypoxanthine and thymidine to ensure an adequate supply of substrates for the salvage pathway (HAT medium) is added, which kills myeloma cells but allows hybridomas to survive as they inherit HGPRT from the lymphocyte parent. Unfused lymphocytes die after a short period of culture, which results in a pure preparation of hybridomas.

10.5 Immunoprecipitation

Immunoprecipitation (IP) is the small-scale affinity purification of antigens using a specific antibody. It is a widely used method for isolation of proteins and other biomolecules from cell or tissue lysates for the purpose of subsequent detection by western blotting and other assay techniques.

The principle of an IP is very straightforward – an antibody (usually monoclonal) against a specific target antigen forms an immune complex with that target in a sample, such as a cell lysate. The immune complex is then captured, or precipitated, on a beaded support. In general, there are two approaches to capture antigen using antigen-specific antibodies – the *direct capture method* (**using pre-immobilized antibodies**) and the *indirect capture method* (**using free antibodies**).

Direct capture method

An antigen-specific antibody is pre-immobilized onto a solid support, such as agarose or magnetic beads. The beads with bound antibodies are then incubated with a cell lysate that contains the target antigens. The immobilized immune complexes are then collected from the lysate, washed and eluted from the solid support and analyzed based on the nature of the target antigen.

Indirect capture method

Free, unbound antibodies that are specific for a particular protein (i.e. antigen) are added directly to the cell lysate containing mixture of proteins. The antibodies have not been attached to a solid support yet. The antibodies are free to float around the protein mixture and bind their targets. As time passes, the beads coated with either Protein A or Protein G or Protein A/G are added to the mixture of antibody and protein. At this point, the antibodies, which are now bound to their targets, will stick to the beads. Protein A/G is a recombinant fusion protein that contain IgG binding domains of both Protein A and Protein G.

Protein A, Protein G and Protein A/G have different affinities for antibodies from different species, as well as different subclasses of immunoglobulin Gs. An inherent problem of using Protein A, G and A/G in immunoprecipitation reactions is that the antibody is not covalently bound to the antibody binding proteins. This results in the release of the antibody and antigen complex during elution. Protein cross-linking can be used to overcome the issue of the antibodies eluting with the protein of interest. The use of small, amine reactive homobifunctional protein cross-linkers, such as **disuccinimidyl suberate** (DSS) and **Bissulfosuccinimidyl suberate** (BS3), are used to covalently attach the antibody to the Protein A, Protein G and Protein A/G.

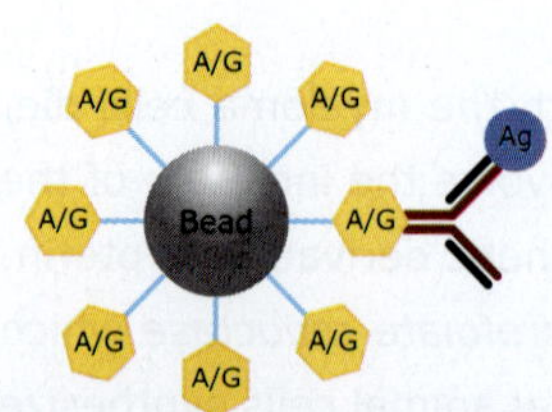

The antibody would first be coupled to the Protein A, G or A/G solid beads and then incubated with an appropriate cross-linker. The antibody would then be covalently coupled to the support and ready for use in the immunoprecipitation.

While the pre-immobilized antibody approach is more commonly used for IP, using free antibody to form immune complexes is beneficial if the target protein is present in low concentrations, the antibody has a weak binding affinity for the antigen or the binding kinetics of the antibody to the antigen are slow.

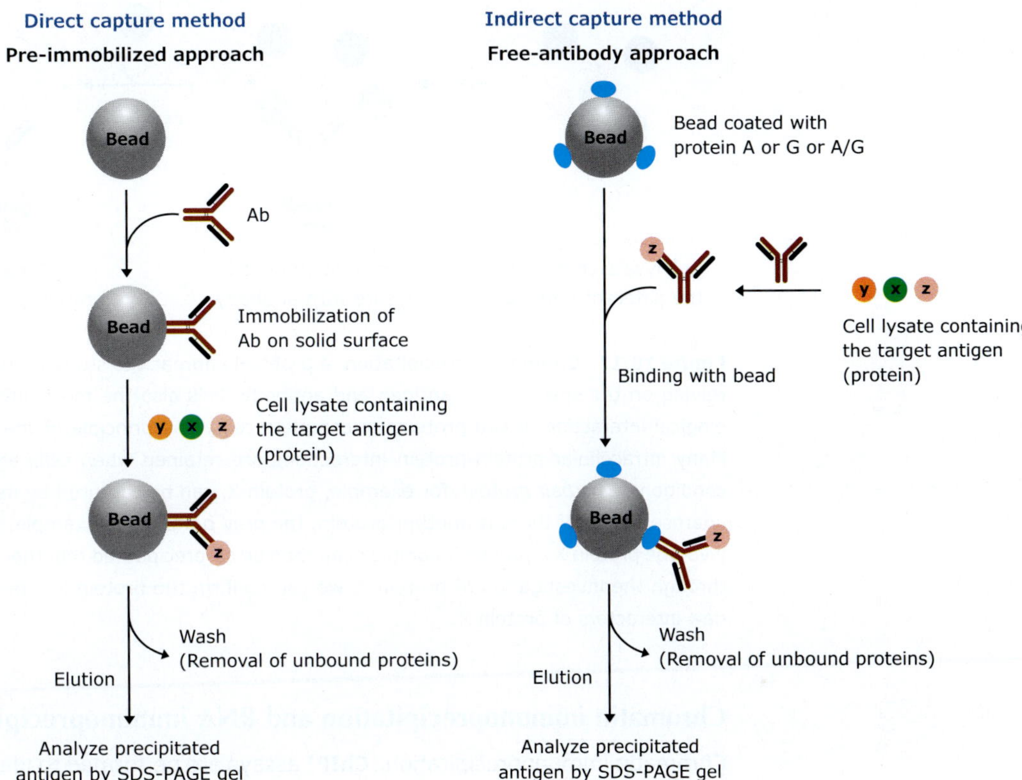

Figure 10.16
Direct capture method: The beads bound to antibodies are added to the protein mixture present in cell lysates and the target proteins are captured by the beads through antibodies, or simply immunoprecipitated. The antibodies are immobilized on solid-phase substrate such as microscopic agarose and superparamagnetic microbeads.
Indirect capture method: Antibodies are added to cell lysates containing protein mixture and bind to target proteins. Beads coated with protein A or G or A/G is added subsequently to target bound proteins in the complex. Beyond this point, the protocols converge and as the end-results are one and the same.

Agarose beads (also known as agarose resins or slurries) are highly-porous sponge-like structures of varying shapes and sizes (50 to 150 μm diameter). **Magnetic beads**, such as Dynabeads, have largely replaced agarose beads as the preferred support for immunoprecipitation. Magnetic beads are solid and spherical, and antibody binding is limited to the surface of each bead. Magnetic beads are also significantly smaller than agarose resin (1 to 4 μm diameter), which collectively increases the surface area for high-capacity antibody binding. The capture or collection of immunoprecipitated complexes in presence of magnetic beads is done with the help of a small micro-centrifuge tube stand that is fitted with a magnet and thus allows the precipitation of complex leaving supernatant behind.

Co-immunoprecipitation

Variations of IP are also used to study the interactions of the primary antigen protein with other proteins (Co-IP) or nucleic acids (ChIP and RIP).

Co-immunoprecipitation (**Co-IP**) helps determine whether two proteins interact or not in physiological conditions *in vitro*. Co-IP is conducted in essentially the same manner as an IP, except that the target antigen precipitated by the antibody is used to co-precipitate its binding partner(s) or associated protein complex from the lysate. Many protein-protein associations that exist within the cell remain intact when a cell is lysed under nondenaturing conditions. Thus, if protein *X* is immunoprecipitated, then protein *Y*, which stably associated with *X*, may also precipitate. Coimmunoprecipitation is most commonly used to test whether two proteins of interest are associated *in vivo*. The target protein is specifically immunoprecipitated from the cell extracts, and the immunoprecipitates are fractionated by SDS-PAGE.

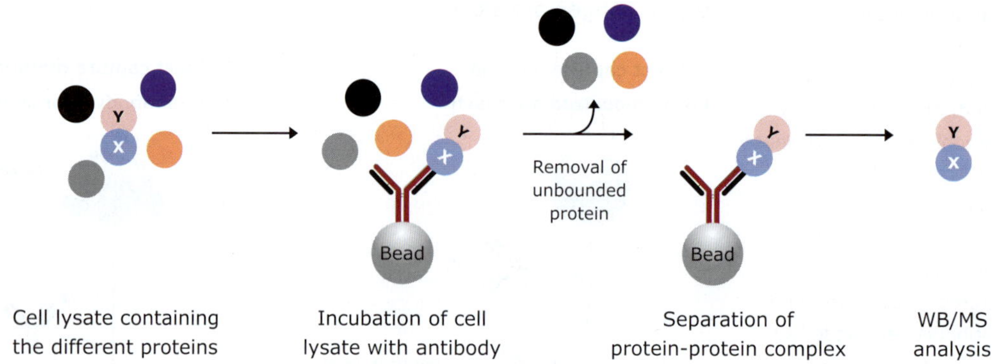

Figure 10.17 Co-immunoprecipitation, a protocol approach to study protein-protein interactions, mainly relying on the specificity of antigen and antibody. It is also the most effective way to detect the physiological interaction of two proteins in complete cells. The principle of the Co-IP technique is as follows: Many intracellular protein-protein interactions are retained when cells are lysed under non-denaturing conditions. The *bait protein*, for example, protein X, can be captured by its specific antibody stabilized to agarose beads. If there is another protein, the *prey protein*, for example, protein Y, binds to protein X *in vivo*, the protein X – protein Y complex can then be co-precipitated together by the antibody. Subsequently, through the investigation of protein Y, we can confirm the protein X – protein Y interaction, or discover new interactors of protein X.

Chromatin immunoprecipitation and RNA immunoprecipitation

Chromatin immunoprecipitation (**ChIP**) assays are performed to identify regions of the genome with which DNA-binding proteins, such as transcription factors and histones, associate. In ChIP assays, proteins bound to DNA are temporarily cross-linked and the DNA is sheared prior to cell lysis. The target proteins are immunoprecipitated along with the cross-linked nucleotide sequences, and DNA is then removed and identified by PCR, sequenced, applied to microarrays or analyzed in some other way. RNA immunoprecipitation (**RIP**) is similar to ChIP, except that RNA-binding proteins are immunoprecipitated instead of DNA-binding proteins. Immunoprecipitated RNAs can then be identified by RT-PCR and cDNA sequencing.

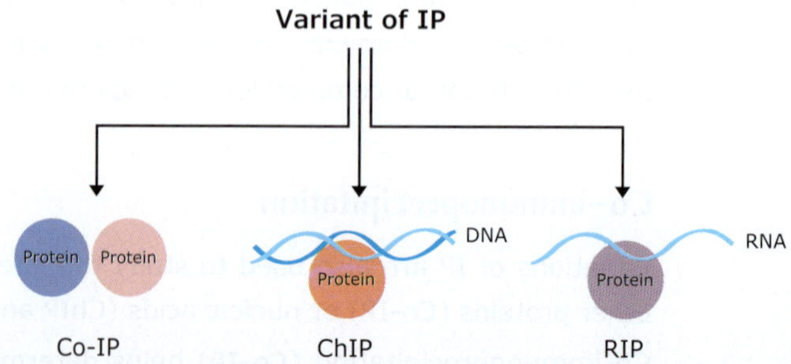

Figure 10.18 There are different types of immunoprecipitation. Variations of IP (e.g. Co-IP) are used to study the interactions between the primary antigen protein and other proteins or nucleic acids (e.g. ChIP or RIP).

The **chromatin immunoprecipitation** (ChIP) assay is a powerful method for analyzing epigenetic modifications and genomic DNA sequences bound to specific regulatory proteins under a particular set of conditions. In this method, protein-DNA complexes are cross-linked, immunoprecipitated, purified, and amplified for gene- and promoter-specific analysis of known targets. When performing the ChIP assay, cells are first fixed with formaldehyde. It acts as a reversible protein-DNA and protein-protein cross-linking agent that serves to fix the protein-DNA interactions occurring in the cell. Cells are then lysed and chromatin is released and fragmented using either sonication or enzymatic digestion. Antibodies directed against a given gene regulatory protein are then used to purify the DNA that are covalently cross-linked to that protein in the cell. After immunoprecipitation, the protein-DNA cross-links are reversed and the DNA is purified. The purified DNA is analyzed by dot blot or Southern blot using a radiolabeled probe derived from the cloned DNA fragment of interest.

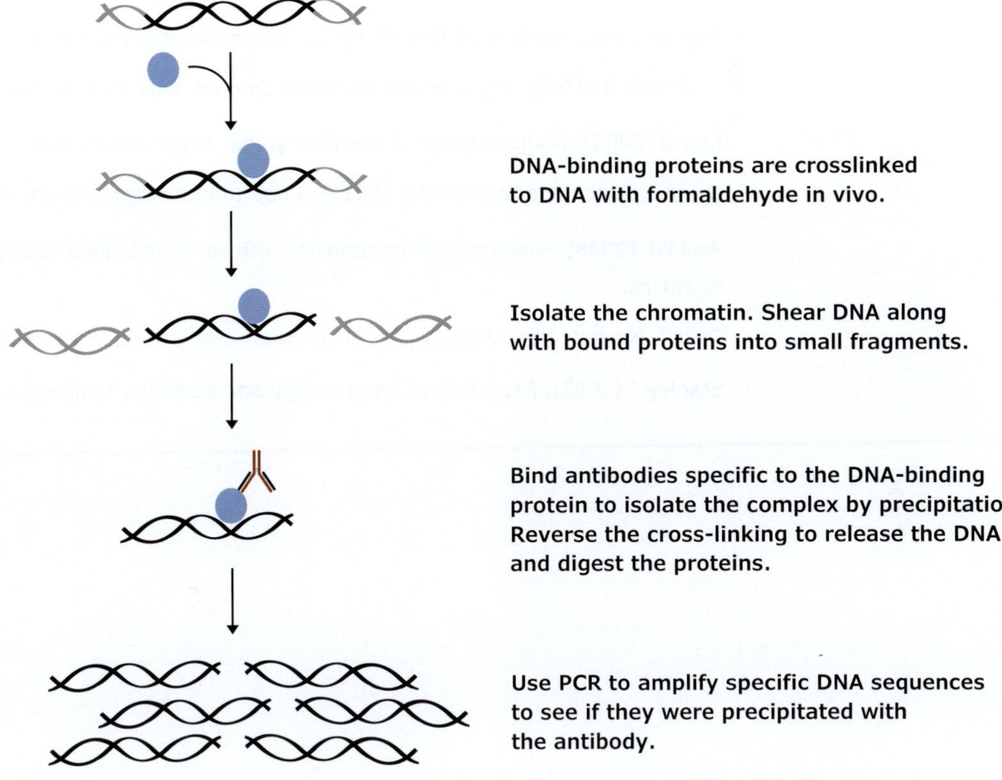

Figure 10.19 Cells grown under the desired experimental condition are fixed with formaldehyde which forms heat-reversible DNA-protein cross-links. After cross-linking, cells were lysed and sonicated to fragment the chromatin. These fragments are purified and then used to perform ChIP. It is performed by incubation of fractionated chromatin with an antibody directed to a protein of interest. The antibody recognizes its targeted protein and precipitates protein-DNA from solution. In this way, only DNA fragments crosslinked to the protein of interest are enriched, while DNA-protein complexes that are not recognized by the antibody are washed away.

As formaldehyde inactivates cellular enzymes immediately upon addition to cells, ChIP provides snapshots of protein-protein and protein-DNA interactions at a particular time point, and hence is useful for kinetic analysis of events occurring on chromosomal sequences *in vivo*. In addition, if the immunoprecipitated DNA is hybridized to microarrays that contain the entire genome displayed as a series of discrete DNA fragments, the precise genomic location of each precipitated DNA fragment can be determined. In this way, all the sites occupied by the gene regulatory protein in the original cells can be mapped on a genome-wide basis. ChIP has been widely used to map the localization of post-translationally modified histones, histone variants, transcription factors, or chromatin modifying enzymes on the genome or on a given locus.

References

Abbas AK, Lichtman AH and Pillai S (2010), *Cellular and Molecular Immunology*, 6th ed. Saunders Elsevier.

Berg JM, Tymoczko JL and Stryer L (2007), *Biochemistry*, 6th ed. W.H. Freeman and Company.

Coico R, Sunshine G and Benjamini E (2003), *Immunology, A short course*, 5th ed. John Wiley and Sons, Inc.

Decker J and Reischl U (2004), *Molecular diagnosis of infectious diseases*, 2nd ed. Humana Press.

Freifelder D (1982), *Physical Biochemistry*, 2nd ed. W.H. Freeman and Company.

Law B (2005), *Immunoassay: A practical guide*, Taylor and Francis.

Owen J, Punt J and Stranford S (2013), *Kuby Immunology*, 7th ed. Macmillan Publishers.

Paul WE (2008), *Fundamental immunology*, 6th ed. Philadelphia, Wolters Kluwer/Lippincott Williams & Wilkins.

Roitt I, Brostoff J and Male D (2012), *Immunology,* 8th ed. Saunders Elsevier.

Stanley J (2002), *Essentials of immunology and serology*, Cengage Learning.

11

FRET and FRAP

11.1 FRET

FRET (Fluorescence Resonance Energy Transfer) is a phenomenon in which an excited donor molecule transfers energy (not an electron) to an acceptor molecule through a non-radiative process. It is a highly distance-dependent radiationless energy transfer process. In this energy transfer process, two molecules interact with each other in which one acts as donor and other as acceptor. The donor is a *fluorophore* that initially absorbs the energy and the acceptor is the *fluorophore* or *chromophore* to which the energy is subsequently transferred. A pair of molecules that interact in such a manner that FRET occurs is often referred to as a donor-acceptor pair.

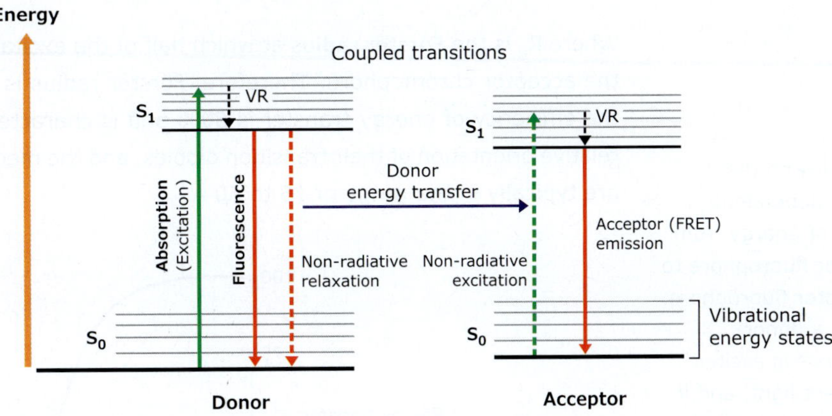

Figure 11.1 A *Jablonski diagram* illustrating the coupled transitions involved between the donor emission and acceptor absorbance in fluorescence resonance energy transfer. In this diagram, the donor molecule is excited by a photon. Under typical fluorescence conditions, relaxation of the electron energy would result in the emission of a photon (*fluorescence*) by the donor fluorophore. However, if a suitable acceptor molecule is within a certain distance (*Förster distance*) then donor energy can experience the transfer of energy to the acceptor, resulting in acceptor fluorescence. The transfer of energy between molecules is nonradiative, and that the distances involved are much less than the wavelength of light. In the presence of a suitable acceptor, the donor fluorophore can transfer excited state energy directly to the acceptor without emitting a photon.

Resonance energy transfer, a photophysical process in which the excited state energy from a donor fluorophore is transferred via a non-radiative mechanism to a ground state acceptor chromophore via weak long-range dipole–dipole coupling.

FRET is a non-radiative quantum mechanical process that does not require a collision and does not involve the production of heat. When energy transfer occurs, the acceptor molecule quenches the donor molecule fluorescence, and if the acceptor is itself a fluorophore, increased or sensitized fluorescence emission is observed.

There are some criteria that must be satisfied in order for FRET to occur. The process of resonance energy transfer can take place when a donor fluorophore in an electronically excited state transfers its excitation energy to a nearby fluorophore, the acceptor. The primary condition is the distance between the donor and the acceptor. The donor and acceptor molecules must be in close proximity to one another for FRET to occur.

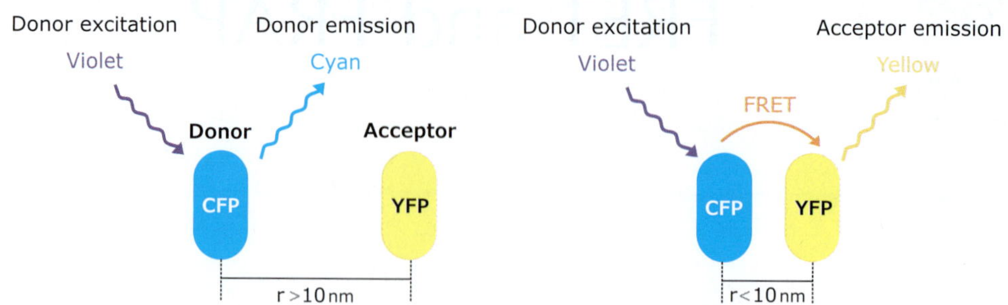

Figure 11.2 The acceptor and donor fluorophores must be very close for the FRET to occur. FRET is a process by which radiationless transfer of energy occurs from an excited state donor fluorophore to an acceptor fluorophore in close proximity. Because the range over which the energy transfer can take place is limited to approximately 10 nanometers, and the efficiency of transfer is extremely sensitive to the separation distance between fluorophores. In this example, CFP (Cyan fluorescent protein) is a donor fluorophore and YFP (Yellow fluorescent protein) is a acceptor fluorophore.

The extent of energy transfer for a single donor-acceptor pair at a fixed distance depends on the inverse sixth power of the distance between the donor and acceptor pair (r) and is given by:

$$E = \frac{R_0^6}{R_0^6 + r^6}$$

Where R_0 is the **Förster radius** at which half of the excitation energy of donor is transferred to the acceptor chromophore. Therefore, Förster radius is referred to as the distance at which the efficiency of energy transfer is 50% and is characteristic of the donor-acceptor pair, the relative orientation of their transition dipoles, and the medium between them. Förster distances are typically in the range of 15 to 60 Å.

FRET relies on the distance dependent transfer of energy from the donor fluorophore to an acceptor fluorophore. In FRET, a donor fluorophore is excited by incident light, and if an acceptor is in close proximity, the excited-state energy from the donor can be transferred. The process is non-radiative (not mediated by a photon). The donor molecule must have an emission spectrum that overlaps the absorption spectrum of the acceptor molecule. In most applications, both donor and acceptor are fluorophores.

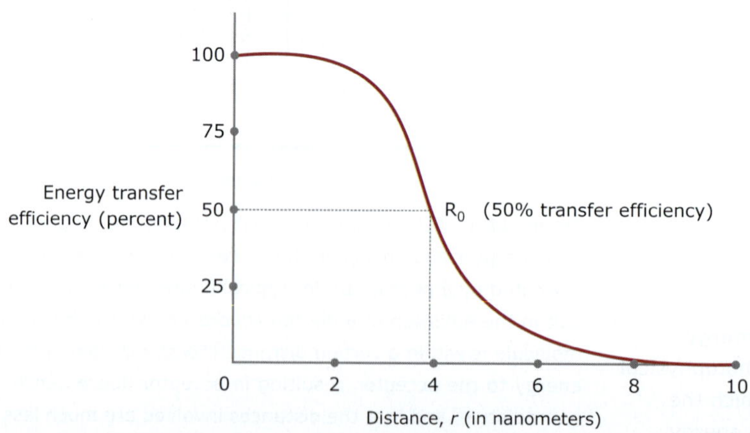

Figure 11.3 Distance and energy transfer efficiency. Energy transfer efficiency is most sensitive to distance between the donor-acceptor molecules. The graph illustrates the exponential relationship between transfer efficiency and the distance separating the donor and the acceptor. The efficiency rapidly increases to 100% as the separation distance decreases below R_0 and conversely, decreases to zero when r is greater than R_0. Because of the strong (sixth-power) dependence of transfer efficiency on distance, measurements of the donor-acceptor separation distance are only reliable when the donor and acceptor radius lies within the Förster distance by a factor of two.

The resonance energy transfer rate depends upon the extent of spectral overlap between the donor emission and acceptor absorption spectra, the quantum yield of the donor, the relative orientation of the donor and acceptor transition dipole moments, and the distance separating the donor and acceptor molecules.

Secondly, for FRET to happen the absorption or excitation spectrum of the acceptor must overlap the fluorescence emission spectrum of the donor. The degree of overlap is referred to as **spectral overlap integral**. The overlap of emission spectrum of the donor and absorption spectrum of the acceptor means that the energy lost from excited donor to ground state could excite the acceptor group. The more overlap of spectra, the better a donor can transfer energy to the acceptor.

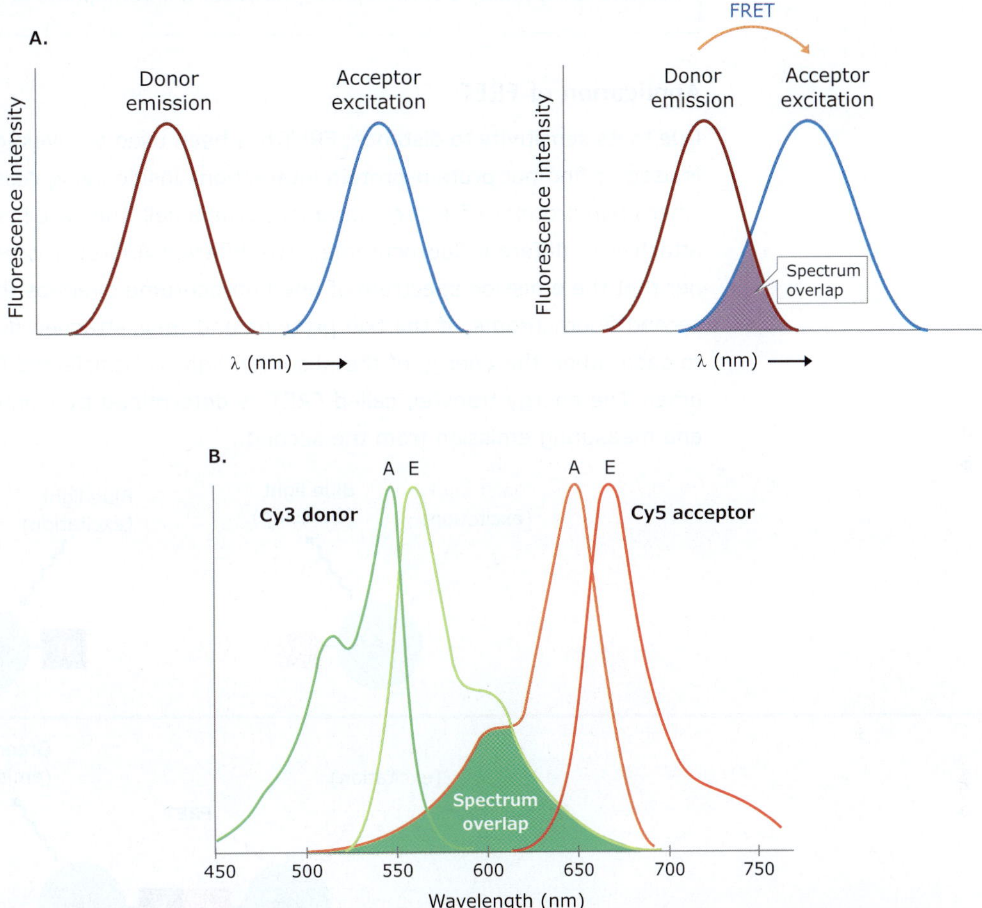

Figure 11.4 Spectral overlap and FRET. **A**. For FRET to happen the absorption or excitation spectrum of the acceptor must overlap the fluorescence emission spectrum of the donor. **B**. FRET between two different cyanine dyes (Cy3 and Cy5). Cy3 is the donor fluorophore and Cy5 is the acceptor fluorophore. The donor (Cy3) is excited at a maximal wavelength of 550 nm, and emits with light at 565 nm. The acceptor (Cy5) absorbs light at 652 nm and emits at 670 nm. Energy transfer occurs due to the spectral overlap between the emission of the donor and excitation spectrum of the acceptor (shown in green). Crosstalk occurs when donor emission overlaps with acceptor emission, such that the donor emission is counted as FRET; similarly, the acceptor can be directly excited during donor excitation if their absorbance spectra overlap.

Table 11.1 Examples for common FRET Donor-Acceptor pairs:

Donor (*Emission*)	**Acceptor** (*Excitation*)
FITC (520 nm)	TRITC (550 nm)
Cy3 (566 nm)	Cy5 (649 nm)
EGFP (508 nm)	Cy3 (554 nm)
CFP (477 nm)	YFP (514 nm)
EGFP (508 nm)	YFP (514 nm)

FITC – Fluorescein isothiocyanate; TRITC – Isothiocyanate derivative of rhodamine; Cy3 – Cyanine

Chromophore and fluorophore

The structural component of a molecule responsible for the absorption of UV or visible light is called **chromophore**. A chromophore that re-emits an absorbed light at a longer wavelength (i.e. exhibits fluorescence) is called **fluorophore**. The amount and wavelength of the emitted light depend on both the fluorophore and the chemical environment of the fluorophore. A fluorophore is inevitably a chromophore, however a chromophore is not necessarily a fluorophore.

Application of FRET

Due to its sensitivity to distance, FRET has been used to investigate molecular interactions. It is used to find out **protein-protein interactions** inside living cells. To determine whether (and when) two proteins of interest interact inside a cell, they are first produced as fusion proteins attached to different fluorochrome. Two different fluorochromes are selected in such a manner that the emission spectrum of one fluorochrome overlaps the absorption spectrum of the second fluorochrome. If the two proteins and their attached fluorochromes come very close to each other, the energy of the absorbed light is transferred from one fluorochrome to the other. The energy transfer, called FRET, is determined by illuminating the first fluorochrome and measuring emission from the second.

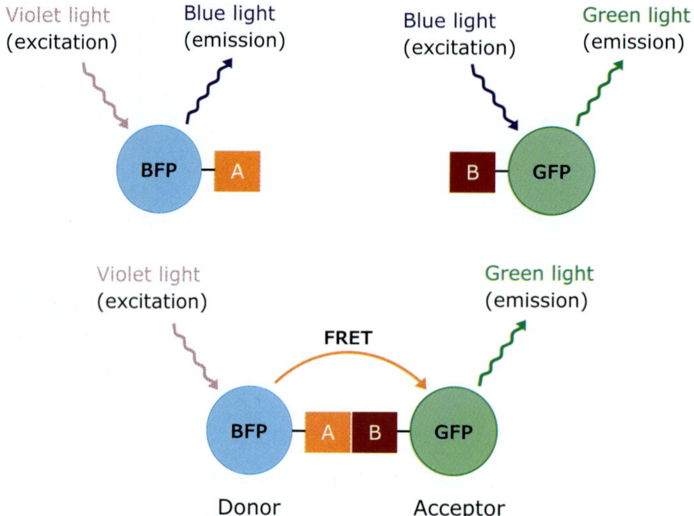

Figure 11.5 Schematic representation of the interaction of two different fusion proteins. In this example, protein A is coupled to a BFP, which is excited by violet light (370–440 nm) and emits blue light (440–480 nm); protein B is coupled to a GFP, which is excited by blue light and emits green light (510 nm). If proteins A and B do not interact, illuminating the sample with violet light yields fluorescence from the BFP only. When proteins A and B interact, FRET can occur. Illuminating the sample with violet light excites the BFP, whose emission, in turn, excites the GFP, resulting in an emission of green light.

FRET is also used to assess nucleic acid annealing. In the figure 11.6, two complementary RNA oligonucleotides are labeled with Cy3 and Cy5 respectively. When these labeled molecules are not annealed, excitation of an RNA oligonucleotide labeled with Cy3 with light at 550 nm results only in the emission of light by Cy3 at 565 nm, while the complementary RNA oligonucleotide labeled with Cy5 does not emit any light at 565 nm or 670 nm. However, when the two oligonucleotides are allowed to anneal, the close proximity of the molecules allows for FRET transfer to occur. This results in the emission of light at 670 nm; when the annealed molecule is excited with 550 nm light.

Figure 11.6 Schematic representation of FRET occurring between Cy3 and Cy5 fluorescent moieties when labeled oligonucleotides are annealed.

Green fluorescence protein, GFP

GFP (discovered by Osamu Shimomura) is an autofluorescent protein (238 amino acid residues) of bioluminescent jellyfish *Aequorea victoria*. In GFP, eleven β-strands make up the β-barrel and an α-helix runs through the center. The chromophore is located in the middle of the β-barrel. The chromophore is *p*-hydroxybenzylidene imidazolinone formed from the spontaneous cyclization and oxidation of the amino acid residues, Ser^{65} (or Thr^{65}) - Tyr^{66} - Gly^{67}. The chromophore of GFP is responsible for its fluorescence.

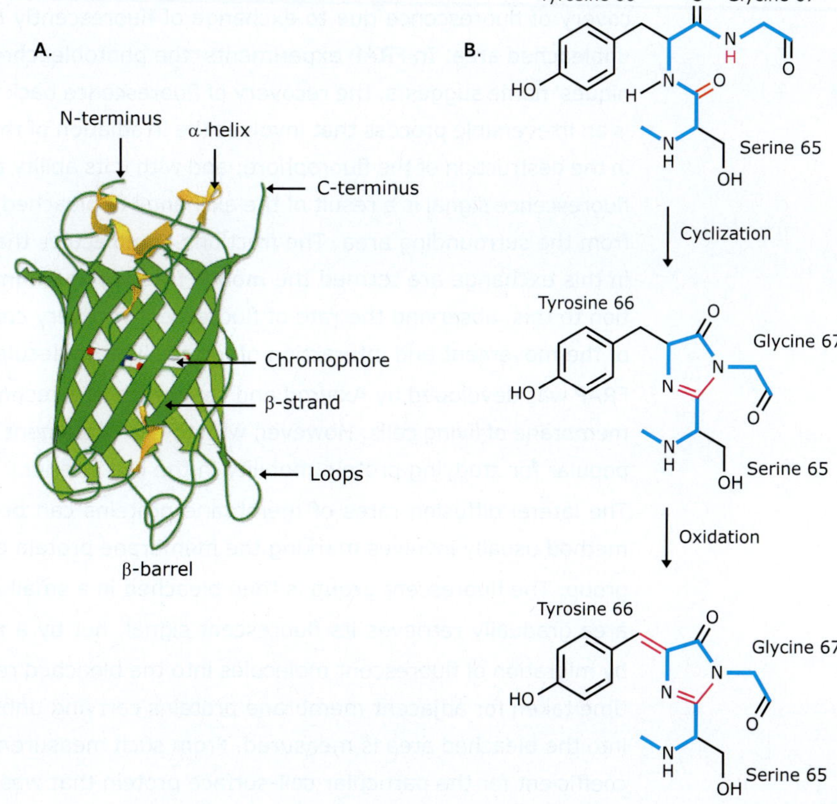

Figure 11.7
A. Structure of green fluorescence protein.
B. The chromophore of GFP is formed from a sequence of three adjacent amino acids—serine, tyrosine and glycine. In the presence of oxygen, the backbone cyclizes to form a five-membered ring whose double bonds are conjugated with those of the phenol ring of tyrosine.

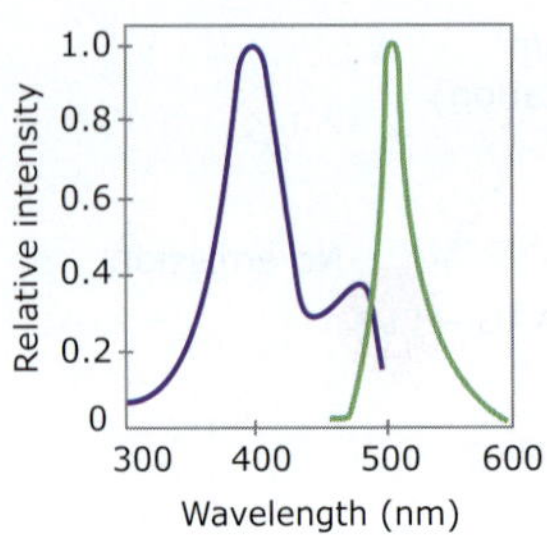

Figure 11.8 The excitation spectrum of native GFP from *A. victoria* (blue) has two excitation maxima at 395 nm and at 470 nm. The fluorescence emission spectrum (green) has a peak at 509 nm.

Wild-type GFP from jellyfish has two excitation peaks, a major one at 395 nm and a minor one at 475 nm. Its emission peak is at 509 nm in the lower green portion of the visible spectrum. Several spectral variants of the original wild-type green fluorescent protein (wtGFP) have been developed. Examples of these variants include: a blue fluorescent protein known as **BFP**; a cyan fluorescent variant known as **CFP**; a yellow fluorescent variant known as **YFP**; a violet-excitable green fluorescent variant known as **Sapphire** and a cyan-excitable green fluorescing variant known as enhanced green fluorescent protein or **EGFP**.

Application

The most popular applications of GFP involve exploiting them for imaging of the localization and dynamics of specific organelles or recombinant proteins in live cells. It serves as a unique reporter and is used as a fusion tag for monitoring protein localization. GFP is used as a tag in a fusion protein, where it is coupled with the protein whose expression is to be tracked. In such cases, the principal aim is to investigate the subcellular localization of the protein under investigation. Genetic engineering can be used to produce vectors containing a GFP coding sequence into which a coding sequence for an uncharacterized protein, X, can be cloned. The resulting GFP-X fusion construct can be transfected into suitable target cells and expression of the GFP-X fusion protein can be monitored to track the subcellular location of the protein.

11.2 FRAP

FRAP (fluorescence recovery after photobleaching) is used to measure the dynamics of two or three dimensional movement of fluorescently labeled molecules within or between cells. The study of molecular mobility is an important parameter in the understanding of cell physiology.

The principle of FRAP is to photobleach the fluorescently labeled molecules in a small region of the sample. Then the mobility of the fluorescently labeled molecules is evaluated from recovery of fluorescence due to exchange of fluorescently labeled molecules from surrounding unbleached area. In FRAP experiments, the photobleached area is restricted and as the techniques' name suggests, the recovery of fluorescence back into it, is monitored. Photobleaching is an irreversible process that involves the irradiation of the fluorophore with light. This results in the destruction of the fluorophore; and with it its ability to emit fluorescence. Recovery of the fluorescence signal is a result of the exchange of bleached fluorophores with those unbleached from the surrounding area. The fractions of molecules that are able and unable to participate in this exchange are termed the **mobile fraction** and **immobile fraction** respectively. In addition to this, observing the rate of fluorescent recovery can provide important understandings of the movement and interaction of intracellular molecules.

FRAP was developed by Axelrod and coworkers as a technique to study protein mobility in the membrane of living cells. However, with the advancement of technique, this technique became popular for studying protein mobility in the cell interior.

The lateral diffusion rates of membrane proteins can be measured by using the *FRAP.* The method usually involves marking the membrane protein of interest with a specific fluorescent group. The fluorescent group is then bleached in a small area by a laser beam. This bleached area gradually retrieves its fluorescent signal, not by a reversal of the bleaching effect, but by migration of fluorescent molecules into the bleached region from the unbleached area. The time taken for adjacent membrane proteins carrying unbleached fluorescent group to diffuse into the bleached area is measured. From such measurement, one can calculate the diffusion coefficient for the particular cell-surface protein that was marked.

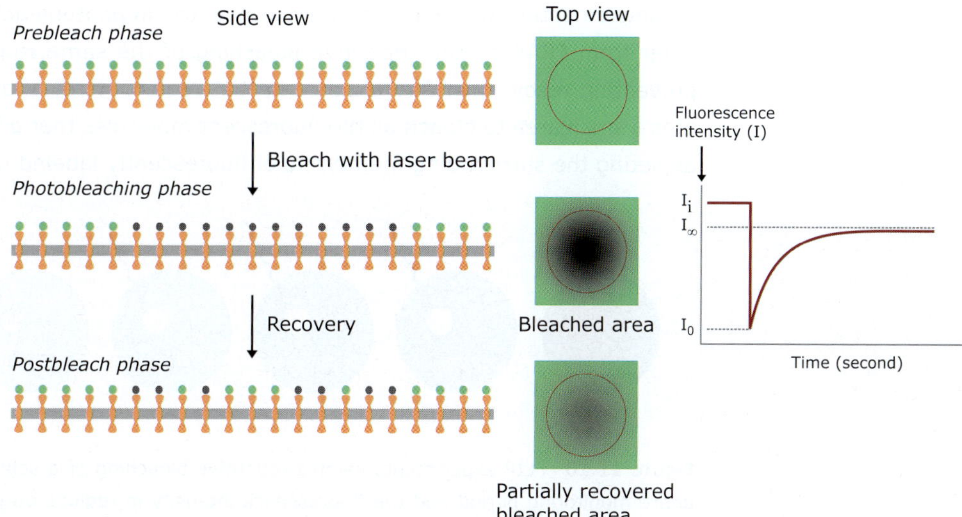

Figure 11.9 Measuring the rate of lateral diffusion of a membrane protein by FRAP technique. This technique is composed of three phases – prebleach phase, photobleaching phase and postbleach phase. A specific protein of interest can be labeled with a fluorescent molecule. Fluorescent molecules are bleached in a small area using a laser beam. The fluorescence intensity recovers as the bleached molecules diffuse away and unbleached molecules diffuse into the irradiated area. The diffusion coefficient is calculated from a graph of the rate of recovery: the greater the diffusion coefficient of the membrane protein, the faster the recovery. From this plot, the mobile and immobile fractions can be determined by calculating the ratios of the final to the initial fluorescence intensity. In the graph, I_0 is the fluorescence intensity immediately after the photobleaching, I_∞ the fluorescence intensity after full recovery and I_i the initial fluorescence intensity before photobleaching.

Analysis of typical FRAP curve

From the initial (prebleach) fluorescence intensity (I_i), the signal drops to a particular low value (I_0) as the high intensity laser beam bleaches fluorochromes in the region of interest. Over time, the signal recovers from the post-bleach intensity (I_0) to a maximal plateau value I_∞. From this plot, the mobile fraction (M_f) and immobile fraction (IM_f) can be calculated. The information from the recovery curve (from I_0 to I_∞) can be used to determine the diffusion constant and the binding dynamics of fluorescently labeled proteins.

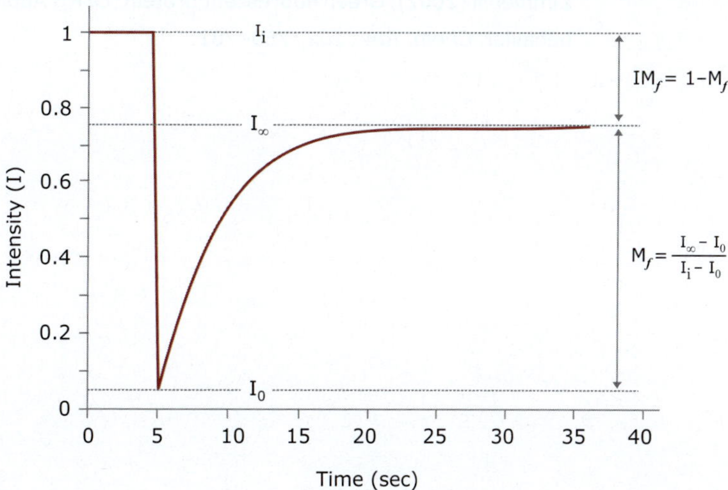

A complementary technique is **fluorescence loss in photobleaching** (FLIP). FLIP experiments differ from FRAP by the repetitive bleaching of the same region in the specimen, thereby preventing recovery of fluorescence in that region. Here, a laser beam continuously irradiates a small area to bleach all the fluorescent molecules that diffuse into it, thereby gradually depleting the surrounding membrane of fluorescently labeled molecules.

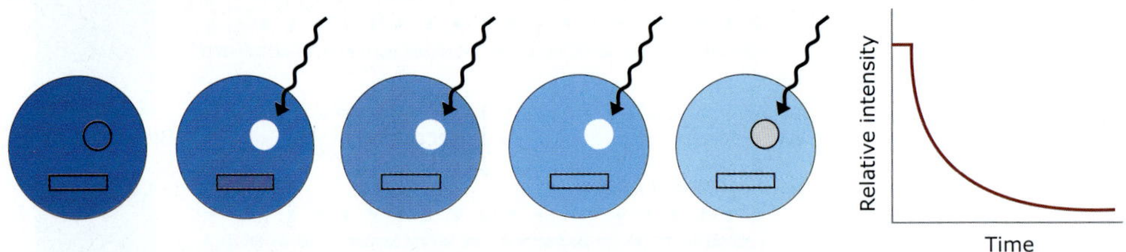

Figure 11.10 FLIP experiments involve repetitive bleaching of a selected region of interest during the entire monitoring period and the fluorescence intensity in regions outside the selected bleached area is measured. The decline in fluorescence intensity in the surrounding regions is due to bleaching of fluorochromes that move through the region of interest during the repetitive bleaching process. The drop in fluorescence intensity outside the bleached region is caused by a steadily increasing population of bleached, non-fluorescent molecules within the cell and thus provides quantitative data on their molecular mobility.

References

Alberts B, et al (2008), *Molecular Biology of the Cell*, 5th ed. Garland Science Publishing.

Ankerhold HC, Ankerhold R and Drummen GP (2012), Advanced Fluorescence Microscopy Techniques—FRAP, FLIP, FLAP, FRET and FLIM, *Molecules* 17(4), 4047-4132.

Chang R (2005), *Physical Chemistry for the Biosciences*, 1st ed. University Science Books.

Lakowicz JR (1999), *Principles of Fluorescence Spectroscopy*, 2nd Ed. Kluwer, New York.

Tsien RY (1998), The green fluorescent protein. *Annu. Rev. Biochem. 67*, 509–544.

van Royen ME, Dinant C, Farla P, Trapman J and Houtsmuller AB (2009), FRAP and FRET methods to study nuclear receptors in living cells. *Methods Mol. Biol. 505*, 69–96.

Zimmer M (2002), Green fluorescent protein (GFP): Applications, structure and related photophysical behavior. *Chem. Rev. 102*, 759–781.

12 Molecular biology techniques

12.1 Nucleic acid extraction

The extraction of nucleic acid is a crucial step for biochemical and diagnostic processes. It is a series of steps to obtain pure nucleic acid samples that are suitable for different downstream applications. Different extraction methods result in the difference in yield, purity and extraction time.

DNA extraction

The choice of the DNA extraction method depends on the types of DNA (chromosomal and plasmid), source organism (bacteria, fungi, plant or animal), starting material (organ, tissue, cell, etc.) and desired results (molecular weight of the desired DNA, purity, extraction time required, etc.).

The extraction of DNA from biological material requires: cell lysis (cell disruption), inactivation of cellular nucleases, removal of biomolecules other than DNA, purification and quantification of the DNA. Common lysis procedures include **mechanical disruption** (for example, grinding, ultrasonication hypotonic lysis, osmotic lysis) and **chemical lysis** (for example, detergent lysis, alkali treatment). Lytic enzymes, chaotropic agents, and different types of detergents are the main components of chemical lysis.

Common DNA extraction methods

Nucleic acid extraction methods can be widely characterized into two different types: solution-based methods and solid-phase based methods. In **solution-based extraction methods**, cell extracts are mixed with chemical solutions devised to purify nucleic acid. **Solid-phase extraction methods** work by causing nucleic acids to bind to solid supports, such as magnetic beads coated with silica or other materials. Different extraction methods result in different yields and purity of DNA.

Organic extraction

A traditional method that can be used to obtain highly pure DNA. This method involves organic extraction (e.g. phenol:chloroform) followed by ethanol precipitation. In this method, cells are lysed and cell debris is usually removed by centrifugation. The remaining soluble material is then mixed with organic solvents such as phenol, or 1:1 mixture of phenol and chloroform.

A phenol-chloroform organic solvent is used to concentrate DNA in a hydrophilic phase. Phenol dissociates proteins bound to DNA while chloroform denatures proteins and lipids. Three distinct phases will form: the *aqueous phase*, the *interphase* and the *organic phase*. Of these, the aqueous phase contains the DNA, whereas the proteins and lipids remain in the other two phases. The aqueous phase is then separated and treated with ethanol to precipitate the DNA (termed **ethanol precipitation**). The precipitated DNA can then be pelleted by centrifugation and dissolved in a buffer of choice for use in downstream reactions.

If the protein content in cell extract is very high, then proteases such as pronase or proteinase K are used to degrade proteins before phenol extraction. RNA is digested from the preparation by treatment with RNase.

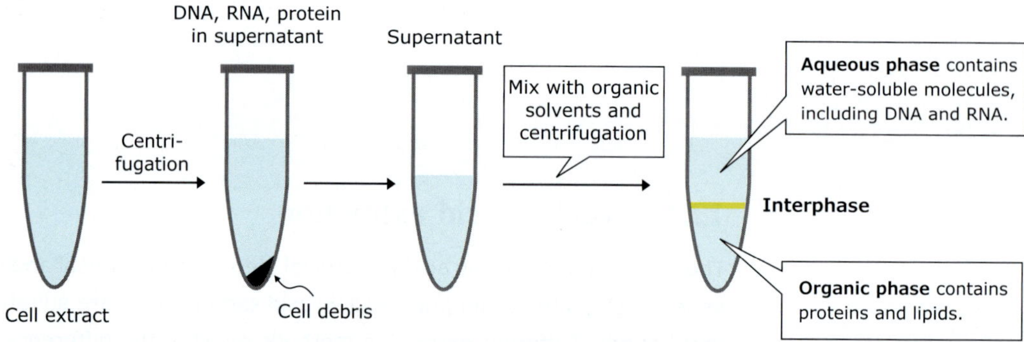

Figure 12.1 Organic extraction method.

Silica-based method

It is a solid-phase DNA extraction method. Silicates have a high binding affinity for DNA under alkaline conditions and high salt concentration. The mechanism involved in this technique is the affinity between negatively charged DNA and positively charged silica material, resulting in selective binding of nucleic acids to the silica matrices (consist of silica material, in the form of either gel or glass particle). As a final step, DNA can be eluted from the silica matrix by any hypoosmotic solution, such as nuclease-free water or buffers such as alkaline Tris-EDTA.

Magnetic bead-based method

Magnetic beads are paramagnetic materials with a uniform particle sizes that have been coated with a functional group that interacts with DNA. Under optimized conditions, DNA selectively binds to an appropriately-coated bead surface, leaving contaminants in solution. A magnetic field is used to separate magnetic beads from a suspension. Hence after DNA binding, beads are separated from other contaminating cellular components by placing a magnet outside of the tube to create a strong external magnetic field. This approach removes the need for vacuum or centrifugation, which minimizes stress or shearing forces on the target molecules.

Anion exchange method

Just like silica matrices, anion exchange resins are also widely used in DNA extraction. DNA extraction by anion exchange chromatography is based on the specific interaction between negatively charged phosphates of the nucleic acid and positively charged surface molecules on the substrate. Unlike silicate negative charge, anion exchange resin makes use of the positively charged diethylaminoethyl (DEAE) cellulose to attract the negatively charged phosphate of nucleic acid. DNA binds specifically to the substrate in the presence of low salt, contaminants are removed by wash steps using a low or medium salt buffer, and purified DNA is eluted using a high salt buffer. Anion exchange has the advantage of extracting very pure DNA as compared to silica and the ability to reuse the resin upon renaturation. However, this method used high-salt concentration in the elution step, thus requiring desalting for downstream applications.

CTAB-based method

Cetyltrimethylammonium bromide (CTAB) extraction method is mainly used for plant samples. The basic composition of CTAB extraction buffer includes 2% CTAB at alkaline pH. CTAB works by precipitating nucleic acids and acidic polysaccharides in low ionic strength solutions, while proteins and neutral polysaccharides remain in solution. Next, the CTAB-nucleic acid precipitated complex is solubilized at high-salt concentrations, leaving the acid polysaccharides in the precipitate. During the precipitation and washing steps, CTAB method uses various organic solvents and alcohols such as phenol, chloroform, isoamyl alcohol and mercaptoethanol.

Plasmid DNA isolation

In a plasmid DNA purification, it is always necessary to separate the plasmid DNA from the large bacterial chromosomal DNA. The separation of plasmid DNA from chromosome DNA is based on the physical differences (size and conformation) between plasmid DNA and chromosomal DNA. Plasmid DNA and the bacterial chromosome are mostly circular, but during the preparation of the cell extract, the large chromosomal DNA is always broken into linear fragments with relaxed conformation. Due to smaller size, plasmids remain as covalently closed circular DNA in supercoiled conformation.

Supercoiled DNA molecules can be easily separated from non-supercoiled DNA. One method of separation is through **alkaline denaturation**. It is based on the differential denaturation of chromosomal and plasmid DNA. The addition of a high concentration of sodium hydroxide denatures the chromosomal and plasmid DNA. In the case of linear chromosomal DNA fragments, strands are separated, whereas the plasmid DNA remains topologically constrained. A neutralization buffer of potassium acetate is added to neutralize the strong alkaline conditions. Subsequent neutralization allows only the covalently closed plasmid DNA to reanneal and to stay solubilized. Denatured chromosomal DNA reaggregate into a tangled mass. The insoluble network can be pelleted by centrifugation, leaving pure plasmid DNA in the supernatant.

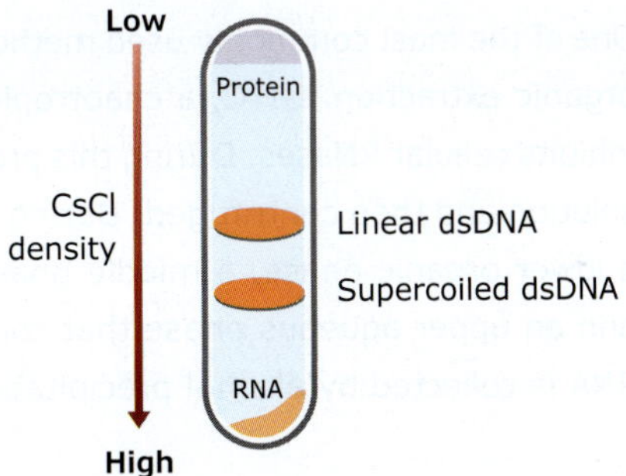

Figure 12.2 Separation of supercoiled DNA from non-supercoiled DNA molecules by density gradient centrifugation in the presence of EtBr. EtBr binds to DNA molecules by intercalating between adjacent base pairs, causing partial unwinding of the double helix. This unwinding results in a decrease in the buoyant density. However, supercoiled DNA, with no free ends, has very little freedom to unwind, and can only bind a less amount of EtBr as compared to linear and nicked circular dsDNA. The decrease in the buoyant density of a supercoiled molecule is therefore much less. As a consequence, supercoiled molecules form a band in an EtBr-CsCl gradient at a different position to linear and nicked-circular DNA.

Another method of separation is through **CsCl buoyant density gradient centrifugation** in the presence of DNA intercalating dye ethidium bromide (EtBr). EtBr intercalation causes local unwinding of the DNA helix, which reduces the molecular density of DNA. Large linear chromosomal DNA (non-supercoiled) molecules can bind more EtBr than supercoiled DNA owing to the difference in their topology. Because supercoiled plasmid DNA cannot be easily unwound, much less EtBr can intercalate. In contrast, linear DNA molecules are not as topologically constrained and can, therefore, allow more EtBr molecules to intercalate, resulting in an overall decrease in molecular density.

RNA extraction

DNA extraction methods cannot be directly applied to RNA extraction as RNA is structurally and chemically different from DNA. RNA is single-stranded, while DNA is mostly double-stranded. Apart from that, RNA is not as biochemically stable as DNA. Two factors contribute to the biochemical instability of RNA. First, RNases, a group of enzymes that degrade RNA molecules, are abundant in the environment and it is difficult to remove RNases completely. Second, RNA is less stable than DNA because of the 2′ hydroxyl group on the ribose ring that promotes hydrophilic attack on the 5′-3′ phosphodiester bond to form a 2′-3′ cyclic phosphate. Therefore, even if all RNases are eliminated or inhibited during RNA purification, RNA spontaneously degrades while in an alkaline solution. To circumvent this biological decay of RNA, purified samples are stored at –20°C.

Isolating intact RNA requires– 1. Disruption of cells or tissue; 2. Inactivation of endogenous RNase activity; 3. Removal of contaminating DNA and proteins. The most important step is the immediate inactivation of endogenous RNases that are released from membrane-bound organelles when cells are disrupted.

There are three major techniques extensively used for RNA extraction: organic extraction, such as phenol-guanidine isothiocyanate-based solutions, silica-membrane based spin-column technology, and paramagnetic particle technology. Silica column and paramagnetic particle based RNA isolation systems do not require the use of toxic organic solvents, are relatively simple, efficient, low cost, and yield total intact RNA with low levels of contamination from proteins and other cellular materials. However, these methods can often result in significant levels of genomic DNA contamination.

One of the most commonly used methods is the phenol- guanidine isothiocyanate (GITC)-based organic extraction. GITC, a chaotropic agent is a protein denaturant that lyses the cells and inhibits cellular RNases. During this process, the sample is homogenized in a phenol-containing solution and then centrifuged. During centrifugation, the sample separates into three phases: a lower organic phase, a middle phase that contains denatured proteins and genomic DNA, and an upper aqueous phase that contains RNA. The upper aqueous phase is recovered and RNA is collected by alcohol precipitation and rehydration.

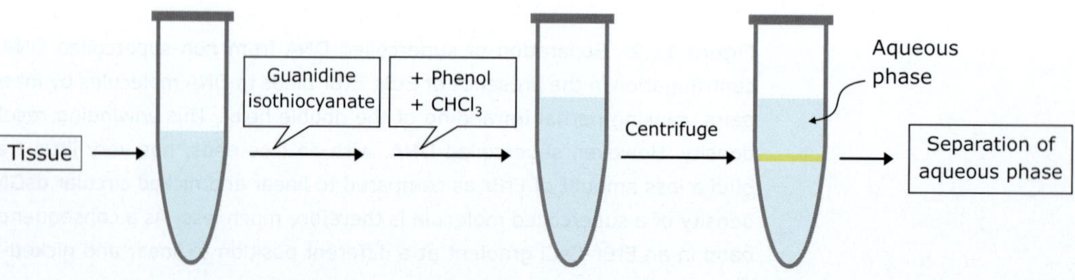

Figure 12.3 Isolation of RNA from tissue culture cells using phenol-guanidine isothiocyanate-based organic extraction.

Qualitative and quantitative analysis of nucleic acid

Nucleic acid quantitation is an important and necessary step prior to most nucleic acid analysis methods. Common methods used for determining quality and quantity of nucleic acid include absorbance, fluorescence and qPCR.

Absorbance based quantification method is performed at 260 nm (based on the fact that the heterocyclic rings of nucleotides absorb UV light with an absorption maximum at around 260 nm). Absorbance method is simple, quick and has a wide detection range. This method does not require any reagents. But this method is less sensitive than fluorescence or qPCR based methods and also cannot distinguish between ssDNA, dsDNA or RNA.

Nucleic acid can be quantified by measuring the absorption at 260 nm. For dsDNA, 1 absorbance unit is equivalent to 50 µg DNA (i.e. a solution of dsDNA with a concentration of 50 µg/mL will have an absorbance at 260 nm equal to 1); for ssDNA, it is equivalent to 33 µg DNA; for ssRNA, it is equivalent to 40 µg RNA.

Absorbance is also used to measure purity of nucleic acids present in the sample. The ratio of absorbance at 260 nm and absorbance at 280 nm (A_{260}/A_{280}) and the ratio of absorbance at 260 nm and absorbance at 230 nm (A_{260}/A_{230}) are used to check nucleic acids purity. For a pure DNA sample, A_{260}/A_{280} ratios is ~1.8 and for a pure RNA sample, the ratio of A_{260}/A_{280} is ~2. A ratio of < 1.8 indicates the DNA sample is contaminated with protein or an organic solvent such as phenol. Typically, protein contamination can be detected by a reduction of this ratio; RNA contamination can be detected by an increase of this ratio.

The A_{260}/A_{230} values for 'pure' nucleic acid are commonly in the range of 2.0–2.2. The A_{260}/A_{230} is a sensitive indicator of contaminants that absorb at 230 nm. These contaminants are significantly more numerous than those absorbing at 280 nm and include chaotropic salts such as guanidine thiocyanate, EDTA, non-ionic detergents like Triton X-100 and phenol. Substances like polysaccharides also show absorbance at this wavelength, but will have a weaker effect.

Fluorescence-based quantification methods use fluorescence dyes that preferentially bind a given species of nucleic acid (e.g. dsDNA, ssDNA or RNA). When a dsDNA binding dye, for example, is excited by a given wavelength of light, only dye in the dsDNA-bound state will fluoresce. As the dye binds to the target nucleic acid, fluorescent quantum yield increases as a function of shift in fluorophore molecular geometry. The intensity of fluorescent signal relates the amount of nucleic acid present. Fluorescence detection methods are comparatively more sensitive than absorbance.

12.2 Polymerase chain reaction

Polymerase chain reaction (PCR) is a rapid and versatile *in vitro* method for amplifying defined target DNA sequences present within the source of DNA. This technique was formulated by *Kary Mullis* in 1983. Usually, the method is designed to permit *selective amplification* of a specific target DNA sequence(s) within a heterogeneous collection of DNA molecules (e.g. total genomic DNA or a complex cDNA population). To permit such selective amplification, some prior DNA sequence information from the target sequences is required. This information is used to design two oligonucleotide primers (amplimers), which are specific to the target sequence and are often about 15–25 nucleotides long. After the primers are added to denatured template DNA, they bind specifically to complementary DNA sequences at the target site. In the presence of a suitably heat-stable DNA polymerase and DNA precursors (the four deoxynucleoside triphosphates, dATP, dCTP, dGTP and dTTP), primer initiates the synthesis of new DNA strands which are complementary to the individual DNA strands of the target DNA segment and will overlap each other.

Essential components of PCRs:

- Thermostable DNA polymerase
- Primers
- dNTP (dATP, dCTP, dGTP and dTTP)
- Divalent cations (usually Mg^{2+}) as a cofactor for activity of DNA polymerase
- Buffer, provides a suitable chemical environment for activity of DNA polymerase
- Template DNA

Primer designing

In PCR, primer designing is the most important aspect for selective amplification. For primer design, some prior DNA sequence information from the target DNA is required. The information is used to design two primers, which are specific to sequences flanking the target DNA sequence. So, for most PCR reactions, it is very important to reduce the chance of the primers binding to other locations in the DNA than the desired one. Hence, certain rules for primer design are important to consider. These rules follow:

Primer length

Length of the primers should not be very short or long. If the primers are too short, they might hybridize to non-target sites and give undesired amplification products.

To illustrate this point, imagine a DNA molecule of 10 Mbp is used in a PCR experiment with a pair of primers eight nucleotides in length. The attachment sites for these primers are expected to occur, on average, once every 4^8 = 65,536 bp. This means that it would be very unlikely that a pair of 8 nucleotides long primers would give a single, specific amplification product. If the primer length will be 15 nucleotides, then the expected frequency of attachment site for the primer will be once every 4^{15} = 1073741824 bp. This figure is higher than the length of the DNA molecule, so 15 nucleotides long primer would be expected to have just one attachment site. But we cannot take very long primer because the long primer influences the rate at which it hybridizes to the template DNA; long primers hybridize at a slower rate. It is generally accepted that the optimal length of primers is 18–20 nucleotides. This length is long enough for adequate specificity and short enough for primers to bind easily to the template at the annealing temperature.

Calculation of length of primers: The longer the primer, the higher its specificity for a particular target. The following equation can be used to calculate the probability that a sequence exactly complementary to the primer sequence will occur by chance in a DNA molecule that consists of a random sequence of nucleotides.

$$K = [g/2]^{G+C} \times [(1-g)/2]^{A+T}$$

where, K is the expected frequency of occurrence of target sequence in a DNA molecule,

g is the relative G + C content of the DNA molecule and

G, C, A and T are the number of specific nucleotides in the primer.

For a double-stranded genome of size N (in nucleotides), the expected number (n) of sites complementary to the primer is n = 2NK.

General recommendations on designing PCR primers:

- 15–30 nucleotides long
- T_m 55–70°C (within 5°C, for two primers)
- 40–60% GC (with uniform distribution)
- One C or G at 3′ end

Nature of sequences

Inverted repeats or any self-complementary sequences >3bp in length should not be present. Sequence of this type tends to form hairpin structure and prevents the primer annealing to its target. There should be no complementarity between the two primers. Because primers are present at high concentration in PCR, even small complementarity between them leads to hybrid formation.

T_m values

T_m (*melting temperature*) is the temperature at which half of the DNA strands are in double-stranded state. T_m values for two primers used together should not differ by >5°C and the T_m of the amplification product should not differ from those of the primers by >10°C.

Base composition

The GC content should be between 40 to 60% with an even distribution of all four nucleotides.

Degenerate primers

In order to make the PCR primers, some sequence information is required. Degenerate primers are used when partial sequence information is available, but the complete sequence is unknown. A *degenerate primer* is a mixture of oligonucleotide sequences in which some positions contain a number of possible bases, giving a population of primers with similar sequences that cover all possible nucleotide combinations for a given protein sequence. Degenerate DNA primers are generally used if only a protein sequence is available. In this case, the protein sequence is translated backward to give the corresponding DNA sequence. Due to the degeneracy of the genetic code, several possibilities will exist for the sequence of DNA that corresponds to any particular polypeptide sequence. Again, most of the ambiguity is in the third codon position. This ambiguous sequence may be used to make degenerate primers.

Partial sequence of polypeptide:

Met — Tyr — Cys — Asn — Thr — Arg — Pro — Gly

Possible codons in DNA:

Met	Tyr	Cys	Asn	Thr	Arg	Pro	Gly
ATG	TAC	TGT	AAT	ACT	AGA	GCT	GGT
	TAT	TGC	AAC	ACC	AGG	GCC	GGC
				ACA		GCA	GGA
				ACG		GCG	GGG

Corresponding degenerate primer:

Met	Tyr	Cys	Asn	Thr	Arg	Pro	Gly
ATG	TAC	TGT	AAT	ACT	AGA	GCT	GGT
	T	C	C	C	G	C	C
				A		A	A
				G		G	G

Bases in the third codon position are shown in red. The degenerate primer consists of a mixture of primers with these bases varied as shown.

Figure 12.4 Degenerate DNA primers are used if only partial DNA sequence information is available. Often, as here, a short amino acid sequence from a protein is known. Because many amino acids are encoded by several alternative codons, the deduced DNA coding sequence is ambiguous. For example, the amino acid tyrosine is encoded by TAC or TAT. Hence, the third base is ambiguous and when the primer is synthesized, a 50:50 mixture of C and T will be inserted at this position. This ambiguity occurs for all the bases shown in red letters, resulting in a pool of primers with different, but related sequences. Hopefully, one of these primers will have enough complementary bases to anneal to the target sequence that is to be amplified.

Reaction cycle

The PCR is a chain reaction because newly synthesized DNA strands will act as templates for further DNA synthesis in subsequent cycles. It consists of a series of cycles of three successive reactions:

- **Denaturation**: Performs typically at about 93–95°C for human genomic DNA.
- **Primer annealing**: Annealing temperature must be low enough to enable hybridization between primer and template, but high enough to prevent mismatched hybrids from forming. The annealing temperature is typically about 5°C below the calculated T_m of primers. It is usually kept between 50°C to 70°C depending on the T_m of the expected duplex.

  Several equations are used to calculate the T_m value of duplex formed between an oligonucleotide primer and its complementary target sequence. One common equation, known as **Wallace rule**, can be used to calculate the T_m for perfect duplex 15-20 nucleotides in length in solvents of high ionic strength (e.g. 1M NaCl):

  T_m (in °C) = (4×[G + C]) + (2×[A + T])

  A + T is the sum of A and T containing nucleotides and G + C is the sum of G and C containing nucleotides in the oligonucleotide primer.
- **DNA synthesis** (or **primer extension**): After primer annealing, the next step in PCR is to extend the 3′ end of primers, complementary to the template. In this step, 5′ to 3′ polymerase activity of the DNA polymerase incorporates dNTPs and synthesizes the complementary strands. The reaction temperature is raised to the optimal temperature of the enzyme for its maximal activity, which is generally 70–75°C for thermostable DNA polymerases. If the primer annealing temperature is within 3°C of the extension temperature, both annealing and extension temperatures can be combined into a single step called **two-step PCR**, instead of conventional **three-step PCR**. Two-step PCR shortens the time taken for the PCR process as there is no need for switching and stabilizing temperatures between annealing and extension.

These three steps constitute one cycle of the PCR amplification and can be carried out repetitively just by changing the temperature of the reaction mixture. The thermostability of the polymerase makes it feasible to carry out PCR. Suitable heat-stable DNA polymerases have been obtained from microorganisms whose natural habitat is hot springs.

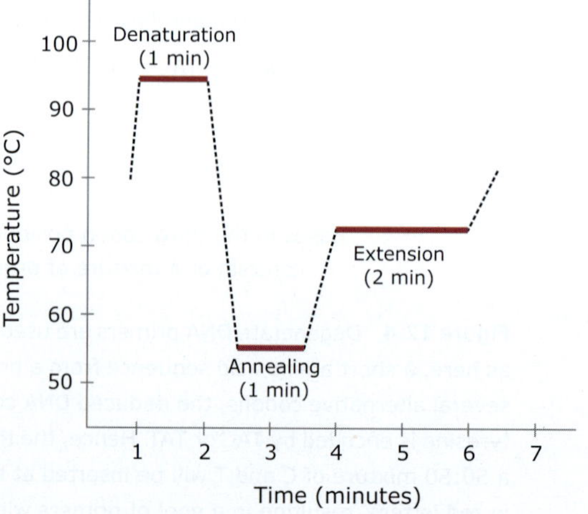

Figure 12.5
In each cycle of the reaction, the strands of the duplex DNA are separated by heat denaturation, the preparation is cooled to such level that synthetic DNA primers anneal to a complementary segment on each strand, and the primers are extended by DNA polymerase. The process is then repeated for numerous cycles. The number of 'unit-length' strands doubles with every cycle after the second cycle.

For example, the widely used *Taq* DNA polymerase is obtained from *Thermus aquaticus* and is thermostable up to 94°C, with an optimum working temperature of 75–80°C. It has an extension rate of 35 to 100 nucleotides per second at 72°C. *Taq* polymerase lacks a 3′ to 5′ exonuclease (proofreading) activity. When the *Taq* polymerase incorporates a wrong dNTP, subsequent extension of the strand either proceeds very slowly or stops completely. This problem can be overcome by using other thermostable DNA polymerases with 3′ to 5′ proofreading exonuclease activity.

DNA Pol	5′ to 3′ exonuclease	3′ to 5′ exonuclease	Source
Tli (or *Vent*)	No	Yes	*Thermococcus litoralis*
Pfu	?	Yes	*Pyrococcus furiosus*
Pwo	No	Yes	*Pyrococcus woesei*
Tth	?	No	*Thermus thermophilus*

After each cycle, the number of templates doubles, so that if one starts with a single dsDNA molecule, after 20 cycles, the number of molecules synthesized by the PCR is about 1×10^6, and after 30 cycles the number increases to about 1×10^9. This number can be calculated by applying the following formula:

$$N_f = N_i \times 2^n$$

where, N_f is the final number of DNA molecules produced by the PCR,

N_i is the initial number of molecules (template), and

n is the number of cycles performed.

When amplifying a small segment of a long double-stranded DNA template, the desired blunt-ended target fragments first appear in the *third cycle* of PCR. The number of blunt-ended target products synthesized after 25 cycles of PCR with different numbers of starting molecule (assuming that amplification is 100% efficient) is given below:

Initial number of molecules	Number of blunt-ended target products
01	4,194,304
02	8,388,608
05	20,971,520
10	41,943,040

In theory, each amplification cycle should double the number of target molecules, resulting in an exponential increase in PCR product. However, even before substrate or enzyme becomes limiting, the efficiency of exponential amplification is less than 100% due to sub-optimal DNA polymerase activity, poor primer annealing and incomplete denaturation of the templates. So, the actual amount of PCR product can be calculated by considering efficiency term. This PCR efficiency formula can be expressed as:

PCR product = (Initial input amount) × (1 + % efficiency)$^{\text{cycle number}}$

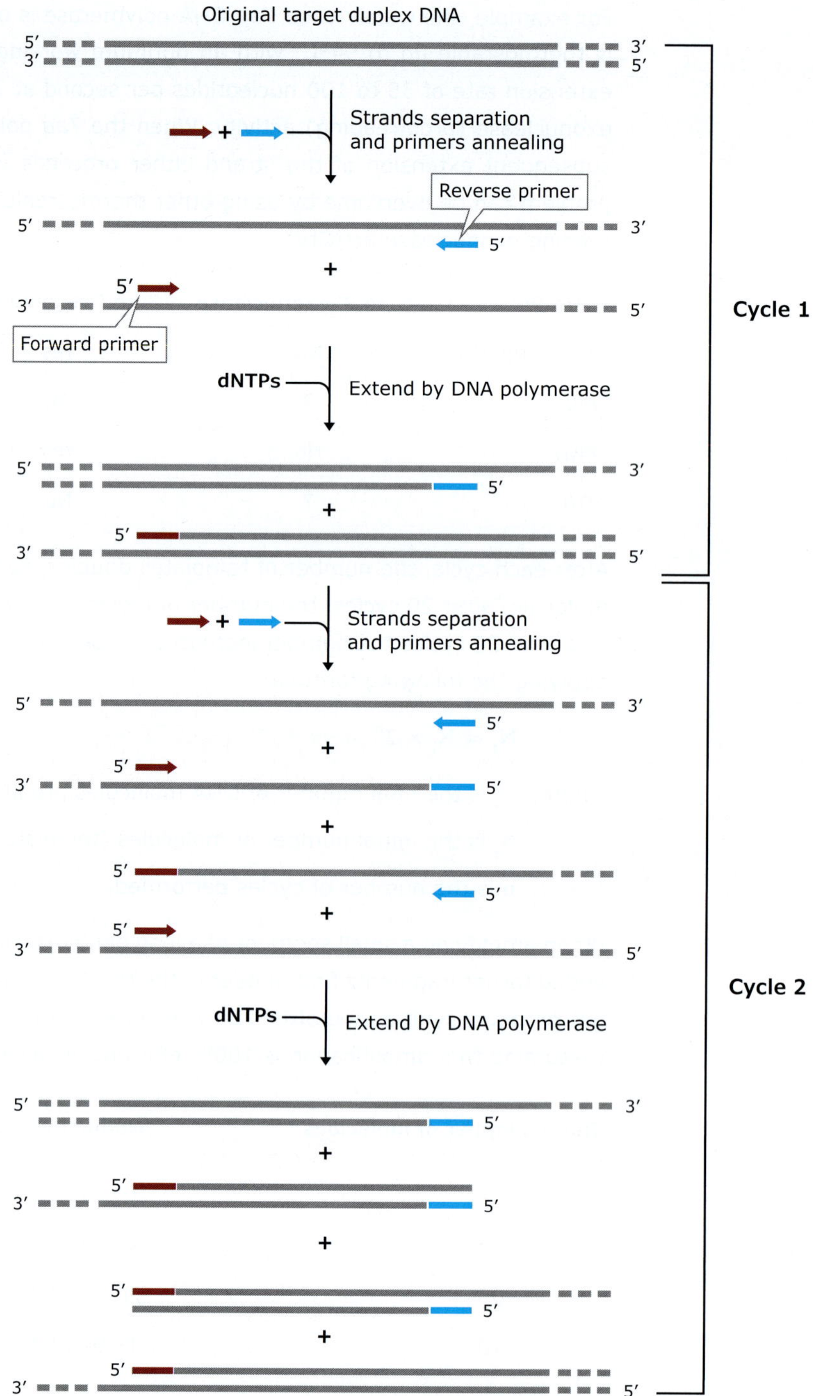

Figure 12.6 A typical temperature profile for a PCR. The denaturation temperature is usually 94°C, which denatures the dsDNA and releases single-stranded DNA to act as templates in the next round of DNA synthesis. The annealing temperature, at which the primers attach to the templates, is dependent on the nucleotide composition of the primers. The extension temperature, at which the bulk of DNA synthesis occurs, is usually set at 74°C, optimum for *Taq* polymerase.

Modifications to the PCR techniques

Nested PCR

Nested PCR increases the specificity of DNA amplification, by reducing non-specific amplification of DNA. Two sets of primers are being used in two successive PCR reactions. In the first reaction, one pair of primers is used to generate DNA products, which besides the intended target, may still consist of non-specifically amplified DNA fragments. The product(s) is, then, used in a second PCR reaction with a set of primers whose binding sites are just downstream of the first primer or *nested* between the original set of primers. Binding sites are completely or partially different from the primer pair used in the first reaction, but are completely within the DNA target fragment.

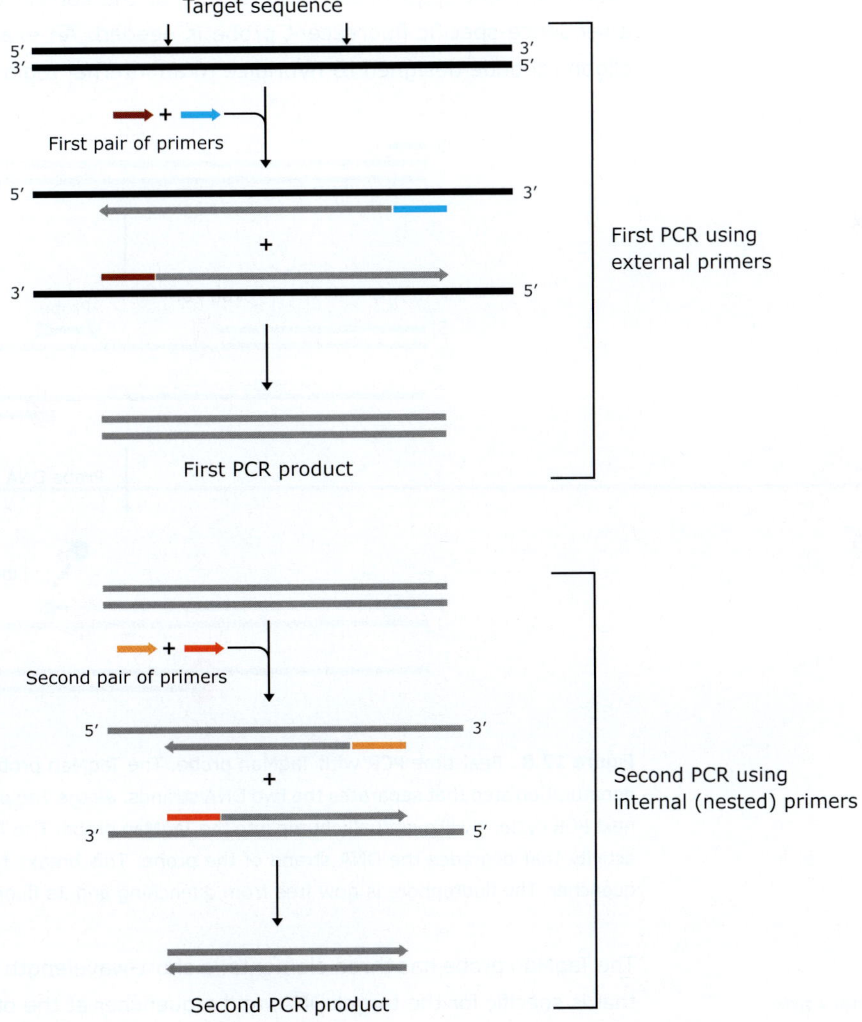

Figure 12.7 Nested PCR. It involves using two sets of primers. The first external set generates a normal PCR product. Primers that lie inside the first set are, then, used for a second PCR reaction. These internal or nested primers generate a shorter product. It is used to increase the specificity and fidelity of the PCR.

Quantitative Real–Time PCR

Quantitative Real-Time PCR is based on the general principle of PCR, which is used to amplify and simultaneously quantify a target DNA molecule. This is called Real-Time PCR because it allows the scientist to actually view the increase in the amount of DNA as it is amplified. Real-Time PCR systems rely upon the detection and quantitation of a fluorescent reporter, whose signal increases in direct proportion to the amount of PCR product in a reaction. These fluorescent reporter molecules include dyes that bind to the dsDNA (i.e. SYBR Green) or sequence-specific probes. **SYBR Green** binds to the minor groove of the dsDNA only. In the solution, the unbound dye exhibits very little fluorescence. This fluorescence is substantially enhanced when the dye is bound to dsDNA. SYBR Green has, however, a limitation that includes preferential binding to G.C rich sequences.

SYBR Green monitors the total amount of double-stranded DNA, but cannot distinguish between different sequences. To be sure that the correct target sequence is being amplified, a sequence-specific fluorescent probe is needed. An example is the **TaqMan probe** which is oligonucleotide designed to hybridize to an internal region of a PCR product.

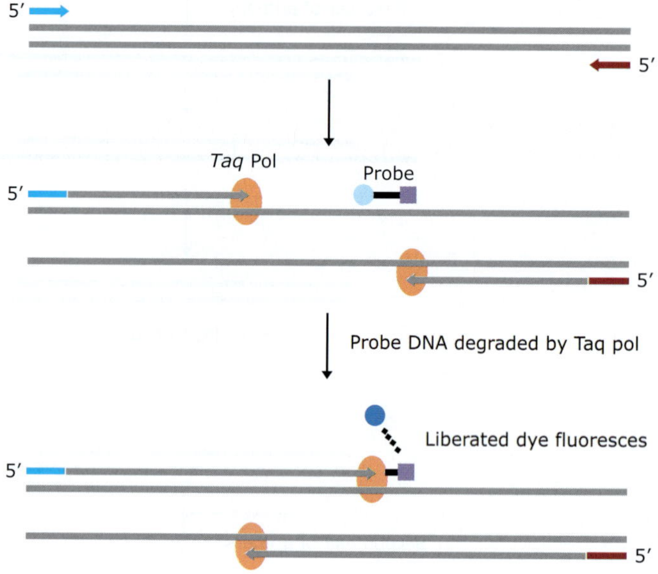

Figure 12.8 Real-time PCR with TaqMan probe. The TaqMan probe binds to the target sequence after the denaturation step that separates the two DNA strands. As the *Taq* polymerase extends the primer during the next PCR cycle, it will eventually bump into the TaqMan probe. The *Taq* polymerase has a 5′ to 3′ exonuclease activity that degrades the DNA strand of the probe. This breaks the linkage between the fluorophore and quencher. The fluorophore is now free from quenching and its fluorescence increases.

Quenchers are substances capable of absorbing radiation from a fluorophore and re-emitting much of that energy as either heat (in case of **dark quenchers**) or visible light (in case of **fluorescence quenchers**).

The TaqMan probe has three elements: a short-wavelength fluorophore on one end, a sequence that is specific for the target DNA, and a quencher at the other end. As long as the fluorophore and the quencher are close to each other, fluorescence is quenched and no fluorescent light is emitted. This probe is designed to anneal to the center of the target DNA. When *Taq* polymerase elongates the second complementary strand during PCR, its 5′ to 3′ exonuclease activity cuts the probe into single nucleotides. This removes the close proximity between the fluorophore and the quencher and abolishes quenching. The short-wavelength fluorophore can now fluoresce and a signal will be detected that is proportional to the number of newly synthesized strands.

RT–PCR

RT-PCR (Reverse Transcription PCR) is a method used to amplify, isolate or identify a known sequence from a cellular or tissue RNA. The PCR reaction is preceded by a reaction using reverse transcriptase to convert RNA to cDNA. Some thermostable DNA polymerase can use RNA templates as substrate. An example of this is the recombinant form of **Tth polymerase** from *Thermus thermophilus*, which can catalyze high-temperature reverse transcription of RNA in the presence of $MnCl_2$. It has both intrinsic reverse transcriptase and thermostable DNA-dependent DNA polymerase activities. This has led to the development of protocols for single-enzyme reverse transcription and PCR amplification.

Inverse PCR

Standard PCR is used to amplify a segment of DNA that lies between two inward-pointing primers. In contrast, inverse PCR (also known as inverted or inside-out PCR) is used to amplify unknown DNA sequences that flank one end of a known DNA sequence and for which no primers are available. The inverse PCR method involves a series of restriction digestion and ligation, resulting in a circular DNA that can be primed for PCR from a single section of known sequence. It involves isolating a restriction fragment that contains the known sequence plus flanking sequences. The restriction fragments circularize to form circular DNA under very low concentrations in the presence of DNA ligase. To perform PCR, two primers that bind specifically to the known sequence but they are oriented in opposite directions. Successful PCR with these primers produces a linear product in which central unknown region remains flanked by two short known sequences.

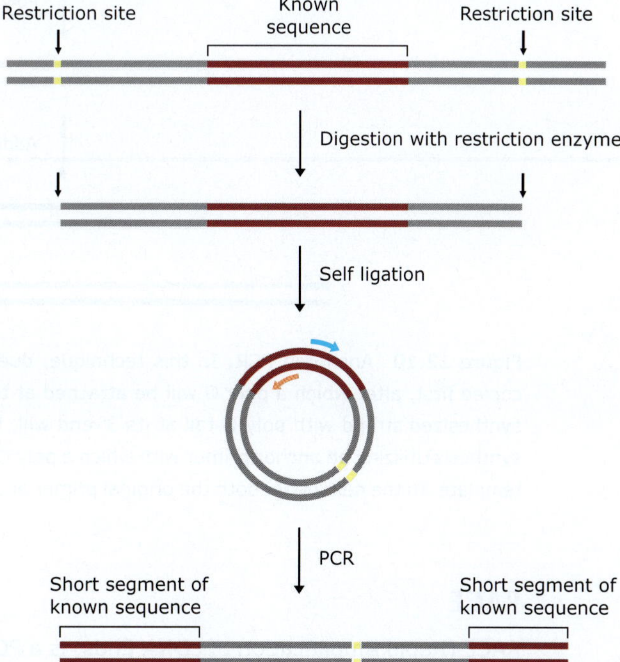

Figure 12.9 Inverse PCR. A region of DNA in which part of the sequence is known. If the areas of interest lie outside the known region, inverse PCR can be used to amplify the flanking region. It involves the digestion of genomic DNA with appropriate restriction endonucleases, intramolecular ligation to circularize the DNA fragments and PCR amplification. First, the DNA is cut with a restriction enzyme that does not cut within the region of known sequence. This generates a fragment of DNA containing the known sequence flanked by two regions of unknown sequence. Under low DNA concentrations, self-ligation is induced to give a circular DNA product. Finally, PCR is performed on the circular fragments of DNA. Two primers are used that face outwards from the known DNA sequence. PCR amplification gives linear products.

Anchored PCR

In the basic PCR technique and the inverse PCR, one has to use two primers representing the sequences lying at both ends of sequence to be amplified. But sometimes, we may have knowledge about the sequence at only one of the two ends of the DNA sequence to be amplified. In such cases anchored PCR may be used, which will utilize only one primer instead of two primers. In this technique, due to the use of one primer, only one strand will be copied first, after which a poly G tail will be attached at the end of the newly synthesized strand. This newly synthesized strand with poly G tail at its 3′-end will, then, become the template for the daughter strand synthesis utilizing an anchor primer with which a poly C sequence is linked to complement with poly G of the template. In the next cycle, both the original primer and anchored primer will be used for gene amplification.

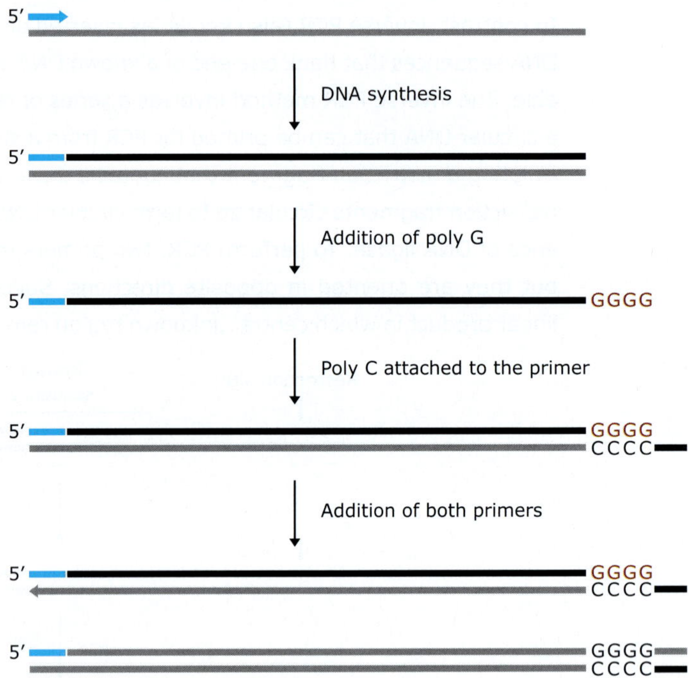

Figure 12.10 Anchored PCR. In this technique, due to the use of one primer, only one strand will be copied first, after which a poly G will be attached at the end of the newly synthesized strand. This newly synthesized strand with poly G tail at its 3′-end will, then, become the template for the daughter strand synthesis utilizing an anchor primer with which a poly C sequence is linked to complement with poly G of the template. In the next cycle, both the original primer and anchored primer will be used for gene amplification.

RACE

RACE (Rapid Amplification of cDNA Ends) is a PCR-based method for locating the precise start and end points of gene transcripts. It is of two types– 5′-RACE and 3′-RACE.

In **5′-RACE**, the first step is the conversion of mRNA into cDNA with enzyme reverse transcriptase. The primer used during reverse transcription is specific for an internal region (gene specific primer) close to 5′-end of the gene under study. Since only a small segment of mRNA is copied, the cDNA so generated will correspond exactly with the start of the mRNA. After synthesis of above cDNA, a short poly(A) tail is added to its 3′-end using enzyme terminal deoxynucleotidyl transferase and it is subjected to normal PCR. The second primer will anneal to this poly(A) sequence and convert the single-stranded cDNA into a double-stranded molecule, which is subsequently amplified as the PCR proceeds. The sequence of the final PCR product will reveal the precise position of the start or 5′-end of the transcript. This is known as 5′-RACE, because it results in amplification of the 5′-end of the starting RNA.

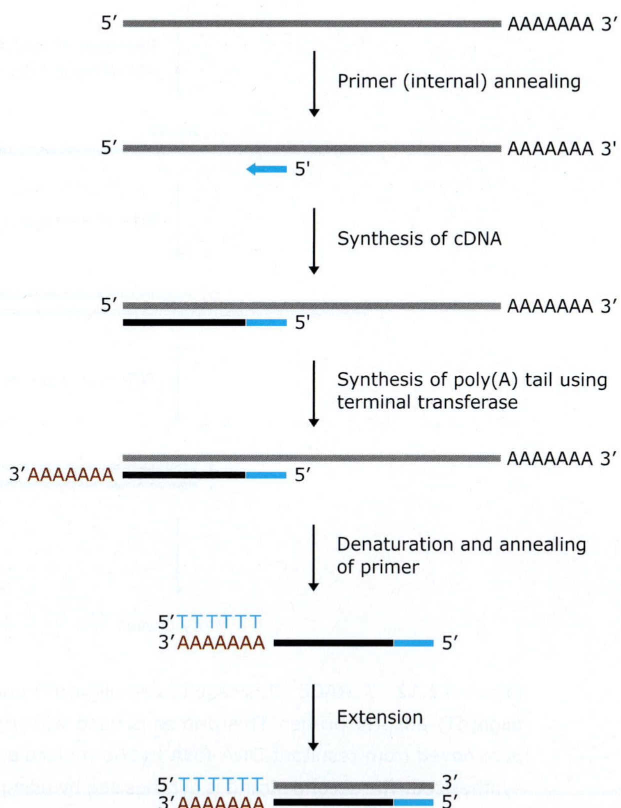

Figure 12.11 5′-RACE PCR begins using mRNA as a template for the first round of cDNA synthesis (or reverse transcription) reaction using an antisense primer that recognizes a known sequence in the gene of interest; the primer is called a gene-specific primer, and it copies the mRNA template in the 3′ to the 5′ direction to generate a specific single-stranded cDNA product. Following cDNA synthesis, the enzyme terminal deoxynucleotidyl transferase is used to add homopolymer poly(A) tail to the 3′-end of the cDNA. A PCR reaction is then carried out, by using primers that hybridize to the 3′ poly(A) tail of the cDNA and convert single-stranded cDNA into a double-stranded molecule.

3′-RACE like 5′-RACE, requires knowledge of a sequence within the target RNA. A population of mRNAs is transcribed into cDNA with an adapter-primer consisting oligo(dT) primer-linked with adapter sequence. It uses the natural polyA tail that exists at the 3′-end of all eukaryotic mRNAs for primer formation during reverse transcription. After first strand cDNA synthesis, the original mRNA template is destroyed with RNase H, which is specific for RNA-DNA hybrid molecules. Since the internal sequence is known, an internal primer (*gene specific primer*) is used for the synthesis of second strand. Further, the same internal primer and a primer corresponding to the adapter sequence are used in a standard PCR reaction to amplify just the 3′-end of the cDNA.

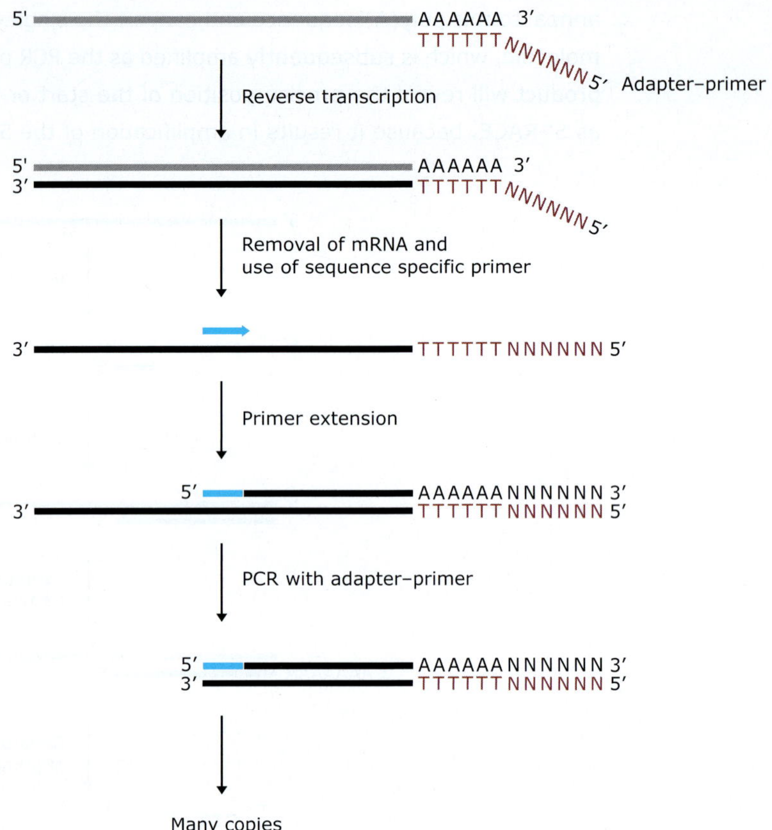

Figure 12.12 3′-RACE. This requires an oligo(dT) primer that has an adapter sequence at the 5′-end i.e. oligo(dT) adapter-primer. This primer is used with reverse transcriptase to make the cDNA. The mRNA is removed from resultant DNA-RNA hybrid molecule. After removal of mRNA, a second strand of DNA is synthesized. The second strand is synthesized by using internal sequence-specific primer. The same internal primer and a primer corresponding to the adapter sequence are then used in a standard PCR reaction to amplify just the 3′-end of the cDNA.

Touchdown PCR

The specificity of conventional PCRs decreases as the sequence complexity of the template DNA increases. The greater the complexity, the greater is the chance that the primers will bind promiscuously to sequences other than the intended target. The number of mispriming events can be reduced by optimizing the concentrations of the components of the PCR, in particular, the concentrations of Mg^{2+}, primers, dNTPs and template. Further minimization of off-target amplification can be obtained by using stringent annealing temperature.

Touchdown PCR is a method used to increase specificity without compromising the yield. The principle is to initiate synthesis at very high annealing temperatures which permit only perfectly matched primer-template hybrids to form. The annealing temperature is dropped in a stepwise fashion with each cycle (1-2°C/every second cycle). Once copies of the target sequence have begun to accumulate over the first few cycles, high-temperature annealing becomes much less critical for specificity, as it's the previous products which form the major template. These products are unlikely to have any sites for mispriming. The benefit of decreasing the annealing temperature is to increase the probability of stable primer-target interaction.

RAPD

RAPD stands for Random Amplification of Polymorphic DNA. It is a type of PCR, but the segments of DNA that are amplified are random. Unlike traditional PCR analysis, RAPD does not require any specific knowledge of the DNA sequence of the target organism. In RAPD, several arbitrary, short primers (8–12 nucleotides) and a large template of genomic DNA are used. No knowledge of the DNA sequence for the targeted gene is required, as the primers will bind somewhere in the sequence, but it is not certain exactly where. This makes the method popular for comparing the DNA of biological systems that have not had the attention of the scientific community, or in a system in which relatively few DNA sequences are compared. In recent years, RAPD has been used to characterize and trace, the phylogeny of diverse plant and animal species.

RAPD involves amplification of DNA fragments from any species by use of a single arbitrary oligonucleotide primer without prior sequence information. As the approach requires no prior knowledge of the genome that is being analyzed, it can be employed across species using universal primers. The major limitation of the RAPD method is the reproducibility and dominant inheritance. Several factors influence the reproducibility of RAPD reactions such as quality and quantity of template DNA, PCR buffer, concentration of magnesium chloride, primer to template ratio and annealing temperature. RAPD markers are dominant markers, and hence do not distinguish dominant homozygotes from heterozygotes.

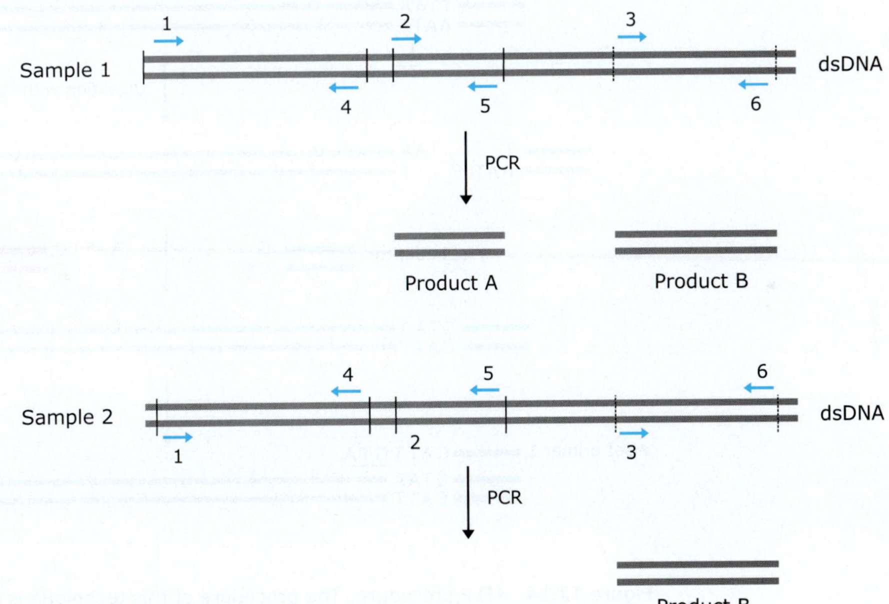

Figure 12.13 Schematic drawing of reaction conditions for RAPD. The primers must anneal in a particular orientation (such that they point towards each other) and within a reasonable distance of one another. The arrows represent multiple copies of a single primer and the direction of the arrow indicates the direction in which DNA synthesis will occur. The numbers represent primer annealing sites on the DNA template. For sample 1, primers anneal to sites 1, 2, and 3 on the top strand of the DNA template and to sites 4, 5, and 6 on the bottom strand of the DNA template. In this example, only 2 RAPD products are formed. For sample 1: (i) product A is produced by PCR amplification of the DNA sequence which lies in between the primers bound at positions 2 and 5; and (ii) product B is produced by PCR amplification of the DNA sequence which lies in between the primers bound at positions 3 and 6. No PCR product is produced by the primers bound at positions 1 and 4 because these primers are too far apart to allow completion of the PCR reaction. No PCR products are also produced by the primers bound at positions 4 and 2 or positions 5 and 3 because these primer pairs are not oriented towards each other. For sample 2, the primer failed to anneal at position 2 and PCR product was obtained only for primers bound at position 3 and 6.

AFLP

To overcome the limitation of reproducibility associated with RAPD, AFLP (Amplified Fragment Length Polymorphism) technology was developed. Like RAPD, AFLP does not require any DNA sequence information from the organism under study and also a dominant marker. However, it is highly reliable and reproducible. It combines the power of RFLP with the flexibility of PCR-based technology by ligating primer recognition sequences (adaptors) to the restricted DNA and selective PCR amplification of restriction fragments using a limited set of primers. The first step in AFLP analysis involves restriction digestion of genomic DNA with a combination of rare cutter (EcoRI or PstI) and frequent cutter (MseI or TaqI) restriction enzymes. Double-stranded oligonucleotide adaptors are, then, designed in such a way that the initial restriction site is not restored after ligation. Such adaptors are ligated to both ends of the fragments to provide known sequences for PCR amplification. PCR amplification will only occur where the primers are able to anneal to fragments which have the adaptor sequence plus the complementary base pairs to the additional nucleotides called selective nucleotides. An aliquot is then subjected to two subsequent PCR amplifications under highly stringent conditions with primers complementary to the adaptors, and possessing 3′ selective nucleotides of 1–3 bases.

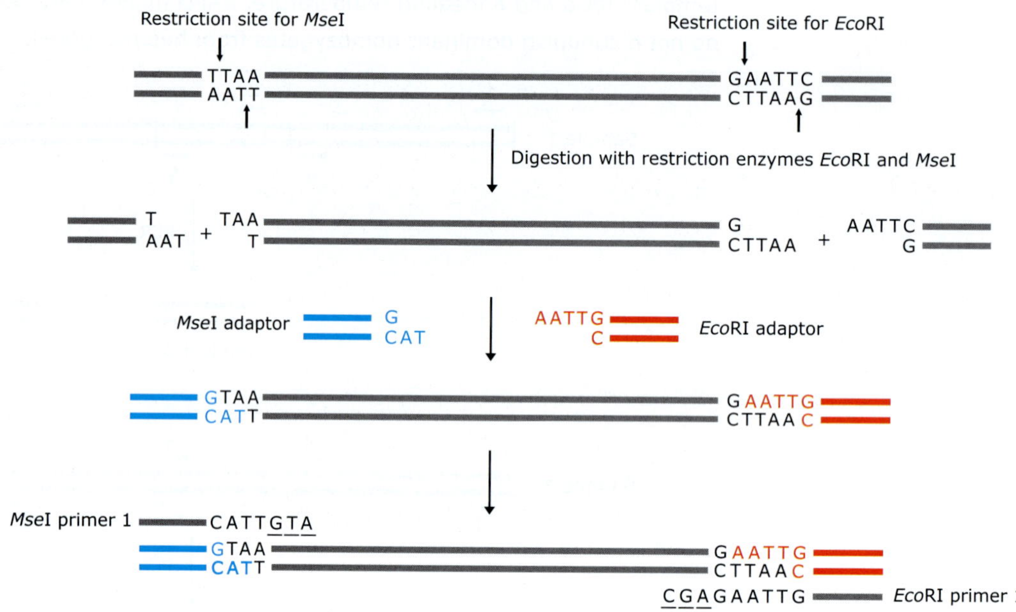

Figure 12.14 AFLP procedure. The procedure of this technique is divided into two steps:

1. Digestion of cellular DNA with two restriction enzymes and ligation of restriction half-site specific adaptors. The adaptor is designed in such a way that ligation of a fragment to an adaptor does not reconstitute the reaction site.
2. Selective amplification of some of these fragments with two PCR primers that have corresponding adaptor and restriction site-specific sequences.
   To achieve selective amplification of a subset of these fragments, primers are extended into the unknown part of the fragments [underlined base], usually one to three arbitrarily chosen bases beyond the restriction site. The first is performed with a single-bp extension, followed by a more selective primer with up to a 3-bp extension. Because of the high selectivity, primers differing by only a single base in the AFLP extension amplify a different subset of fragments.

The first PCR (preamplification) is performed with primer combinations containing a single bp extension, while final (selective) amplification is performed using primer pairs with up to 3-bp extension. Because of the high selectivity, primers differing by only a single base in the AFLP extension amplify a different subset of fragments. A primer extension of one, two or three bases reduces the number of amplified fragments by factors of 4, 16 and 64, respectively. Ideal primer extension lengths will vary with genome size of the species and result in an optimal number of bands: not too many bands to cause smears or high levels of band co-migration during electrophoresis, but sufficient to provide adequate polymorphism. AFLP fragments are visualized either on agarose gel or on denaturing polyacrylamide gels with autoradiography.

12.3 Nucleic acid hybridization

Nucleic acid hybridization is a fundamental tool in molecular biology which takes advantage of the ability of individual single-stranded nucleic acid molecules to form double-stranded molecules by standard base pairing. For this to happen, the interacting single-stranded molecules must have a sufficiently high degree of base complementarity. Nucleic acid hybrids can be formed between two strands of DNA, two strands of RNA, or one strand of DNA and one of RNA.

Nucleic acid hybridization is used to identify related or identical molecules on the basis of base complementarity. The hybridization assays involve a labeled nucleic acid *probe* to identify complementary DNA or RNA molecules present within a complex mixture of nucleic acid molecules, the *target* nucleic acid. It can be used to detect homologous DNA or RNA sequences not only in cell extracts but also in chromosomes or intact cells - a procedure called *in situ* hybridization.

Stringency of a hybridization reaction

Stringency is a term that describes the number of mismatched bases that are allowed in a hybridization reaction and still have a double-stranded hybrid. The more stringent the hybridization conditions are, the fewer the mismatched bases that are allowed before the two strands come apart.

Stringency is mainly influenced by the *temperature* and the *salt concentration* of the hybridization mixture. All these conditions affect the stability of the double-stranded molecule. High stringency conditions are usually achieved with high temperatures (approaching the melting temperature of the DNA-DNA hybrid) and low salt concentrations (<0.1 M). At high temperature and low salt concentration, only closely matched sequences will hybridize.

The optimum hybridization temperature is experimentally determined, starting with temperatures 5°C below the melting temperature (T_m). The T_m is the temperature at which the probe and target are 50% dissociated. A variety of equations have been derived that can be used to estimate the T_m under a range of conditions.

Factors influencing the hybridization reaction of nucleic acid in solution:

- Temperature
- Ionic strength
- Destabilizing agents
- Mismatched base pairs
- Duplex length
- Viscosity

Salt concentration also influences the stringency of hybridization reaction. The phosphate groups are moderately strong acids, and thus they are always ionized at physiological pH and bear the negative charges. These negative charges tend to repel the strands in double helical structure apart. Binding of positive ions (cations) with phosphates neutralize the negative charge of phosphates and thus screens the repulsive force between the sugar-phosphate backbones in dsDNA. Lowering the salt concentration of a DNA solution promotes denaturation by removing the cations that shield the negative charges on the two strands from each other. At low ionic strength, the mutually repulsive forces of these negative charges from the phosphoryl groups are enough to denature the DNA, even at a relatively low temperature. So, at low salt concentrations, the two strands will only stay double-stranded if the two strands are well matched (complementarity). High salt concentrations tend to stabilize the double-stranded molecule and allow more mismatches to take place when two strands hybridize.

The pH of the solution in which hybridization takes place also affects the stability of hybridizing molecules. Under slightly acidic conditions (pH 6–7), the double-stranded DNA molecule is more stable because the increased H^+ concentration also helps to reduce the repulsive forces between the two backbones of the DNA molecule.

Nucleic acid probe

Nucleic acid hybridization with a labeled probe is the only practical way to detect a complementary target sequence in a complex nucleic acid mixture. Nucleic acid probes are oligonucleotides or polynucleotides that can bind with high specificity to complementary sequences.

Probes can be complementary to either DNA or RNA and can be from as few as 20 nucleotides to hundred of nucleotides long. A nucleic acid probe may be DNA (**DNA probes**), RNA (**RNA probes**) and **synthetic oligonucleotide**. Probes are usually labeled (isotopically or non-isotopically) to aid their easy detection when bound to the target nucleic acids.

Oligonucleotide probes are synthesized chemically and end-labeled. DNA probes, which are cloned DNAs and may either be end-labeled or internally labeled during *in vitro* replication. RNA probes are internally labeled during *in vitro* transcription from cloned DNA templates. DNA probes can be double-stranded or single-stranded. Prior to hybridization, the double-stranded probe will be denatured. RNA probes and oligonucleotide probes are generally single-stranded. DNA probes, RNA probes and oligonucleotide may be *heterologous* and *homologous*.

A **heterologous probe** is a probe that is similar to, but not exactly the same as, the nucleic acid sequence of interest. If the gene being sought is known to have a similar nucleotide sequence to a second gene that has already been cloned, then it is possible to use this known sequence as a probe. For example, a mouse probe could be used to search a human genomic library. A **homologous probe** is a probe that is exactly complementary to the nucleic acid sequence of interest.

12.4 Labeling of nucleic acids

Nucleic acids may be modified with tags that enable detection or purification. The resulting nucleic acid can be used to identify or recover other interacting molecules. Nucleic acids can be labeled by *isotopic* and *non-isotopic labeling* methods:

Isotopic labeling

Isotopic labeling involves the replacement of specific atoms by their isotopes. The isotopes may be stable or radioisotopes. The labeling with radioisotopes or radioactive isotopes is called *radiolabeling*.

Isotopic labeling of nucleic acids has been conducted by incorporating nucleotides containing radioisotopes (radiolabeling). Such radiolabeled probes contain nucleotides with a radioisotope (often ^{32}P, ^{33}P, ^{35}S or ^{3}H), which can be detected specifically in solution or, much more commonly, within a solid specimen. In molecular biology, two are especially important: the radioactive isotopes of phosphorus, ^{32}P, and sulfur, ^{35}S. Since sulfur is not a normal component of DNA or RNA, we use *phosphorothioate* derivatives. A normal phosphate group has four oxygen atoms around the central phosphorus. In a phosphorothioate, one of these is replaced by sulfur. To introduce ^{35}S into DNA or RNA, phosphorothioate groups containing radioactive sulfur atoms are used to link together the nucleotides.

Atoms that have the same atomic number, but have different masses are known as isotopes. Some isotopes of an atom have unstable nuclei which after nuclear reaction emit characteristic radiation. These isotopes are called **radioisotopes**, or more commonly *radionuclides*.

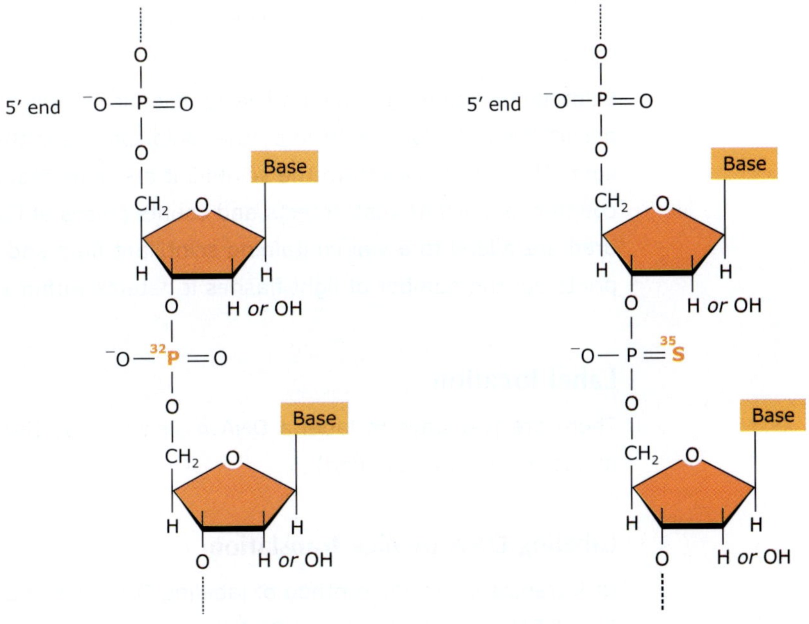

5′ end $^{-}O-P=O$ ... ^{32}P labeled

5′ end $^{-}O-P=O$ ... ^{35}S labeled

Table 12.1 Radioisotopes which are commonly used in biological research

Isotope	Half-life
^{32}P	14 days
^{131}I	8.1 days
^{35}S	87 days
^{14}C	5570 years
^{45}Ca	164 days
^{3}H	12.3 years

The intensity of signal produced by radioisotopes is dependent on the intensity of the radiation emitted by the radioisotopes, and the time of exposure, which may often be long (one or more days, or even weeks in some applications). ^{32}P has been used widely in Southern blot hybridization, dot-blot hybridization, colony and plaque hybridization because it emits high energy β-particles which provide a high degree of sensitivity of detection.

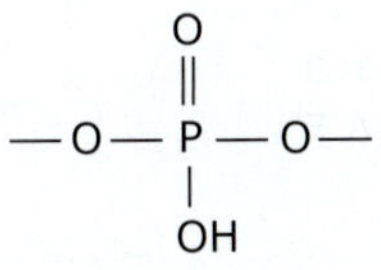

The incorporation of radioisotopes in a sample can be detected by two common methods. In *autoradiography,* radiolabeled material is allowed to expose a photographic emulsion. Development of the emulsion reveals the distribution of labeled material. In the second detection method, the amount of radioactivity in radiolabeled samples is directly measured, either by a *Geiger counter* or by a *scintillation counter*. In a **Geiger counter**, emission from radioisotopes is detected by the ionization they produce in gas. In **scintillation counter**, the sample is mixed with a material that will fluoresce upon interaction with a radiation emitted by radioactive decay. The scintillation counter quantifies the resulting flashes of light.

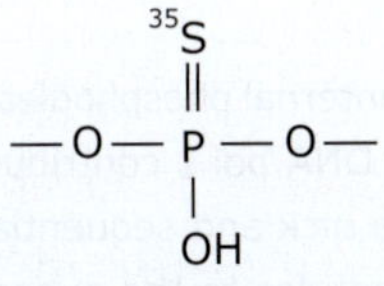

Phosphate group

Phosphorothioate group

Autoradiography is a technique which is used for detecting radioisotopes present in a solid sample on gels or membranes. It involves the production of an image in a photographic emulsion. Such emulsions consist of silver halide crystals in gelatin base. When a β-particle or γ-ray from a radionuclide passes through the emulsion, the silver ions are converted to silver atoms. The resulting latent image can then be converted to a visible image once the image is developed; an amplification process in which entire silver halide crystals are reduced to give metallic silver.

Scintillation counting relies on special chemicals called **scintillants**. The scintillant molecules absorb the radiations emitted by the radioisotopes in the samples, and in turn emit a flash of light. The light pulses from the scintillant are detected by a photocell. To use the scintillation counter (a machine that detects and counts pulses of light), radioactive samples to be measured are added to a vial containing scintillant fluid and loaded into the counter. The counter prints out the number of light flashes it detects within a designated time.

Label location

There are two ways to label a DNA molecule – by the ends (*end labeling*) or all along the molecule (*uniform labeling*).

Labeling DNA by nick translation

Nick translation is one method of labeling DNA, which uses the enzymes pancreatic DNase I, *E. coli* DNA polymerase I and DNA ligase. The endonuclease DNase I is used to create nicks at random sites in both strands of double-stranded target DNA. Following DNase I treatment, DNA polymerase I is used to add nucleotide residues to the free 3′-hydroxyl ends created during the DNase I nicking process. As the DNA polymerase I extends the 3′-ends, the 5′- to 3′ exonuclease activity of the enzyme simultaneously removes bases from the 5′-end of the nick. The sequential addition of base onto the 3′-end with the simultaneous removal of bases from the 5′-end results in translation of the nick along the DNA molecule. When performed in the presence of a radioactive deoxynucleoside triphosphate such as ([α-^{32}P] dCTP), the newly synthesized strand becomes radioactively labeled. For nonradioactive labeling procedures, a digoxigenin or a biotin moiety attached to a dNTP analog is used.

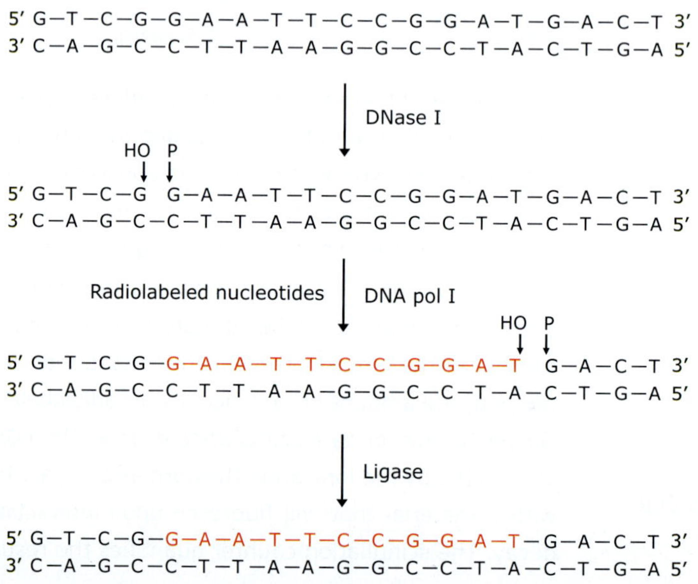

5′ G–T–C–G–G–A–A–T–T–C–C–G–G–A–T–G–A–C–T 3′
3′ C–A–G–C–C–T–T–A–A–G–G–C–C–T–A–C–T–G–A 5′

DNase I

HO P

5′ G–T–C–G G–A–A–T–T–C–C–G–G–A–T–G–A–C–T 3′
3′ C–A–G–C–C–T–T–A–A–G–G–C–C–T–A–C–T–G–A 5′

Radiolabeled nucleotides DNA pol I

HO P

5′ G–T–C–G–G–A–A–T–T–C–C–G–G–A–T G–A–C–T 3′
3′ C–A–G–C–C–T–T–A–A–G–G–C–C–T–A–C–T–G–A 5′

Ligase

5′ G–T–C–G–G–A–A–T–T–C–C–G–G–A–T–G–A–C–T 3′
3′ C–A–G–C–C–T–T–A–A–G–G–C–C–T–A–C–T–G–A 5′

Figure 12.15 Nick translation. DNase I introduces single-stranded nicks by cleaving internal phosphodiester bonds, generating a 5′ phosphate group and a 3′ hydroxyl terminus. Addition of DNA pol I contributes two enzyme activities — a 5′–3′ exonuclease attacks the exposed 5′ termini of the nick and sequentially removes nucleotides in the 5′–3′ direction and a DNA polymerase adds new nucleotides to the exposed 3′ hydroxyl group in the 5′–3′ direction.

Random priming

An alternative method for preparing uniformly labeled DNA is by oligonucleotide-primed DNA synthesis with hexanucleotides (or longer oligomers) of random sequence. Oligonucleotides of random sequence will anneal to a variety of homologous locations on a single-stranded DNA. Once annealed, they serve as primers for DNA synthesis by DNA polymerases. The Klenow fragment is used as this enzyme lacks the 5′–3′ exonuclease activity of DNA polymerase I; and so only fills in the gaps between adjacent primers. Labeled nucleotides are incorporated into the new DNA that is synthesized.

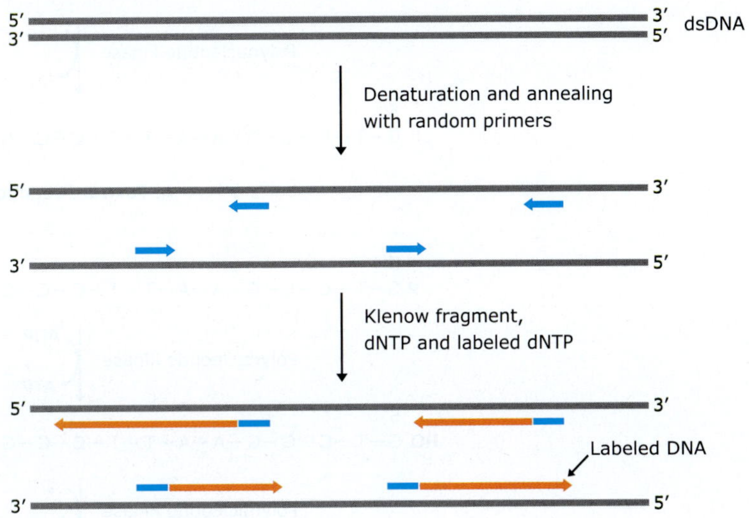

Figure 12.16 Labeling by random priming.

End-labeling of DNA

End labeling can be performed at the 3′- or 5′-end.

3′-end labeling

Template-independent polymerization of [α^{32}P] NTP to the 3′ terminus of DNA is catalyzed by calf thymus terminal deoxynucleotidyl transferase. Terminal deoxynucleotidyl transferase (TdT) is a template-independent DNA polymerase that incorporates dNTPs to the 3′-OH end of single or double-stranded DNA and RNA in an irreversible manner. This enzyme is used for the formation of homo or heteropolymeric tail at the 3′-end and also for incorporating a single nucleotide analog such as [α-^{32}P] cordycepin-5′-triphosphate. Terminal dideoxynucleotidyl transferase labels 3′-protruding ends more efficiently than blunt ends.

5′-end labeling

5′-end labeling is performed by enzymatic methods (T4 polynucleotide kinase), by chemical modification of sensitized oligonucleotides with phosphoramidite, or by combined methods. 5′-end labeling is usually performed using polynucleotide kinase (*kinase end-labeling*). The polynucleotide kinase utilizes two types of reactions: forward reaction and exchange reaction. In the *forward reaction*, a hydroxyl group is first created by removing the unlabeled phosphate residue from the 5′-end of the DNA with alkaline phosphatase. T4 polynucleotide kinase is then used to transfer the labeled gamma phosphate from ATP to the 5′-end of DNA.

In the *exchange reaction*, polynucleotide kinase first transfers the phosphate from the 5′-end of DNA to ADP, forming ATP and leaving a dephosphorylated target. Then enzyme performs a forward reaction and transfers a labeled gamma phosphate from ATP onto the target DNA.

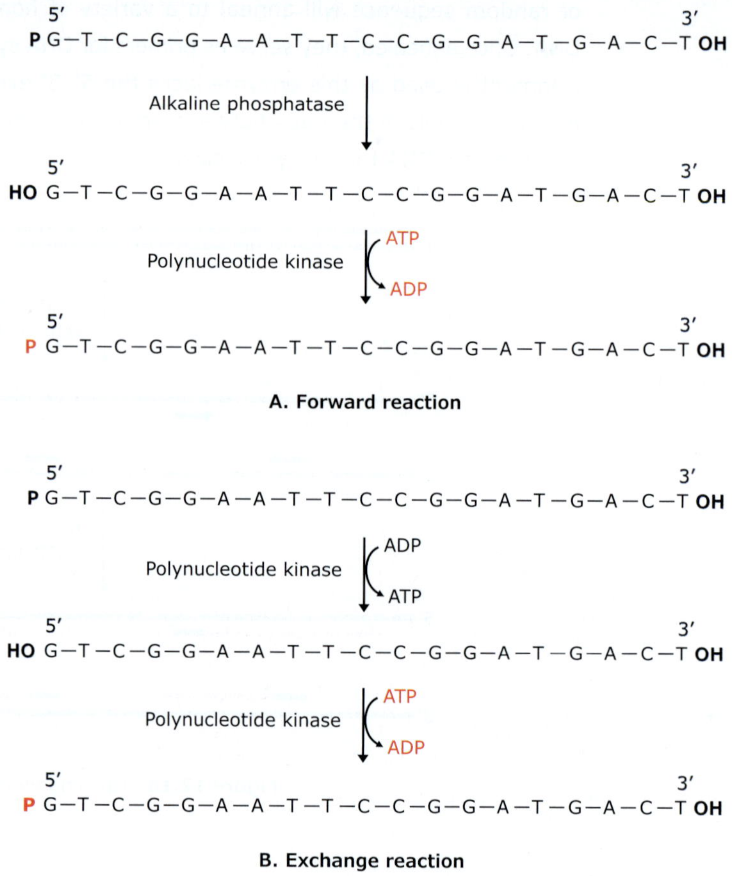

5′ 3′
P G—T—C—G—G—A—A—T—T—C—C—G—G—A—T—G—A—C—T **OH**

Alkaline phosphatase

5′ 3′
HO G—T—C—G—G—A—A—T—T—C—C—G—G—A—T—G—A—C—T **OH**

Polynucleotide kinase ATP ADP

5′ 3′
P G—T—C—G—G—A—A—T—T—C—C—G—G—A—T—G—A—C—T **OH**

A. Forward reaction

5′ 3′
P G—T—C—G—G—A—A—T—T—C—C—G—G—A—T—G—A—C—T **OH**

Polynucleotide kinase ADP ATP

5′ 3′
HO G—T—C—G—G—A—A—T—T—C—C—G—G—A—T—G—A—C—T **OH**

Polynucleotide kinase ATP ADP

5′ 3′
P G—T—C—G—G—A—A—T—T—C—C—G—G—A—T—G—A—C—T **OH**

B. Exchange reaction

Figure 12.17 5′-end labeling. **A.** In the *forward reaction*, polynucleotide kinase transfers the gamma phosphate from ATP to the 5′-end of a polynucleotide (DNA or RNA). **B.** In the *exchange reaction*, target DNA or RNA that has a 5′ phosphate is incubated with an excess of ADP. Polynucleotide kinase first transfers the phosphate from the nucleic acid onto an ADP, forming ATP and leaving a dephosphorylated target. Polynucleotide kinase then performs a forward reaction and transfers a phosphate from ATP onto the target nucleic acid.

Nonisotopic labeling

Compared to radioactive labels, the use of nonradioactive labels have several advantages:

- Safety.
- Higher stability of probe.
- Efficiency of the labeling reaction.
- *In situ* detection.
- Less time taken to detect the signal.

Nonisotopic labeling systems involve the use of non-radioactive probes. Two types of non-radioactive labeling are conducted – direct and indirect. Direct labeling strategies utilize probes that are directly conjugated to a dye or an enzyme, which generates the detection signal. Indirect labeling systems utilize probes that contain a hapten that will bind to a secondary agent generating the detection signal; the probe itself does not generate signal.

Direct nonisotopic labeling, where a nucleotide which contains the label that will be detected is incorporated. Often such systems involve incorporation of modified nucleotides containing a fluorophore, a chemical group which can fluoresce when exposed to light of a certain wavelength. Most commonly used fluorophores for direct labeling are **fluorescein**, a pale green fluorescent dye and **rhodamine**, a red fluorescent dye.

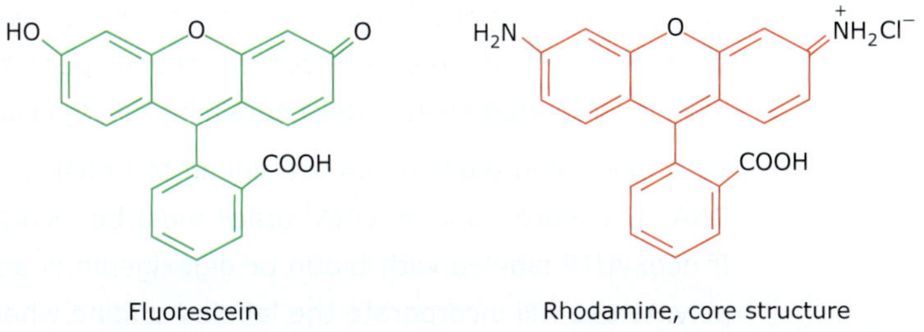

Fluorescein

Rhodamine, core structure

Indirect nonisotopic labeling, usually featuring the chemical coupling of a modified **reporter molecule** to a nucleotide precursor. After incorporation into DNA, the reporter groups can be specifically bound by an **affinity molecule**, a protein or other ligand which has a very high affinity for the reporter group. Conjugated to the latter is a marker molecule or group which can be detected in a suitable assay. The reporter molecules on modified nucleotides need to protrude sufficiently far from the nucleic acid backbone to facilitate their detection by the affinity molecule and so a **spacer** of 4-16 carbon atoms long is required to separate the nucleotide from the reporter group.

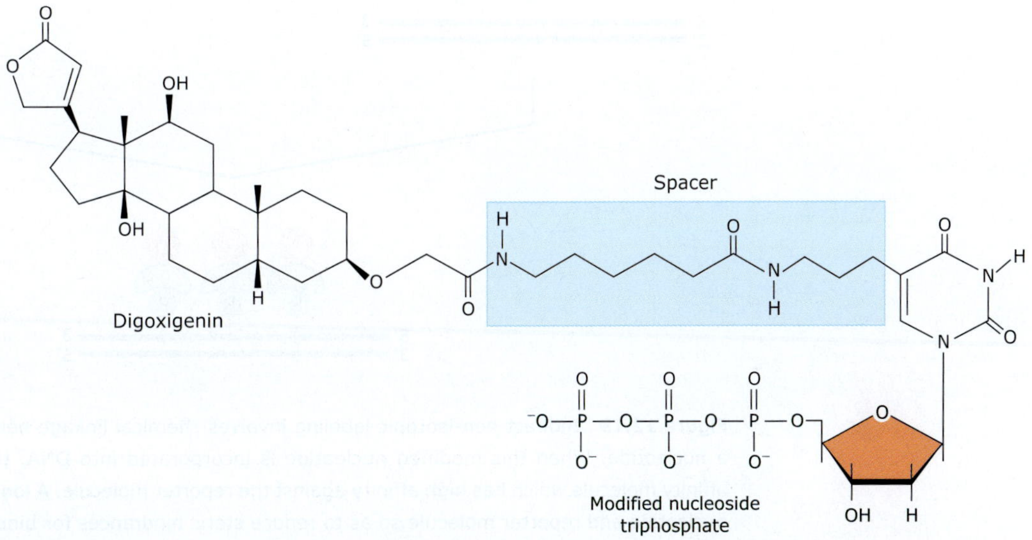

Spacer

Digoxigenin

Modified nucleoside triphosphate

Figure 12.18 The base of the nucleoside triphosphate depicted is an analog of thymine in which the methyl group has been replaced by a spacer arm linked to the plant steroid digoxigenin. The digoxigenin is detected by a specific antibody coupled to a visible marker such as a fluorescent dye.

Two indirect nonisotopic labeling systems are widely used:

The **biotin-streptavidin system** utilizes the extremely high affinity of two ligands – **biotin** (vitamin H or B7) which acts as the *reporter*; and the bacterial protein **streptavidin** which is the *affinity molecule*. Biotin and streptavidin bind together extremely tightly with a dissociation constant in the order of 10^{-14} mol/litre, one of the strongest known in biology. Although both **avidin** and **streptavidin** bind to biotin with very high affinity, but avidin in some applications show high non-specific binding due to the presence of the sugars and high pI. The *streptavidin-biotin* system can be incorporated into virtually every immunoassay, whereby an antibody is conjugated to biotin (**biotinylated antibodies**) and then detected with streptavidin.

Digoxigenin is a plant steroid (obtained from *Digitalis* species) to which a specific antibody has been raised. The digoxigenin-specific antibody permits detection of nucleic acid molecules which have incorporated nucleotides containing the digoxigenin *reporter molecule*.

Both biotin and digoxigenin are linked to uracil, which is normally a component of RNA not DNA. Therefore, to label DNA, uracil must be incorporated into the DNA instead of thymine. If deoxyUTP labeled with biotin or digoxigenin is added to the polymerization reaction, DNA polymerase will incorporate the labeled uridine where thymidine would normally be inserted. The biotin or digoxigenin tags stick out from the DNA without disrupting its structure.

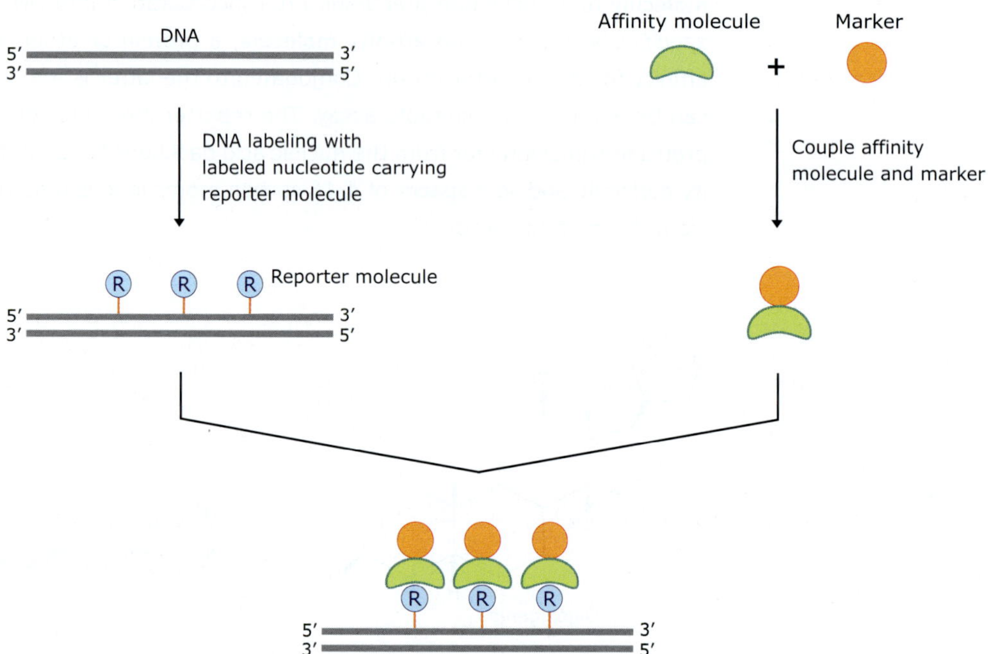

Figure 12.19 Indirect non-isotopic labeling involves chemical linkage between a reporter molecule and a nucleotide. When this modified nucleotide is incorporated into DNA, then it binds specifically to an affinity molecule which has high affinity against the reporter molecule. A long spacer is introduced between nucleotide and reporter molecule so as to reduce steric hindrances for binding of affinity molecule.

Molecular beacons

The *molecular beacon* is a structured fluorescent probe. A structured probe contains stem-loop structure regions that confer enhanced target specificity when compared with a traditional linear probe. Molecular beacons form a stem-loop structure, where the central-loop-sequence is complementary to the target of interest and the stem arms are complementary to each other. It is an oligonucleotide (about 25 nucleotides long) that contains both a fluorophore and a quenching group at opposite ends. Its detection mechanism relies on the principle of FRET, in which a fluorophore in the excited state, can transfer energy to acceptor fluorophore, with subsequent emission of a fluorescent signal from the acceptor, or to a quencher dye, which dissipates the energy without emission of a detectable fluorescence signal. In order for this energy transfer to take place, the donor and acceptor molecules must be situated in close physical proximity.

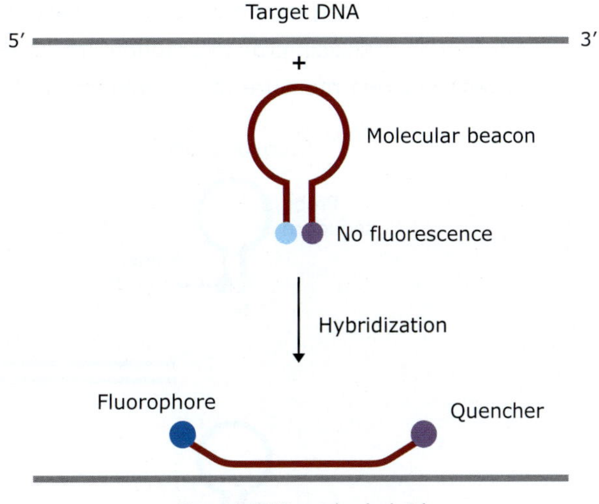

Figure 12.20 Structure and operation of molecular beacons. A typical molecular beacon probe is about 25 nucleotides long. About 15 nucleotides in the middle of probe are complementary to the target DNA or RNA and 5–7 nucleotides at each terminus are complementary to each other; rather than to the target DNA. At the 5′-end of the probe, a fluorescent dye is covalently attached. The quencher dye is covalently attached to the 3′-end. When the beacon is in closed loop shape, the quencher resides in proximity to the fluorophore, which results in quenching the fluorescent emission of the latter. Hence, these molecules are non-fluorescent. But when the probe sequence in the loop hybridizes to its target, forming a double helix, a conformational reorganization occurs that separates the quencher from the fluorophore, restoring fluorescence.

Molecular beacon is a sequence-specific probe designed to fluoresce only when it binds to a specific DNA target sequence. The central region of the probe is complementary to the target sequence. The terminal 5 to 7 nucleotides at each end of the probe are complementary and form a short double-stranded region. In the stem and loop conformation, the quenching group is next to the fluorophore and so prevents fluorescence. When the molecular beacon binds to the target sequence, it is linearized. This separates the quenching group from the fluorophore, which is now free to fluoresce.

Scorpions

Scorpions are bifunctional molecules containing a PCR primer covalently linked to a probe. The fluorophore in the probe interacts with a quencher, which reduces fluorescence. During a PCR reaction, the fluorophore and quencher separates, which leads to an increase in fluorescence.

Scorpion probes are of two basic designs. Both types attach to the 5′-end of a PCR primer through a compound that inhibits PCR extension into the probe. In the first design, the probe consists of a stem-loop structure with a quencher and fluorophore similar to a molecular beacon (unimolecular stem-loop format). In the second design, the fluorophore-containing probe base-pairs with a complementary oligonucleotide that contains the quencher (bimolecular linear scorpion format).

In unimolecular stem-loop format, the probes contain the fluorophore at 5′-end, stem-loop, quencher, a PCR blocker (which prevents read-through by DNA polymerase) and primer.

Scorpions are used as fluorescent reporter molecules that enable the quantification of PCR products in real time. During the extension step of PCR, the primer portion of the probe anneals to the template and *Taq* polymerase makes new DNA. During the next denaturation step, the whole probe plus new DNA strand become single-stranded. During annealing step, the loop section of the scorpion probe is able to base-pair with its complementary sequence within the target DNA, releasing the fluorophore from the quencher. The resultant fluorescence emission gives the direct measure of the amount of PCR product produced.

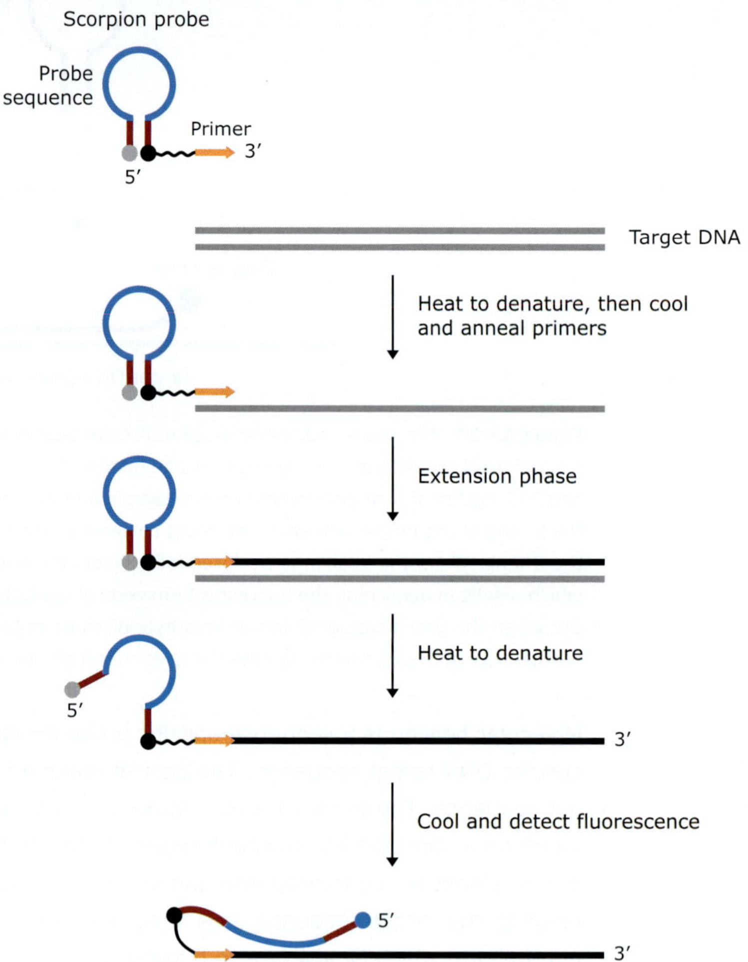

Figure 12.21 Scorpions incorporate two distinct structures - a target-specific DNA probing sequence and a target-specific PCR primer. It contains a stem-loop structure with a fluorophore molecule and quencher. The loop of the Scorpions probe includes a sequence that is complementary to an internal portion of the target sequence. During the first amplification cycle, the Scorpions PCR primer is extended, and the sequence complementary to the loop sequence is generated on the same strand. After subsequent denaturation and annealing, the loop of the Scorpions probe hybridizes to the internal target sequence, and the reporter is separated from the quencher. The resulting fluorescence signal is proportional to the amount of amplified product in the sample. The Scorpion probe contains a PCR blocker, just downstream of the quencher, to prevent read-through during the extension of the opposite strand.

Fluorescence *in situ* hybridization (FISH)

FISH is essentially based upon the same principle as a Southern blot analysis that exploits the ability of single-stranded DNA to anneal to complementary DNA. In the case of FISH, the target is the nuclear DNA of either interphase stage or metaphase stage. It was originally used with metaphase chromosomes. It involves hybridization of a fluorescent-labeled DNA probe to denatured DNA of metaphase chromosomes. In the past, probes were labeled with radio-isotopes. But now-a-days, fluorescent dyes are used for increased sensitivity and resolution.

In FISH, the DNA probe is either labeled directly by incorporation of a fluorescent-labeled nucleotides precursor or indirectly by incorporation of a nucleotides precursor containing a *reporter molecule* which after incorporation into the DNA is then bound by a fluorescently labeled *affinity molecule*. The position at which the probe hybridizes to the chromosomal DNA is visualized by detecting the fluorescent signal emitted by the labeled DNA. If FISH is carried out simultaneously with two DNA probes, each labeled with a different fluorochrome, the relevant positions of the two markers to which probes hybridize on the chromosome can be visualized. Hence, FISH enables the position of a marker on a chromosome or extended DNA molecule to be directly visualized.

FISH involves two main components: the DNA probe and the target DNA to which the probe will be hybridized. One of the most important considerations in FISH analysis is the choice of probe. A wide range of probes can be used, from whole genomes to small cloned probes. There are broadly three types of probes, each with a different range of applications: whole-chromosome painting probes; repetitive sequence probes and locus-specific probes. For the method to work, the DNA in the chromosome must be made single-stranded by denaturation of double helix. Only then will the chromosomal DNA be able to hybridize with the probe. The standard method for denaturing chromosomal DNA without destroying the morphology of the chromosome is to dry the preparation onto a glass microscope slide and then treat with *formamide*.

The most common methods for FISH visualization are: flow cytometry systems and slide-based systems. In **flow systems**, FISH-stained cells are prepared in suspension, and the suspension flows in a narrow stream across a laser beam wherein the detector records their fluorescent intensities. In **slide-based systems**, FISH-stained cells are fixed to a conventional wide-field or a confocal fluorescence microscope slide and observed as a static image. Prior to fixation, the cells can be trapped in the metaphase by colchicine, which interferes with mitosis. The fixed preparations are, then, incubated with various solvents and at elevated temperatures to allow the probe DNA to hybridize with the chromosomes.

Colony hybridization

Colonies of bacteria which contain specific DNA can be selected or identified by colony hybridization. In this hybridization process, bacterial colonies are transferred from the surface of an agar culture plate onto a nylon membrane. The colonies transferred to the membrane are subjected to alkali hydrolysis and detergent treatment to release the DNA content from the bacterial cells, which would then bind to the membrane. The DNA on the membrane is denatured with an alkali to produce single strands which covalently binds with the membrane by UV irradiation. The membrane is, then, immersed in a solution containing a labeled nucleic acid probe and incubated to allow the probe to hybridize to its complementary sequence. After hybridization, the membrane is washed extensively to remove unhybridized probe, and regions where the probe has hybridized are then visualized. By comparing the membrane with the original dish and lining up the regions of hybridization, the original group of colonies can be identified.

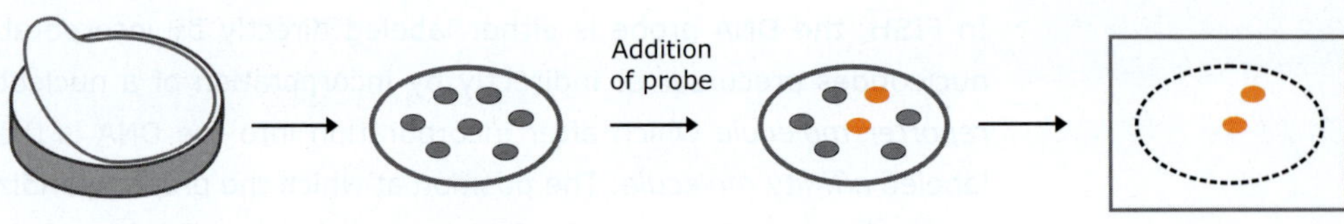

Bacterial colonies are transferred from the surface of an agar culture plate onto nylon membrane.

The membrane is treated with detergent to lyse the bacteria and alkali to denature their DNA.

The membrane is then incubated with nucleic acid labeled probe which hybridize by base pairing to complementary DNA present on the membrane.

Membrane is rinsed and subjected to autoradiography which will make visible only those colonies containing DNA that is complementary in sequence to the probe.

12.5 Blotting

Blotting describes the immobilization of sample nucleic acids/proteins onto a solid support. It involves the transfer of nucleic acids and/or proteins from a gel strip to a specialized, chemically reactive matrix called *blotting membrane* (typically nitrocellulose or activated nylon) on which the nucleic acids/proteins may become immobilized in a pattern similar to that present in the original gel.

Southern blotting

Southern blotting (or Southern blot hybridizations) immobilizes and detects target DNA fragments onto a membrane that have been size-fractionated by gel electrophoresis. This technique was invented in 1975 by E. M. Southern. In the procedure, the DNA fragments separated on an agarose gel are denatured, transferred and immobilized onto a membrane.

Following electrophoresis, the test DNA fragments are denatured in strong alkali. As electrophoretic gels are fragile, and the DNA in them can diffuse within the gel, it is usual to transfer the denatured DNA fragments by blotting onto a durable nitrocellulose paper or nylon membrane, to which single-stranded DNA binds readily. Nowadays, nylon membranes are commonly used. After transfer, the DNA fragments need to be fixed to the membrane so that they cannot detach. In case of nitrocellulose paper, nucleic acid immobilization occurs *non-covalently* after baking for 2 hrs at 80°C. In case of nylon membrane either it is baking for 1 hour at 70°C or UV irradiation at 254 nm. Nucleic acid binds *covalently* with nylon membrane after UV irradiation for 5 minutes. UV irradiated covalent-linking is based on the formation of cross-links between the T-residues in the DNA and the positively charged amino groups on the surface of the nylon membrane. The individual DNA fragments become immobilized on the membrane at positions which are a faithful record of the size separation achieved by gel electrophoresis. Following the fixation step, the membrane is placed in a solution of labeled (radioactive or non-radioactive) RNA, single-stranded DNA or oligodeoxynucleotide which is complementary in sequence to the blot transferred DNA band or bands to be detected. Since this labeled nucleic acid is used to detect and locate the complementary sequence, it is called the **probe**. The probe is allowed to hybridize to its complementary single-stranded target DNA sequences on the membrane. Conditions are chosen which maximize the rate of hybridization, compatible with a low background of non-specific binding on the membrane. After the hybridization, reaction has been carried out, the membrane is then washed extensively to remove non-specifically bound probe.

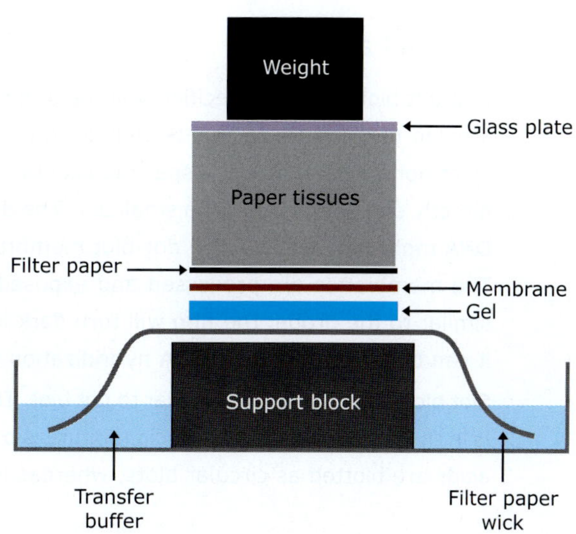

Figure 12.22 Blotting apparatus. The gel is placed on a filter paper wick and a nitrocellulose paper or nylon membrane placed on top. Further, sheets of filter paper and paper tissues complete the setup. Transfer buffer is drawn through the gel by capillary action, and the nucleic acid fragments are transferred out of the gel and onto the membrane.

If the probe is radioisotope labeled, then the membrane is exposed to photographic film. If the probe is non-isotopically labeled with biotin or digoxigenin, the membrane may be treated with chemiluminescent substrate to detect the labeled probe, and then exposed to photographic film. The probe will form a band on the film at a position corresponding to the complementary sequence on the membrane.

Northern blotting

Northern blotting or Northern blot hybridization is a variant of Southern blotting in which the target nucleic acid is RNA instead of DNA. This method is used to measure the amount and size of RNAs transcribed from genes and to estimate their abundance. In this technique, an RNA extract is electrophoresed in an agarose gel, using a denaturing buffer to ensure that the RNAs do not form inter or intra molecular base pairs. After electrophoresis, the gel is blotted onto a reactive DBM (diazobenzyloxymethyl) paper, and hybridized with a labeled probe. RNA bands can also be blotted onto nitrocellulose paper under appropriate conditions and suitable nylon membranes.

Nylon is a generic name for any long-chain synthetic polymer having recurring polyamide (–CONH–) groups. Two types of nylon membranes are available commercially: *unmodified* (or neutral) *nylon* and *charge-modified nylon* which carries amine groups and is, therefore, also known as *positively charged* nylon. Both types of nylon bind single- and double-stranded nucleic acids. Charge-modified nylon has a greater capacity to bind nucleic acids.

Table 12.2 Properties of materials used for blotting of nucleic acids

Materials	Binding capability
Nitrocellulose	ssDNA, RNA
Nylon (Neutral)	ssDNA, dsDNA, RNA
Positively charged nylon	ssDNA, dsDNA, RNA
Activated papers (DBM and DPT)	ssDNA, RNA

DBM–diazobenzyloxymethyl, DPT–Diazophenylthioether

Dot blot assay

In a dot blot assay, a specified volume of nucleic acid mixture is spotted onto a small area of a nylon membrane. In this technique, the nucleic acid molecules are not first separated by electrophoresis. Instead, a specimen containing nucleic acid mixture to be detected is applied directly on a membrane as a small dot. The dot is treated with an alkaline solution to denature DNA molecules. Finally, the dot-blot membrane is allowed to hybridize with a labeled probe. The membranes are processed and exposed to film. If the dot of DNA contains a sequence similar to the probe, the film will turn dark in that area. If no dark spot appears on the film, it can be inferred that no DNA hybridization has occurred.

Slot blot is fundamentally similar to dot blot. The difference between dot and slot blot procedures is in the way that the nucleic acid mixture is blotted onto the membrane. In dot blot, the nucleic acids are blotted as circular blots, whereas in slot blot they are blotted in rectangular slots.

12.6 Phage display

Phage display is a molecular technique that allows expression of foreign polypeptides or peptides on the surface of phage particles. This method was first described by George Smith in 1985. In this technique, the DNA encoding the protein of interest is fused with a gene encoding one of the proteins that forms the viral coat. Genetic engineering techniques are used to insert foreign DNA fragments into a suitable phage coat protein gene. Phage display involved the use of filamentous phages such as fd, f1, M13, where the foreign gene was incorporated into a gene specifying a minor coat protein. Filamentous phage M13 is the most popular choice for phage display. M13 is a filamentous phage contains 6.4 kb single-stranded circular DNA. M13 enters *E. coli* through the bacterial sex pilus, a protein appendage that permits the transfer of DNA between bacteria. The single-stranded DNA in the virus particle [called the (+) strand] is replicated through an intermediate circular double-stranded replicative form containing (+) and (–) strands. Only the (+) strand is packaged into new virus particles.

Protein-protein interaction
Study of interaction between different proteins is very useful as it yields information about the tentative function of a protein through its interaction with well characterized protein with known role or function. Phage display and yeast two hybrid system are two well known and most useful methods for studying protein-protein interactions.

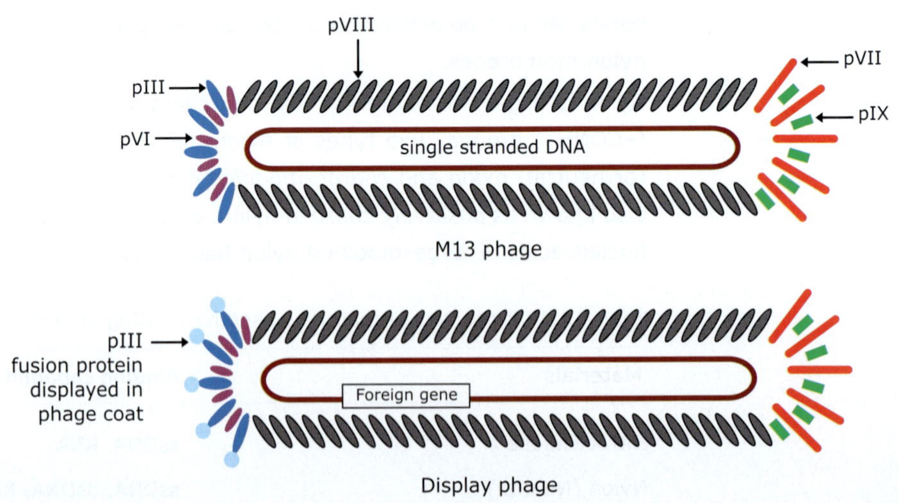

Figure 12.23 Phage display. In order to display a peptide on the surface of a bacteriophage, the DNA sequence encoding the peptide must be fused to the gene for a bacteriophage coat protein (either to the pVIII gene or the pIII gene). The coat protein is expressed as a fusion protein on the virion surface without disturbing the infectivity of the phage.

M13 is preferred since it is non-lytic and does not destroy the host bacteria during phage production. Instead, phage particles are secreted through the bacterial cell envelope. The absence of cell debris simplifies purification of the phage. The M13 phage particle consists of a single-stranded DNA molecule surrounded by a protein coat. The coat consists of major coat protein (pVIII) and minor coat proteins (pIII, pVI, pVII and pIX). At one end of the particle are five copies each of the two minor coat proteins, pIX and pVII and at the other end five copies each of pIII and pVI. The minor coat protein pIII is located on the tip of the phage, which is responsible for attaching the phage to its bacterial host during infection.

Gene encoding the minor coat protein pIII is most commonly used for recombinant formation with the DNA encoding the foreign target peptide or polypeptide. *E. coli* transfected with the recombinant DNA molecules produce phage particles that display peptide or polypeptide as a fusion protein with endogenous pIII coat protein on the surface of the phage particles.

There are two main methods in which phage display can be used to study protein interaction. In one method, test protein is displayed and its interaction sought with a series of purified proteins or protein fragments of known function. In the second method, a phage display library is prepared. The library is made up of many recombinant phages, each displaying a different protein. These libraries can be prepared by cloning a mixture of cDNAs from a particular tissue by cloning genomic DNA fragments. The library consists of the phage displaying a range of different proteins and is used to identify those that interact with a test protein.

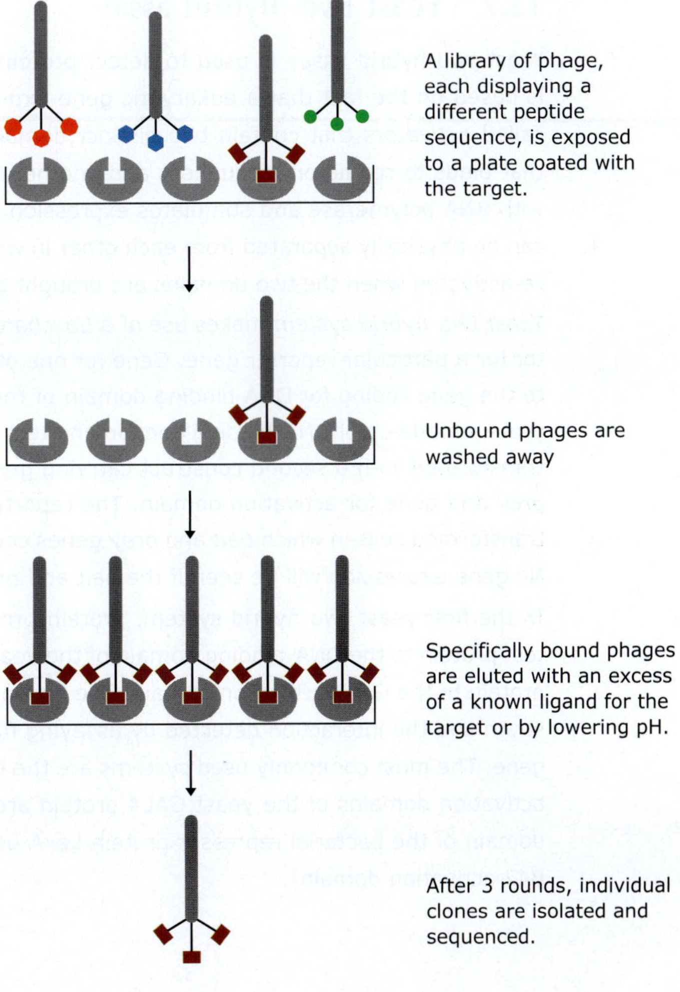

Figure 12.24 Phage display libraries and biopanning. Phage display libraries consist of a large number of modified phages displaying a library of different peptide sequences. They are screened to find peptides that bind to specific target molecules, such as a particular antibody, enzyme or cell surface receptor. The peptide of interest is found by a selection procedure referred to as *biopanning*. The phage display library is incubated with target molecules that are attached to a solid support. Unbound phage is washed away. After several washings, only the phages that contain a protein fragment with high affinity for the ligands will remain bound to the container; the rest will be washed off.

Vector used for phage display

In the **bacteriophage display system**, a segment of foreign DNA is inserted into gene III or gene VIII, a few nucleotides downstream from the cleavage site. *E. coli* transfected with the recombinant viral DNA synthesize and secrete *fusion phage* particles that display on their surface the amino acids encoded by the foreign DNA. Every copy of pIII or pVIII on the surface of an infectious bacteriophage particle carries the sequences encoded by the foreign DNA, which are, therefore, displayed in a densely packed, 'multivalent' fashion.

The **phagemid display system** consists of a plasmid that carries a single copy of gene III or gene VIII and the viral origin of DNA replication. In phagemid display, as in conventional phage display, a segment of foreign DNA is inserted into gene III or gene VIII just downstream from the cleavage site that separates the hydrophobic signal sequence from the mature protein. The recombinant plasmid is then used to transform an appropriate strain of *E. coli.* Bacteriophage particles displaying the amino acid sequences encoded by the segment of foreign DNA are obtained by superinfecting the transformed cells with helper phages. Because replication and packaging of the helper phages are less efficient than that of the phagemid, the population of bacteriophages secreted from the superinfected cells consists mainly of particles that display the cloned target sequence.

12.7 Yeast two-hybrid assay

Yeast two-hybrid assay is used to detect protein interactions inside a yeast cell nucleus. It is based on the fact that a eukaryotic gene expression is promoted by transcription factors called *activators* that contain two distinct domains. One is the **DNA binding domain** (DBD) that binds to regulatory sequences and another is the **activation domain** (AD) that interacts with RNA polymerase and stimulates expression of the associated gene. These two domains can be physically separated from each other in which case the protein loses its activity, but is re-activated when the two domains are brought together.

Yeast two-hybrid system makes use of a *Saccharomyces cerevisiae* strain which lacks activator for a particular reporter gene. Gene for one of the interacting proteins called **bait** is fused to the gene coding for DNA binding domain of the activator and specifies synthesis of fusion protein made up of DNA-binding domain and test protein. The recombinant yeast strain is co-transformed with a second construct carrying gene, for the second interacting protein called **prey** and gene for activation domain. The reporter gene (such as *lac*Z) is expressed only in transformed cells in which *bait* and *prey* genes code for proteins that interact with each other. No gene expression will be seen if the bait and prey do not interact with each other.

In the first yeast two-hybrid system, protein-protein interactions were tested by fusing one test protein to the DNA-binding domain of the yeast GAL4 transcription factor, and the second protein to the GAL4 activation domain. The fusion proteins were expressed in a suitable yeast strain and the interaction detected by assaying for expression of a GAL4 responsive reporter gene. The most commonly used systems are the **GAL4 system** (in which the DNA-binding and activation domains of the yeast GAL4 protein are used) and the **LexA system** (DNA-binding domain of the bacterial repressor protein LexA used in combination with the *Escherichia coli* B42 activation domain).

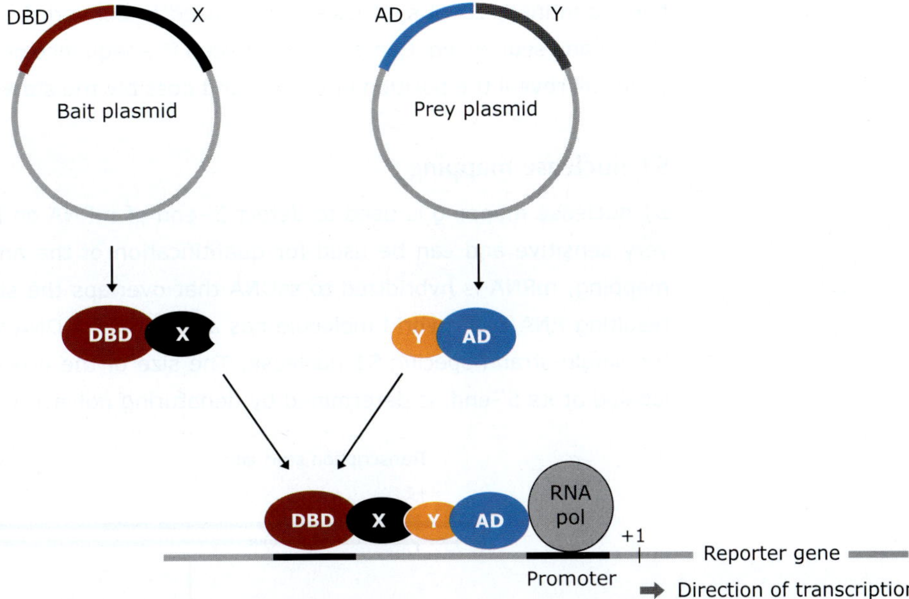

Figure 12.25 Yeast two-hybrid assay. Using the yeast two-hybrid system, the protein of interest (X) is expressed as a fusion protein to the DBD (DBD-X) and the AD is fused to the second protein of interest, Y as AD-Y. The AD-Y fusion vector is introduced into a yeast strain containing the DBD-X fusion partner by transformation. Only if proteins X and Y physically interact with each other, the DBD and AD will be brought together to activate expression of the downstream reporter gene.

A range of modifications to this basic principle have been made to expand the range of molecular interactions that can be identified. The **yeast one-hybrid system** screens for proteins that bind to DNA; for instance, to identify factors that bind specific gene promoter sequences. In addition, a variety of **yeast three-hybrid systems** exist. One such system assesses tertiary protein complexes and can be used to identify proteins that act as a 'bridge' between two other proteins, or that stabilize the interaction of proteins that would otherwise only interact very weakly. Another three-hybrid system assesses RNA–protein interactions, which identifies proteins that bind to a specific RNA species.

The yeast two-hybrid system characterizes the novel protein-protein interaction. The two-hybrid system uses the bi-functional nature of transcription factors to allow protein-protein interactions to be monitored through changes in transcription of reporter genes. Once a positive interaction has been identified, either of the interacting proteins can mutate (either by site-specific or randomly introduced changes) to produce proteins with a decreased ability to interact. Mutants generated using this strategy are very powerful in defining the residues involved in the interaction. Such techniques are termed **reverse two-hybrid system.**

12.8 Transcript analysis

A number of protocols are available for the analysis of mRNA transcripts: Northern blotting, S1 nuclease mapping and prime extension analysis. These protocols vary in the degree of sensitivity of detection and the information generated. In Northern blotting, an RNA extract is electrophoresed in an agarose gel, using denaturing electrophoresis buffer to ensure that the RNAs do not form inter– or intramolecular base pairs (because base pairing affects the rate at which the molecules migrate through the gel). After electrophoresis, the gel is blotted onto a nylon or nitrocellulose membrane and hybridized with a labeled probe. Once a transcript has been identified, cDNA synthesis can be used to convert it into a dsDNA copy, which can be cloned and sequenced. Comparison between the sequence of the cDNA and the sequence of its gene will reveal the position of introns and possibly the start and end point of the transcripts.

S1 nuclease mapping

S1 nuclease mapping is used to detect 5′-end of mRNA on DNA template. This technique is very sensitive and can be used for quantification of the amount of mRNA. In S1 nuclease mapping, mRNA is hybridized to ssDNA that overlaps the start of the target transcript. The resulting RNA/DNA hybrid molecule has an overlap of DNA at the 3′-end that is digested by the single-strand specific S1 nuclease. The size of the processed ssDNA molecule, which is labeled at its 5′-end, is determined by denaturing polyacrylamide gel electrophoresis.

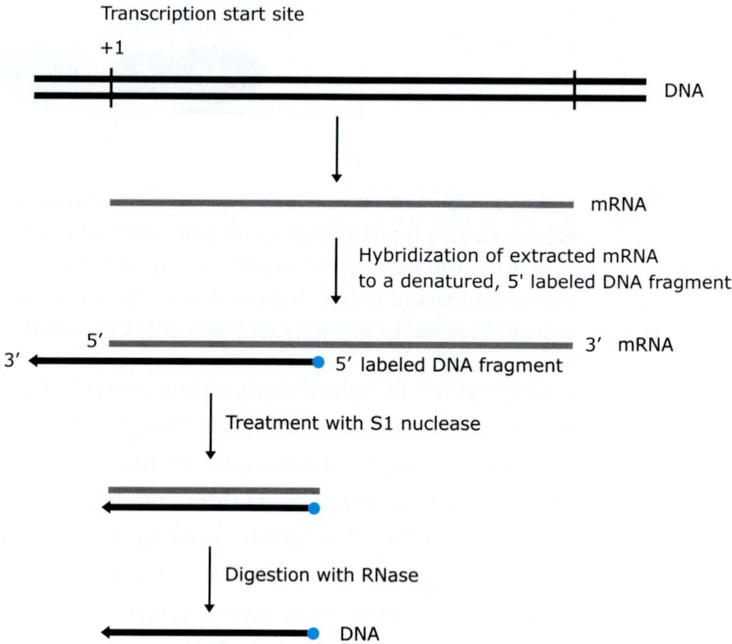

Figure 12.26 In S1 nuclease mapping, RNA is hybridized to the denatured DNA fragment that has been labeled at its 5′-end. The DNA fragment is chosen so that its 5′-end is internal to the target mRNA, while the 3′-end extends beyond the putative mRNA start point. The RNA-DNA hybrid molecule has single-stranded extensions that are degraded by the single-stranded specific S1 nuclease. The 3′-end of the DNA fragment is finally determined.

Primer extension

In the primer extension assay, the transcription start site for a gene is determined experimentally by identifying the 5′-end of the encoded messenger RNA (mRNA).

This approach involves binding of a labeled primer, usually a synthetic oligonucleotide of about 20 residues, that is complementary to an mRNA sequence downstream of the anticipated 5′-end. The primer is then extended by reverse transcriptase, thus synthesizing DNA that is complementary to the mRNA. The 3′-end of this newly synthesized strand of DNA, therefore, corresponds with the 5′ terminus of the transcript. The resulting radiolabeled cDNA products are analyzed by denaturing polyacrylamide gel electrophoresis, followed by autoradiography. The sizes of the bands detected on the gel, as compared to an adjacent sequencing ladder or molecular weight standards, provide a measure of the distance from the 5′-end of the synthetic oligonucleotide to the beginning of the mRNA transcripts. In theory, the 3′-end of the cDNA will coincide with the 5′-end of the mRNA. Thus, the size of the radiolabeled cDNAs should represent the distance from the labeled 5′-end of the primer to the 5′-end of the mRNA (i.e. the 3′-end of the cDNA).

Primer extension analysis has two main applications. First, it is used for mapping the 5′-end of transcripts. This allows one to determine the transcription initiation site. Second, it can be used to quantify the amount of transcript in an *in vitro* transcription system.

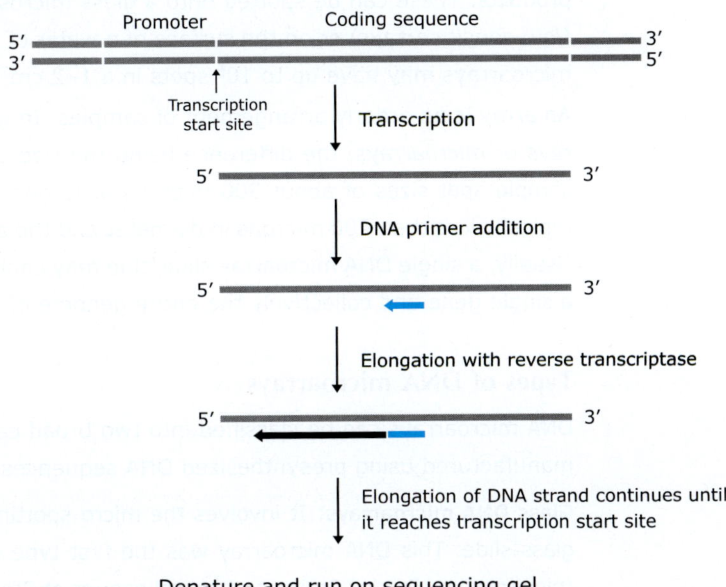

Figure 12.27 A *primer extension assay* involves four steps: First, selection and preparation of a labeled primer complementary to the RNA transcript of interest. Second, hybridization of the primer complementary to a region of the RNA under study. Third, extension from the primer that is catalyzed by reverse transcriptase using RNA as the template to synthesize a cDNA strand. Fourth, analysis of the extended cDNA products on denaturing polyacrylamide gels and autoradiography.

12.9 DNA microarray

Although all of the cells in the human body contain identical genetic material, but the same genes are not active in every cell. Studying which genes are active and which are inactive in different cell types helps scientists to understand both how these cells function normally and how they are affected when various genes do not perform properly. In the past, scientists were able to conduct these genetic analyses on a few genes at once. With the development of DNA microarray technology, however, scientists can now examine how thousands of genes express and their product interact at any given time.

A **DNA microarray** (also commonly known as *gene chip, DNA chip, bioarray* or *gene array*) is an orderly arrangement of thousands of identified genes fixed on a solid support, usually glass, silicon chips or nylon membrane for the purpose of expression profiling, monitoring expression levels for thousands of genes simultaneously. In expression profiling, it detects the presence and abundance of labeled nucleic acids in a biological sample, which will hybridize to the DNA on the array via Watson-Crick base pairing, and which can be detected via the label.

The most well-known use of DNA microarrays is for profiling mRNA levels; however, DNA microarrays have also been used to detect DNA-protein interactions (e.g. transcription factor-binding site and transcription factor), alternatively spliced variants, the epigenetic status of the genome (such as methylation patterns), DNA copy number changes and sequence polymorphisms.

In DNA microarray, a large number of DNA probes, each one with a different sequence, are immobilized at defined positions on a solid surface (such as glass, silicon chips or nylon membrane). The probe can be *synthetic oligonucleotides*, short DNA molecules, such as cDNA or PCR products. These can be spotted onto a glass microscope slide or a piece of nylon membrane (*low-density arrays*) or on the surface of a wafer of silicon (*high-density array*). High-density microarrays may have up to 10^6 spots in a 1–2 cm^2 area.

An *array* is an orderly arrangement of samples. In general, arrays are described as *macroarrays* or *microarrays*, the difference being the size of the sample spots. Macroarrays contain sample spot sizes of about 300 microns or larger. The sample spot sizes in microarray are typically less than 200 microns in diameter and the arrays usually contain thousands of spots. Usually, a single DNA microarray slide/chip may contain thousands of spots each representing a single gene and collectively the entire genome of an organism.

Types of DNA microarrays

DNA microarrays can be classified into two broad categories depending on whether they are manufactured using presynthesized DNA sequences or *in situ* synthesis methods.

Glass DNA microarrays: It involves the micro-spotting of pre-fabricated cDNA fragments on a glass slide. This DNA microarray was the first type of DNA microarray technology developed (pioneered by Patrick Brown and his colleagues at Stanford University). It is fabricated by either pen tip deposition (*spotted arrays*) or inkjet deposition (*sprayed arrays*).

High-density oligonucleotide microarrays often referred to as a 'chip' which involves *in situ* oligonucleotide synthesis. The most prominent microarrays with *in situ* synthesized probes are the 'GeneChips' manufactured by Affymetrix. They are produced by chemical synthesis of the oligonucleotides directly on the coated quartz surface of the array. This technology allows very high spot densities. Therefore, they are called *high-density oligonucleotide arrays*. GeneChips are produced in a unique photolithographic process in combination with chemical reactions developed for combinatorial chemistry. A quartz wafer is coated with a narrow layer of a light-sensitive compound. This light-directed synthesis has enabled the large-scale manufacture of arrays containing hundreds of thousands of oligonucleotide probe sequences on glass slides or 'chips,' less than 2 cm^2 in size.

How does DNA microarray work?

Standard Watson-Crick base-pairing (i.e. A-T and G-C for DNA; A-U and G-C for RNA) between *target* and *probe* is the underlining principle of DNA microarray. Microarray technology evolved from Southern blotting, whereby fragmented DNA is attached to a substrate and then probed with a known gene or fragment.

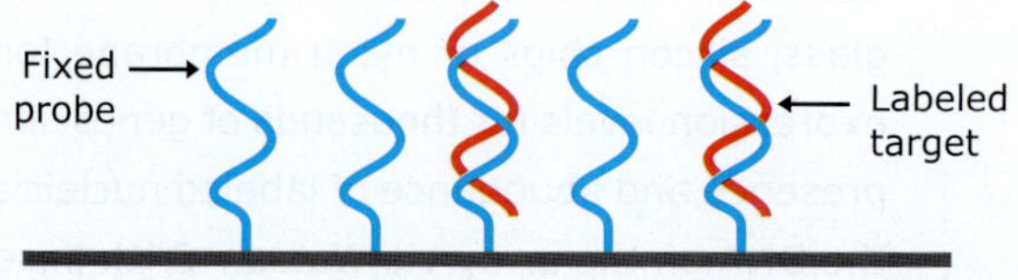

Figure 12.28 Hybridization of the target to the probe.

In a gene expression profiling experiment, the expression levels of thousands of genes are simultaneously monitored. There are four laboratory steps in using a microarray to measure gene expression in a sample:

1. Sample preparation and labeling
2. Hybridization
3. Washing
4. Image acquisition

To determine which genes are turned on and which are turned off in a given cell, a researcher must first collect the mRNA molecules present in that cell. The researcher then labels each mRNA molecule by attaching a fluorescent dye. Next, the researcher places the labeled mRNA onto a DNA microarray slide. The mRNA that was present in the cell will then hybridize - or bind - to its complementary DNA on the microarray, leaving its fluorescent tag. A researcher must then use a special scanner to measure the fluorescent areas on the microarray.

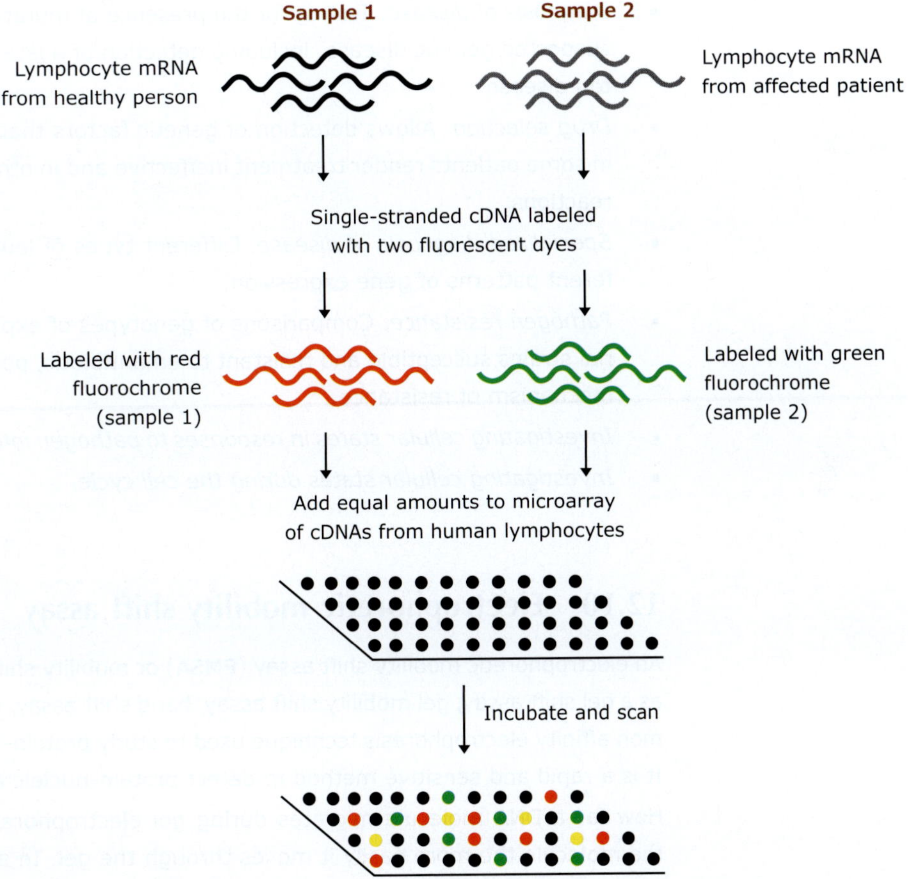

Figure 12.29 DNA microarray. To prepare the microarray, DNA fragments–each corresponding to a gene–are spotted onto a slide. In this example, mRNA is collected from two different lymphocytes for a direct comparison of their relative levels of gene expression. These samples are converted to cDNA and labeled, one with a red fluorochrome, the other with a green fluorochrome. The labeled samples are mixed and then allowed to hybridize on the microarray. After incubation, the array is washed and the fluorescence scanned. Red spots indicate that the gene in sample 1 is expressed at a higher level than the corresponding gene in sample 2. Green spots indicate that expression of the gene is higher in sample 2 than in sample 1. Yellow spots reveal genes that are expressed at equal levels in both samples. Black spots indicate little or no expression in either sample.

If a particular gene is very active, it produces many molecules of mRNA, which hybridize to the DNA on the microarray and generate a very bright fluorescent area. Genes that are comparatively less active produce fewer mRNAs, which result in dimmer fluorescent spots. If there is no fluorescence, none of the messenger molecules have hybridized to the DNA, indicating that the gene is inactive. Researchers frequently use this technique to examine the activity of various genes at different times.

DNA hybridization analysis on microarrays usually involves detecting the signal generated by the binding of a reporter probe (fluorescent, chemiluminescent, colorimetric, radioisotope, etc.) to the target DNA sequence. The microarray is scanned or imaged to obtain the complete hybridization pattern.

Principal applications of microarrays

- *Investigating cellular states and process*. Patterns of expression that change with the cellular state or growth conditions can give clues to the mechanisms of processes such as sporulation, or the change from aerobic to anaerobic metabolism.
- *Diagnosis of disease*. Testing for the presence of mutation can confirm the diagnosis of a suspected genetic disease, including detection of a late-onset condition such as Huntington disease.
- *Drug selection*. Allows detection of genetic factors that govern responses to drugs, that in some patients render treatment ineffective and in other cause unusual serious adverse reactions.
- *Specialized diagnosis of disease*. Different types of leukemia can be identified from different patterns of gene expression.
- *Pathogen resistance*. Comparisons of genotypes of expression patterns, between bacterial strains susceptible and resistant to an antibiotic, point to the proteins involved in the mechanism of resistance.
- *Investigating cellular states in responses to pathogen infection and environmental change.*
- *Investigating cellular states during the cell cycle.*

12.10 Electrophoretic mobility shift assay

An electrophoretic mobility shift assay (**EMSA**) or mobility shift electrophoresis, also referred to as a gel shift assay, gel mobility shift assay, band shift assay, or gel retardation assay, is a common affinity electrophoresis technique used to study protein–DNA or protein–RNA interactions. It is a rapid and sensitive method to detect protein-nucleic acid interactions.

How far a DNA molecule migrates during gel electrophoresis varies with size: the smaller the molecule the more easily it moves through the gel. In addition, if a given DNA molecule has a protein bound to it, migration of that DNA-protein complex through the gel is retarded compared to the migration of the unbound DNA molecule. This forms the basis of an assay to detect specific DNA binding activities. EMSA is based on the observation that the electrophoretic mobility of a protein-nucleic acid complex is typically less than that of the free nucleic acid in the gel matrix. This process tests the ability of a protein to bind a DNA fragment as it migrates through a non-denaturing gel under the influence of an electric current. Binding of the protein will reduce the mobility of the DNA. If the fragment travels further in the absence of a protein than it does in its presence, one can conclude that the protein has an affinity for that piece of DNA. Different proteins may retard a fragment to different extents, depending on their size and shape. If two proteins bind the same piece of DNA, they will reduce its mobility even further.

The mobility shift assay has a number of strengths. The basic technique is simple to perform, yet it is robust enough to accommodate a wide range of binding conditions. Using radioisotope-labeled nucleic acids, the assay is highly sensitive, allowing assays to be performed with small protein and nucleic acid concentrations and small sample volumes. It is very popular because it is easy, quick, versatile and very sensitive.

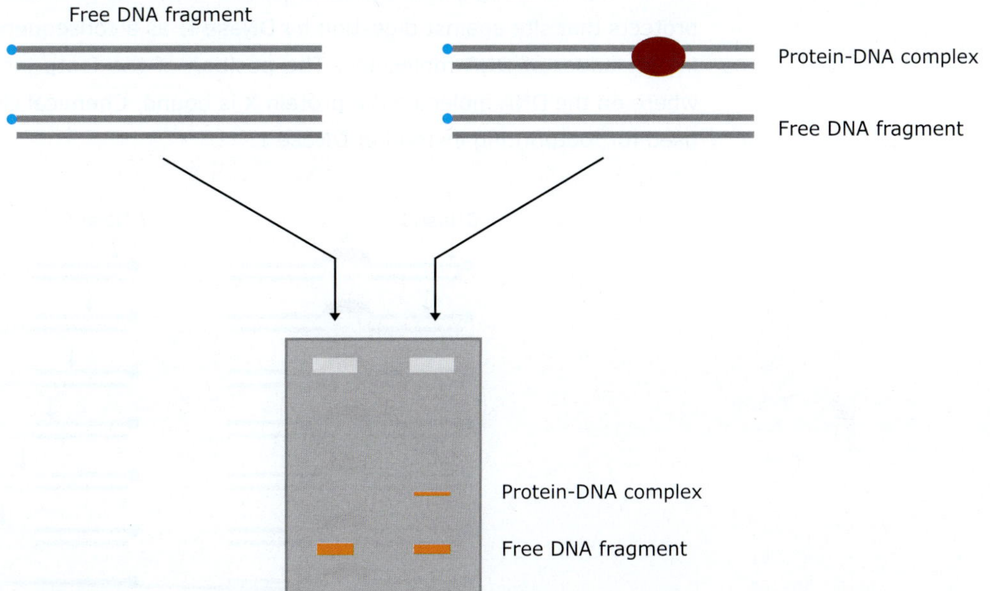

Figure 12.30 Gel mobility shift assay. The effect of the proteins binding on the mobility of the DNA fragment is analyzed by polyacrylamide gel electrophoresis followed by autoradiography. The free DNA fragments migrate rapidly to the bottom of the gel, while those fragments bound to proteins are retarded. In the figure, a protein is mixed with radiolabeled DNA containing a binding site for that protein. The mixture is resolved by polyacrylamide gel electrophoresis and visualized using autoradiography. DNA not mixed with protein runs as a single band corresponding to the size of the DNA fragment (left lane). In the mixture with the protein, a proportion of the DNA molecules (but not all of them at the concentrations used) binds the DNA molecule. Thus, in the right-hand lane, there is a band corresponding to free DNA and another corresponding to the DNA fragment in complex with the protein.

12.11 Footprinting assay

Gel retardation assay gives a general indication about the protein binding with a DNA fragment. However, it does not reveal where a protein is binding within a particular DNA fragment. Footprinting assay allows us to identify the sites, where protein binds. It is of two main types: protection and interference. The **protection footprinting** consists of identifying the sites of protection from – cleavage of phosphodiester bonds or modification of nucleotide – of the DNA by its interaction with the protein, whereas the **interference footprinting** identifies the DNA sites essential to the interaction by the fact that their modification inhibits protein binding.

Nuclease protection footprinting

DNase I footprinting assay is a method of studying *DNA-protein interaction* and identifying the DNA sequence to which a protein binds. It measures the ability of a protein to protect a radiolabeled DNA fragment against digestion by DNase I, which is an endonuclease with little sequence specificity that will cut almost anywhere within a DNA molecule. DNase I treatment is carried out in the presence of a manganese salt, which induces the enzyme to make random, double strand cuts in the target molecules, leaving blunt-ended fragments.

The nuclease treatment is carried out under limiting conditions, such as low temperature and/or very little enzyme, so that on average, each copy of the DNA fragment suffers a single 'hit'—meaning that it is cleaved at just one position along its length. This generates a population of fragments of all possible sizes that appear as a continuous ladder when resolved according to size on a denaturing gel. If a protein (suppose X) binds the DNA at a particular site, then this protects that site against digestion by DNase I; as a consequence, a gap or *footprint* appears in the ladder of DNA molecules. The position of the footprint corresponds to the positions where on the DNA molecule the protein X is bound. Chemical cleavage agents are sometimes used for footprinting instead of DNase I.

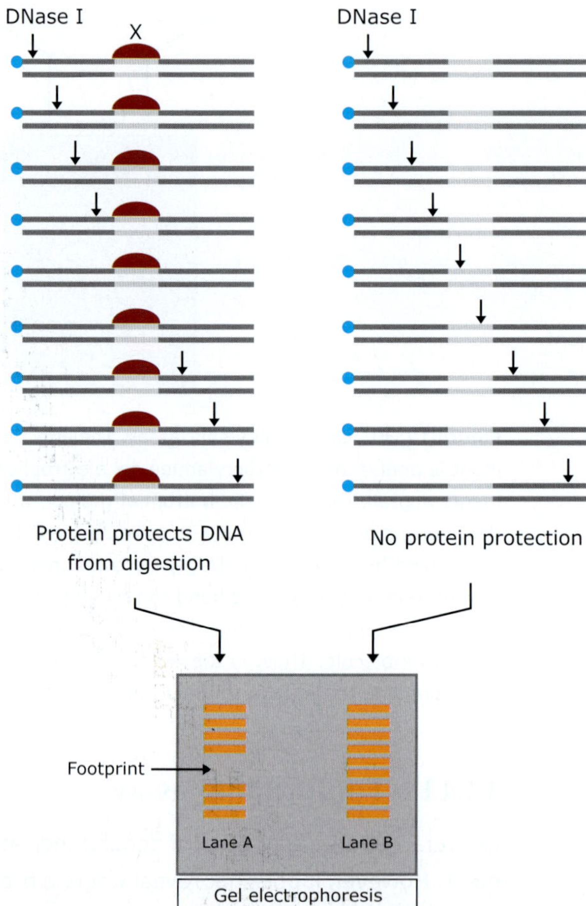

Figure 12.31 DNase I footprinting assay. X represents the protein having a specific binding site on DNA. During the footprinting reaction, a sample of the DNA fragment containing the binding site is mixed with purified protein (X) and DNase I. The DNase I cleaves the DNA randomly. The amount of DNase I is controlled so that each DNA fragment is only digested once. Since DNase does not cut the DNA where the protein is bound, sequence of DNA fragments bound with protein remain protected from DNase I action. The samples are run on a sequencing gel to separate the fragments. When lane B (no protein) and lane A (plus protein) are compared, it can be seen that lane A shows no bands in the boxed region. Therefore, the protein binds to DNA in this region and protects it from cutting by the DNase. Alignment with a sequencing ladder (lane B) allows the precise region of binding to be deduced.

Modification protection footprinting

Modification protection footprinting is similar to DNase I footprinting. The basis of this technique is that if a DNA molecule carries a bound protein, then part of its nucleotide sequence will be protected from modification. In this case instead of DNase I digestion, the fragments are treated with limited amounts of base modifying agent, such as dimethyl sulfate (DMS) which adds methyl groups to G nucleotides. If a protein binds the DNA at a particular site, then this protects guanines at that site from the action from dimethyl sulfate. In DMS modification protection assay, the fragments are treated with limited amounts of DMS so that a single G nucleotide is methylated in each fragment.

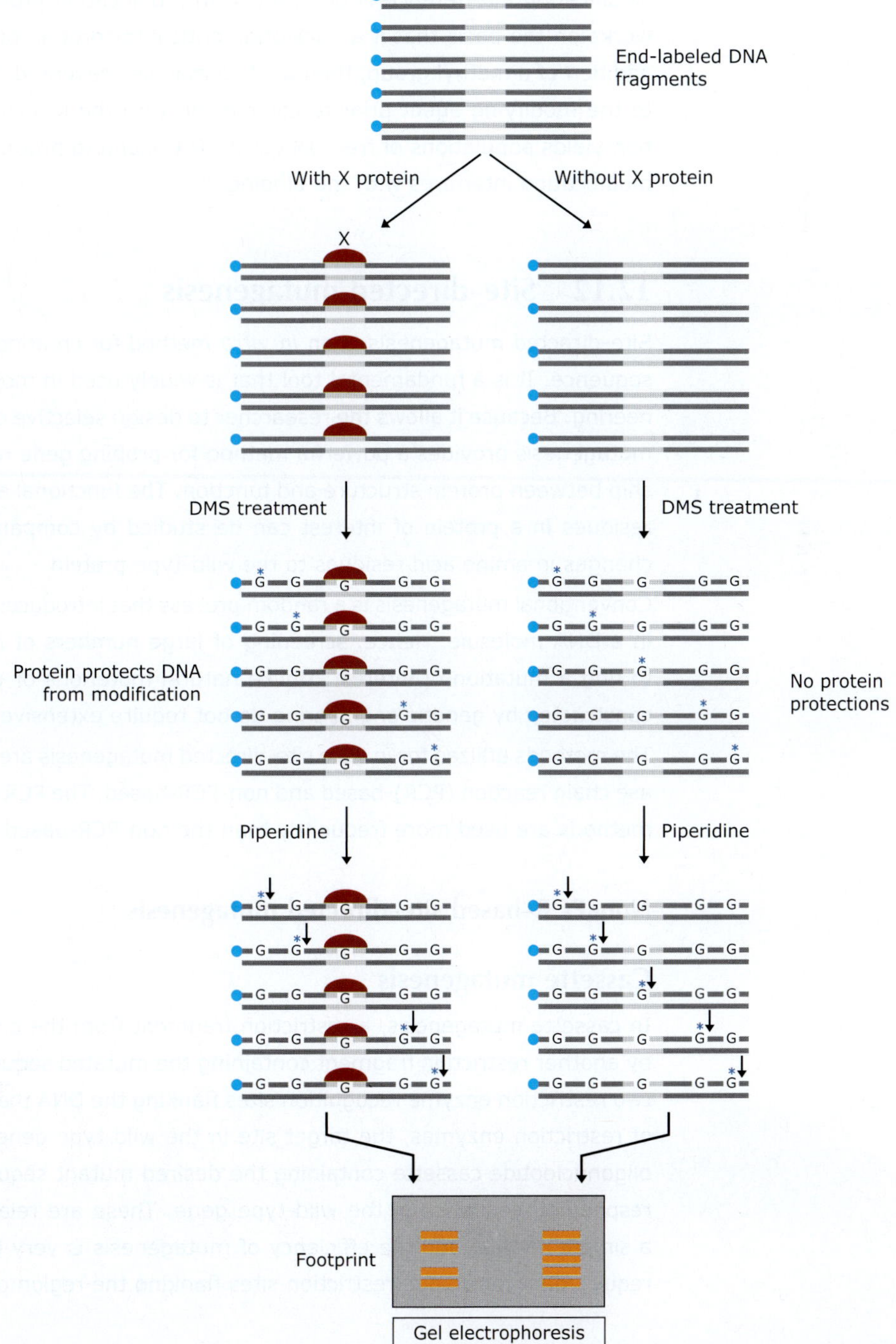

Figure 12.32 The modification protection footprinting experiment with dimethylsulfate (DMS) as the probe. The black circle denotes a radioactive end-label which is required for visualization of the final products on the gel. A portion of the DNA is combined with the protein and other portion is kept free from protein. Both are then treated with the DMS, which modifies G nucleotide on the DNA. The region of the DNA bound with protein remain protected from modification with the DMS. The DNA is, then, cleaved at the sites of modification by treatment with piperidine. The end-labeled fragments of the free and protein-bound DNA are then compared by separation on a denaturing sequencing gel and visualized with an autoradiogram.

After removal of the protein, the DNA is treated with piperidine, which cleaves the phosphodiester bond at the modified nucleotide positions. DNA fragments are then resolved according to size by gel electrophoresis. Piperidine only cuts the strand that is modified, rather than making a double-stranded cut. The samples are, therefore, examined by *denaturing gel* electrophoresis so that the two strands are separated. The resulting autoradiograph shows the sizes of the strands that have one labeled end and one end created by piperidine nicking. The banding pattern for the control DNA strands - those not incubated with the proteins – indicates the positions of G nucleotide in the DNA fragment, and the footprint.

Modification interference assay footprinting

Modification interference is different from modification protection. Modification interference works on the basis that if a nucleotide critical for protein binding is altered, for example by addition of a methyl group, then binding may be prevented. In this case, the DNA is exposed to the modifying agent prior to complexing with the protein. The subsequent binding reaction yields populations of free DNA and DNA bound to protein, defined by whether or not the modification interferes with the binding.

12.12 Site-directed mutagenesis

Site-directed mutagenesis is an *in vitro* method for creating a specific mutation in a known sequence. It is a fundamental tool that is widely used in molecular biology and protein engineering. Because it allows the researcher to design selective changes to the DNA, site-directed mutagenesis provides a powerful method for probing gene regulation as well as the relationship between protein structure and function. The functional and structural roles of amino acid residues in a protein of interest can be studied by comparing the mutant protein carrying changes in amino acid residues to the wild-type protein.

Conventional mutagenesis is a random process that introduces changes at unspecified positions in a DNA molecule. Hence, screening of large numbers of mutated organisms is necessary to find a mutation of interest. Unlike natural mutations or random mutagenesis, mutations constructed by genetic engineering do not require extensive screening.

The methods utilized for *in vitro* site-directed mutagenesis are generally of two types: polymerase chain reaction (PCR)-based and non-PCR-based. The PCR-based site-directed mutagenesis methods are used more frequently than the non-PCR-based methods.

Non-PCR-based site-directed mutagenesis

Cassette mutagenesis

In cassette mutagenesis, a restriction fragment from the cloned DNA of interest is replaced by another restriction fragment containing the mutated sequence. It relies on the presence of two restriction enzyme recognition sites flanking the DNA that is to be mutated. With the help of restriction enzymes, the target site in the wild-type gene is removed. A double-stranded oligonucleotide cassette containing the desired mutant sequence is used to replace the corresponding sequence in the wild-type gene. These are relatively large-scale changes. It is a simple method and the efficiency of mutagenesis is very high. The disadvantages are the requirement for unique restriction sites flanking the region of interest.

Design of mutagenic oligonucleotides

A crucial step in site-directed mutagenesis is the design of the mutagenic oligonucleotide. By definition, mutagenic oligonucleotides must contain at least one base change, but they may incorporate far more complicated mutations including insertions, deletions, and multiple substitutions. The minimum length of the mutagenic oligonucleotide is defined by the complexity of the mutation. Simple single-base substitutions can be accomplished with oligonucleotides ~25 bases in length. More complicated mutations may require oligonucleotides 80 bases or more in length.

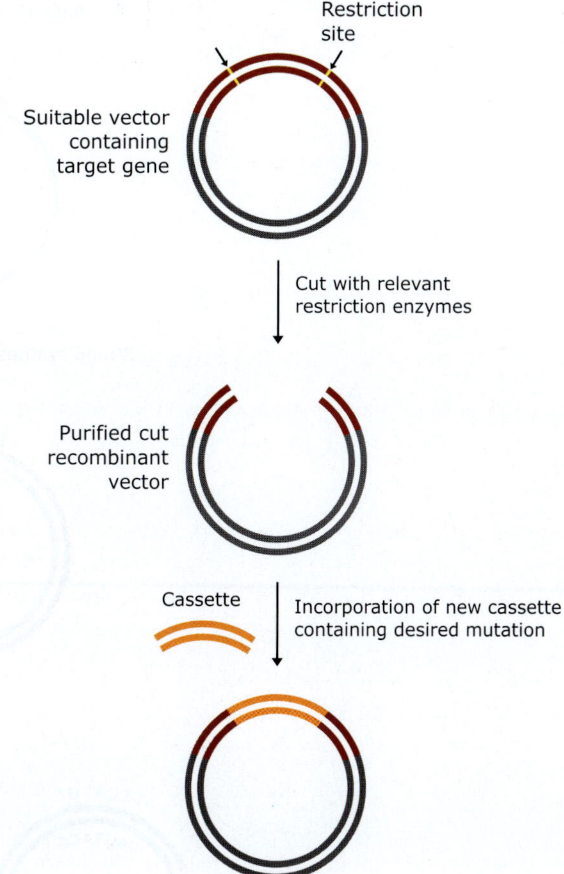

Figure 12.33 Cassette mutagenesis. The gene contained in a suitable vector is cleaved with two restriction enzymes. This releases a small section of DNA from the gene. These restriction enzymes cut at sites that flank the area of DNA to be changed. In this way, only the desired cutting occurs. A synthetic duplex is then ligated in place of the released cassette. This technique can be used to make single or multiple changes. The changes are only limited by the available size of the synthetic DNA cassette.

Primer extension mutagenesis

The simplest method of site-directed mutagenesis is the primer extension mutagenesis (*oligonucleotide-directed mutagenesis*). The method involves DNA synthesis with a chemically synthesized oligonucleotide (7-20 nucleotides long) that carries a base mismatch with the complementary sequence. As shown in figure 12.34, the method requires that the DNA to be mutated is available in single-stranded form. A single-stranded version of the gene is obtained by cloning in an M13 vector. The synthetic oligonucleotide which acts as primer, primes DNA synthesis and is itself incorporated into the resulting heteroduplex molecule.

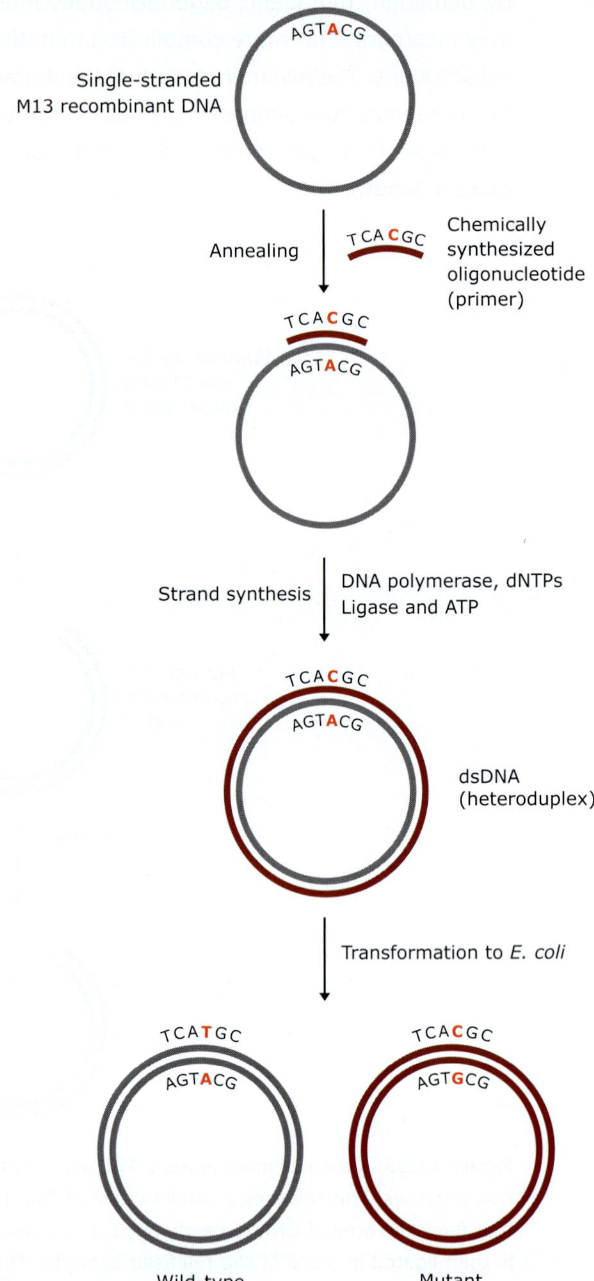

Figure 12.34 Primer extension mutagenesis. *Step* 1, an oligonucleotide primer annealing to template DNA. The gene or the DNA fragment is inserted in an M13 phage vector. *Step* 2, Primer extension by DNA polymerase and synthesis of second strand. *Step* 3, Sealing of nick by ligase. DNA ligase seals the nick and a completely closed, circular double-stranded DNA molecule is produced. One strand of this molecule contains the original sequence while the other strand contains the mutated sequence, that is, a heteroduplex. *Step* 4, Transformation to *E. coli*. The dsDNA is transformed into *E. coli* cells. In *E. coli*, subsequent replication will produce some double-stranded circular DNA molecules with wild-type sequences and some molecules with mutated sequences. *Step* 5, Screening of phages with mutant sequence. Bacteriophages containing either the wild-type or the mutant sequence can be distinguished from each other through hybridization screening.

The newly formed heteroduplex is used to transform cells. After introduction into *Escherichia coli*, DNA replication produces numerous copies of this recombinant DNA molecule. Half of the resulting double-stranded molecules are copies of the original strand of DNA, and the remaining are copies of the strand that contains the mutated sequence.

A variation of the procedure outlined above involves oligonucleotides containing inserted or deleted sequences. As long as stable hybrids are formed with single-stranded wild-type DNA, priming of *in vitro* DNA synthesis can occur, ultimately giving rise to clones corresponding to the inserted or deleted sequence.

PCR based site-directed mutagenesis

PCR approaches have become the method of choice to generate arrays of predefined mutations within the gene of interest. There are numerous PCR-based approaches to DNA mutagenesis. PCR-mediated nucleotide changes, deletions, or insertions can be accomplished by performing PCR synthesis reactions with carefully chosen or modified PCR reaction components.

One strategy for this kind of PCR mutagenesis is based on the principle of 'mispriming'. Because mismatches between templates and primers are tolerated under certain PCR conditions, primers can be designed to include predefined changes (so-called *mutagenic primers*). Other strategies make use of the built-in high error rate of *Taq* DNA polymerase in the PCR reaction. In general, mutagenic primer PCR approaches are used for introducing site-directed mutagenesis into the genes of interest, whereas *Taq* DNA polymerase or base analog PCR approaches are useful in creating random and extensive mutagenesis in the target gene.

Overlap extension method

A specialized technique developed for the introduction of mutations into the center of a PCR fragment is known as the *overlap extension method*. In this technique, primers are designed for the synthesis of two different PCR fragments with a region of common sequence. Each reaction uses one flanking primer and one internal primer containing the desired mutation. The two overlapping pieces of DNA are then annealed to generate a heteroduplex that can be extended by *Taq* polymerase and amplified by subsequent rounds of PCR to produce a full-length mutant segment. By varying the templates and primers, both mutant and chimeric DNA can be generated. The technique is applicable to the introduction of site-specific insertion and deletion mutations. It has also been modified to allow for the introduction of multiple mutations on the same template without loss of efficiency.

The overlap extension method can be modified to simplify the removal of excess primers and wild-type template after each PCR step. Biotinylated primers and streptavidin-coated magnetic beads are employed to purify PCR products from reactants. This approach improves the speed and efficiency of overlap extension mutagenesis.

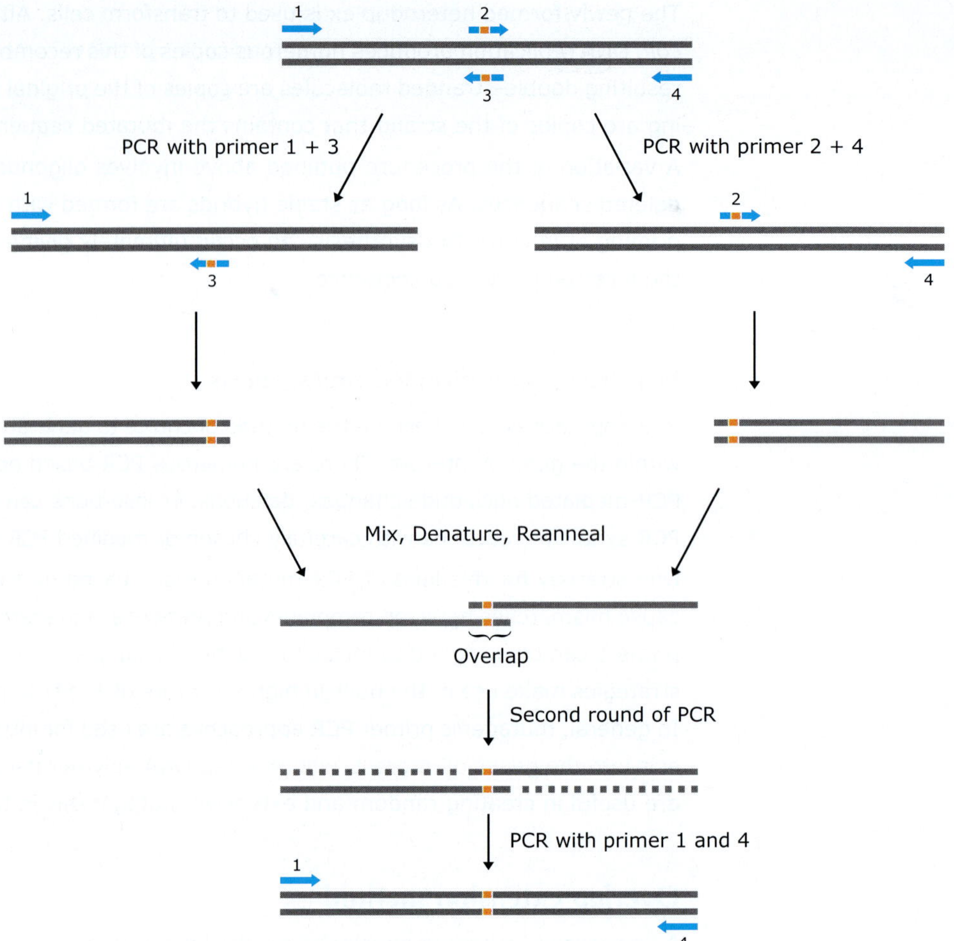

Figure 12.35 Overlap extension method of mutagenesis. Two outermost primers (1 and 4) are designed, which are complementary to the antisense strand and the sense strand of the target DNA, respectively. In addition, two internal primers 2 and 3 containing the desired mutations are taken, so that the required mutation can be introduced to each strand. Two PCR reactions are carried out in two separate tubes, but with the same reaction conditions (as shown in the figure). The products of the two reactions are then mixed, denatured, and re-annealed; some of the strands from the first PCR product will anneal with that obtained from the second reaction especially at the regions of overlap, that is, at the regions corresponding to the sequences of primers 2 and 3. The hybrid molecule containing a short 3′-end can act as a primer and can be extended in the next phase by DNA polymerase to form a complete double-stranded DNA molecule with mutation on both strands. The subsequent and final PCR carried out with flanking primers (primers 1 and 4) will then amplify the full-length product to yield the mutated DNA.

Megaprimer PCR

Although primers are typically short, single-stranded oligonucleotides that a long, double-stranded DNAs could also serve as primers. These larger primers, known as *megaprimers* can be used to introduce any combination of point mutations, deletions or insertions. The megaprimer method of mutagenesis uses three oligonucleotide primers (two flanking and one megaprimer) and two rounds of PCR performed on a DNA template. The first PCR uses one outside primer and the middle mutagenic primer to form a double-stranded product (the megaprimer) containing the desired mutations.

The amplified product is then used in a second round of PCR with wild-type template and the other flanking primer to create a fragment of the same length as the original target DNA containing the desired mutation. The key to this method is that the amplified product from the first round of PCR is used as a primer in the second round of PCR.

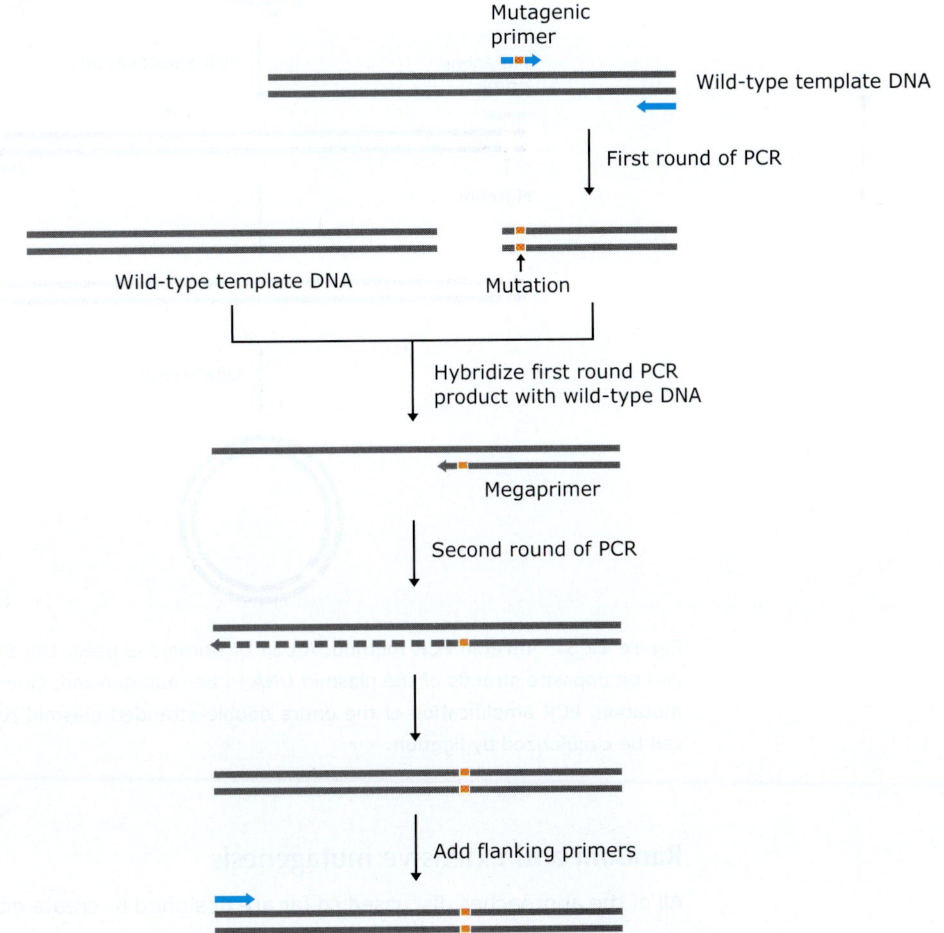

Figure 12.36 Schematic illustration of the megaprimer method for site-directed mutagenesis. The first round of PCR is used to make a fragment of the template DNA containing the desired mutation. This 'megaprimer' is then hybridized to the wild-type template DNA, and a second round of PCR is carried out, to generate the entire molecule with the mutation.

Inverse PCR method

The inverse PCR method uses only two primers to create the desired mutation. The key feature of this method is that in making the mutation, the entire vector is amplified. The two primers, one containing the desired mutation, extend on the circular template DNA in opposite directions. Amplification ultimately yields a linear, double-stranded DNA molecule containing the mutation at one end. Following amplification, the ends are ligated and the resulting circular DNA molecule is transformed into *E. coli*. There are a number of variations of this method that improve the efficiency of mutagenesis.

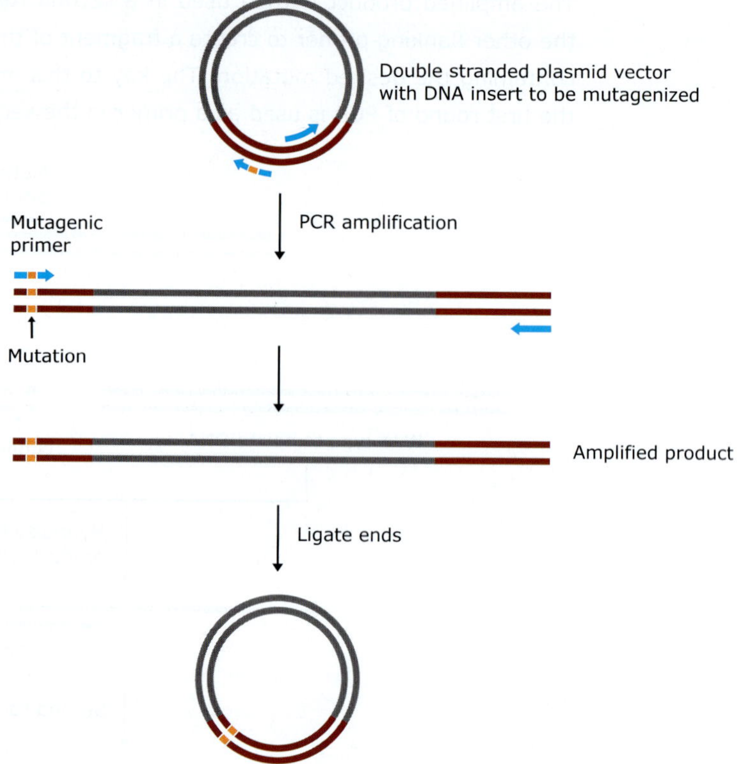

Figure 12.37 Inverse PCR method. A pair of primers is used, the 5′ termini of which hybridize end-to-end on opposite strands of the plasmid DNA to be mutagenized. One of the primers contains the desired mutation. PCR amplification of the entire double-stranded plasmid results in a linear PCR amplicon that can be circularized by ligation.

Random and extensive mutagenesis

All of the approaches discussed so far are designed to create either single or a few site-directed mutations. Sometimes, it is desirable to obtain random and extensive mutations in the gene of interest or to generate a library of such molecules. There are two basic approaches to achieve these goals, and both are based on selected uses of PCR reaction components. The first approach makes use of error-prone DNA polymerases. Certain thermostable DNA polymerases, like *Taq* DNA polymerase, have an intrinsic error rate due to the lack of a 3′-5′ exonuclease activity. Mismatch passes by the polymerase during PCR allows the possibility of mutations. The cumulative error rate can be substantial. This tendency is further enhanced by other factors such as buffer composition (e.g. high-magnesium concentration, or high pH) and other experimental conditions (e.g. a large amount of polymerase, a great number of cycles, a low-annealing temperature).

The second approach to generate random and extensive mutations in the gene of interest is based on the base pairing property of degenerate base analog which can form base pairing with nucleotides A, C, G and T under normal reaction conditions. In the presence of degenerate base analog and a biased ratio of dNTPs, DNA polymerase tends to randomly incorporate a substantial amount of base analog in the newly synthesized DNA strand. This base analog-containing DNA can serve as a template in subsequent PCR amplifications and allows random base insertion sites. As a result, the final PCR products will have base substitution random sites.

12.13 DNA sequencing

The term DNA sequencing encompasses biochemical methods for determining the order of the nucleotide bases, adenine, guanine, cytosine and thymine, in a DNA molecule. The methodologies for DNA sequencing are as follows:

Chain termination method

The enzymatic chain termination method (also dideoxy method) was developed by Frederick Sanger and coworkers in year 1977. The method is based on the DNA polymerase-dependent synthesis of a complementary DNA strand in the presence of natural deoxynucleotides (dNTPs) and dideoxynucleotides (ddNTPs) that serve as terminators. The DNA synthesis reaction is randomly terminated whenever a ddNTP is added to the growing oligonucleotide chain, resulting in truncated products of varying lengths with an appropriate ddNTP at their 3′-terminus. The products are separated by size using polyacrylamide gel electrophoresis and the terminal ddNTPs are used to reveal the DNA sequence of the template strand.

Chemical degradation method

The chemical degradation method was introduced by Allan Maxam and Walter Gilbert in year 1977. In this method, the sequence of a dsDNA molecule is determined by treatment with chemicals that cut the molecule at specific nucleotide positions.

Pyrosequencing method

The pyrosequencing method was developed by Mostafa Ronaghi and Pal Nyrén at the Royal Institute of Technology in Stockholm in 1996. In this method, the addition of a deoxynucleotide to the end of the growing strand is detectable because it is accompanied by the release of a flash of light.

Chain termination method

Chain termination method relies on the use of dideoxyribonucleoside triphosphates (ddNTP), derivatives of the normal deoxyribonucleoside triphosphates that lack the 3′ hydroxyl group. Purified DNA is synthesized in vitro in a mixture that contains single-stranded molecules of the DNA to be sequenced, the enzyme DNA polymerase, a short primer DNA to enable the polymerase to start DNA synthesis, and the four deoxyribonucleoside triphosphates (dATP, dCTP, dGTP, dTTP: A, C, G and T). If a dideoxyribonucleotide analog of one of these nucleotides is also present in the nucleotide mixture, it can become incorporated into a growing DNA chain. Because this chain now lacks a 3′ OH group, the addition of the next nucleotide is blocked, and the DNA chain terminates at that point.

To determine the complete sequence of a DNA fragment, the double-stranded DNA is first separated into its single strands and one of the strands is used as the template for sequencing. Four different chain-terminating dideoxyribonucleoside triphosphates (ddATP, ddCTP, ddGTP and ddTTP) are used in four separate DNA synthesis reactions on copies of the same single-stranded DNA template.

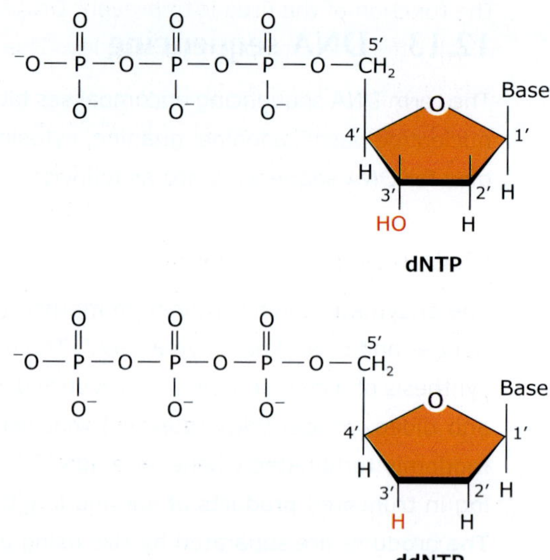

Figure 12.38 The structures of dNTP and ddNTP differ in the number and location of hydroxyl groups on the 2′ and 3′ carbons. DNA polymerase cannot add another nucleotide to a chain ending in dideoxyribose because its 3′ carbon does not have a hydroxyl group.

Each reaction produces a set of DNA copies that terminate at different points in the sequence. The products of these four reactions are separated by electrophoresis in four parallel lanes of a polyacrylamide sequencing gel. Sequencing gels are run in the presence of denaturing agents, urea and formamide. They routinely contain 6–20% polyacrylamide and 7 mol/l urea.

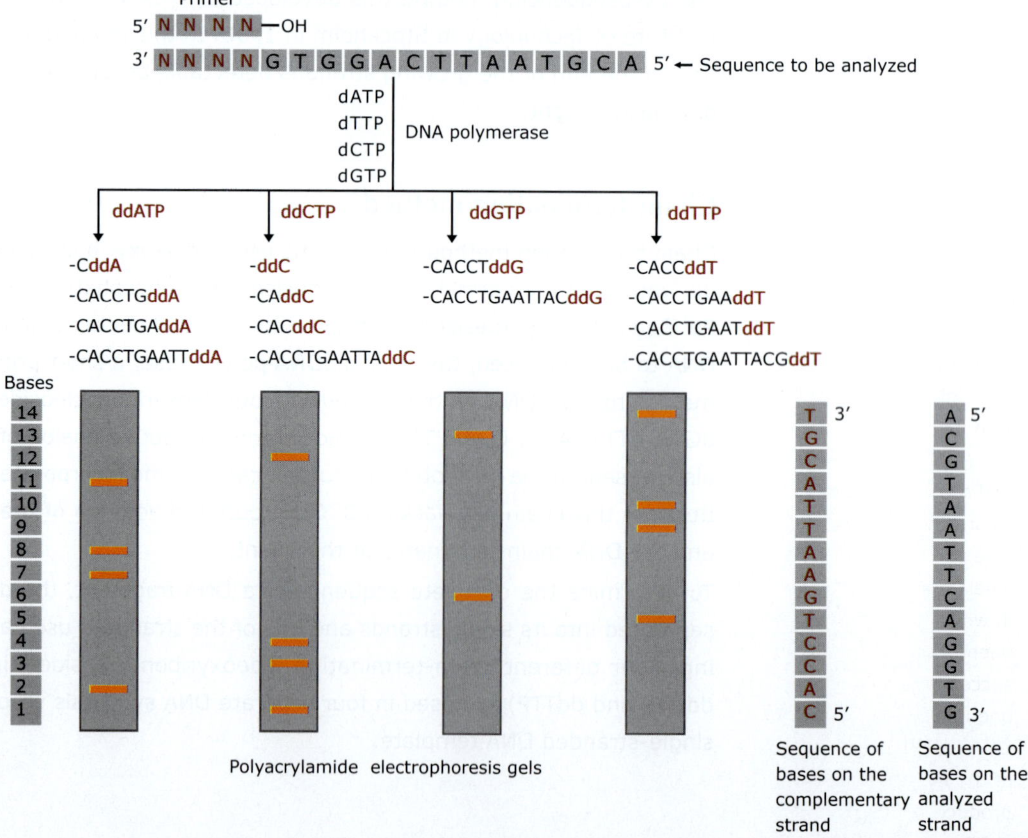

Figure 12.39 DNA sequencing by chain termination method (Sanger method).

The function of the urea is to prevent DNA secondary structure due to intrachain base pairing, which affects electrophoretic mobility. This is important because the change in conformation resulting from base pairing alters the electrophoretic mobility of a single-stranded molecule. DNA sequencing gels tend to be very long (100 cm) to maximize the separation achieved.

The newly synthesized fragments are detected by a label either radioactive or fluorescent that has been incorporated either into the primer or into one of the deoxyribonucleoside triphosphates used to extend the DNA chain. In each lane, the bands represent fragments that have terminated at a given nucleotide, but at different positions in the DNA. By reading off the bands in order, starting at the bottom of the gel and working across all lanes, the DNA sequence of the newly synthesized strand can be determined.

Automated sequencing

The standard chain termination sequencing methodology employs radioactive labels, and the banding pattern in the polyacrylamide gel is visualized by autoradiography. This approach is not well suited to automation. To automate the process, it is desirable to acquire sequence data in real time by detecting the DNA bands within the gel during the electrophoretic separation.

Fluorescent dideoxynucleotides are the basis of automated sequencing. Each of the four dideoxynucleotides carries a spectrally different fluorophore. Chains terminated with A are, therefore, labeled with one fluorophore, chains terminated with C are labeled with a second fluorophore, and so on. Now it is possible to carry out the four sequencing reactions – A, C, G and T – in a single tube and to load all four families of molecules into just one lane of the polyacrylamide gel, because the fluorescent detector can discriminate between the different labels and determine if each band represents an A, C, G or T. The sequence can be read directly as the bands pass in front of the detector; and either printed out in a form readable by eye or sent straight to a computer for storage.

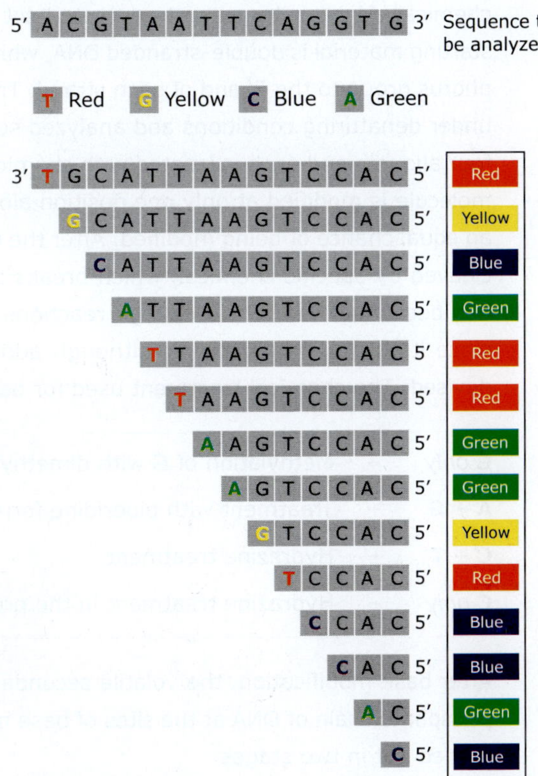

Figure 12.40 Replication of the DNA template strand proceeds with a reaction mixture including the four standard dNTPs and all four ddNTP, each labeled with a different fluorescent dye (ddATP, ddCTP, ddGTP, and ddATP). Random incorporation of the labeled ddNTPs produces a series of DNA fragments in which chain growth has been terminated at each successive position, each one nucleotide longer than the previous. Separation of the fragments by size produces a sequencing ladder as a series of colored bands.

Sequencing enzymes

Naturally occurring polymerases have features that are often not optimal for DNA sequencing. The relevant and desirable properties of sequencing polymerases include:

High processivity

Processivity is the degree to which chain extension continues before the enzyme dissociates from a primer-template annealing complex. T7 Sequenase is the most processive of the current catalog of sequencing enzymes, whereas the Klenow fragment is the least.

Thermostability

Resistance to inactivation or dissociation at high temperatures is the most important factor in the cycle-sequencing reactions that are the basis for modern high-throughput sequencing. *Taq* polymerase or variants are the only feasible options.

Incorporation of nucleotide analogs such as dye terminators

The ability to incorporate analogs is a critical factor for the dideoxy chain-termination method. The efficiency of chain termination with each of the dye-labeled terminators must be similar to avoid low-quality data with uneven peaks.

Exonuclease activities

Polymerases often have 3′-exonuclease 'proofreading' and/or 5′-exonuclease activities that remove RNA primers after DNA replication. Because neither activity is desirable for sequencing, variants of the polymerases should be used that lack these activities (e.g. thermal sequenase).

Chemical degradation method

Chemical degradation method is based on preferential base specific modification followed by chemical cleavage to generate a nested set of end-labeled derivatives. In this method, the starting material is double-stranded DNA, which is first labeled by attaching a radioactive phosphorus group to the 5′ end of each strand. The strands are then separated by electrophoresis under denaturing conditions and analyzed separately. DNA labeled at one end is divided into four aliquots and each is treated with chemicals that act on specific bases. On average, each molecule is modified at only one position along its length; every base in the DNA strand has an equal chance of being modified. After the modification reactions, the separate samples are cleaved by specific chemical, which breaks phosphodiester bonds of nucleotide whose base has been modified. The cleavage reactions in this method have not changed significantly since its initial development, although additional chemical cleavage reactions have been devised. The chemical treatment used for base modification are:

G *only*	–	Methylation of G with dimethyl sulfate (DMS) at pH 8.0.
A + G	–	Treatment with piperidine formate at pH 2.0.
C + T	–	Hydrazine treatment.
C *only*	–	Hydrazine treatment in the presence of 1.5 M NaCl.

After base modification, the volatile secondary amine piperidine is used to cleave the sugar-phosphate chain of DNA at the sites of base modifications. Thus, these cleavage reactions are carried out in two stages:

First, chemical modification of specific bases, and

Second, removal of the modified base from its sugar, and cleavage of the phosphodiester bonds 5′ and 3′ to the modified base.

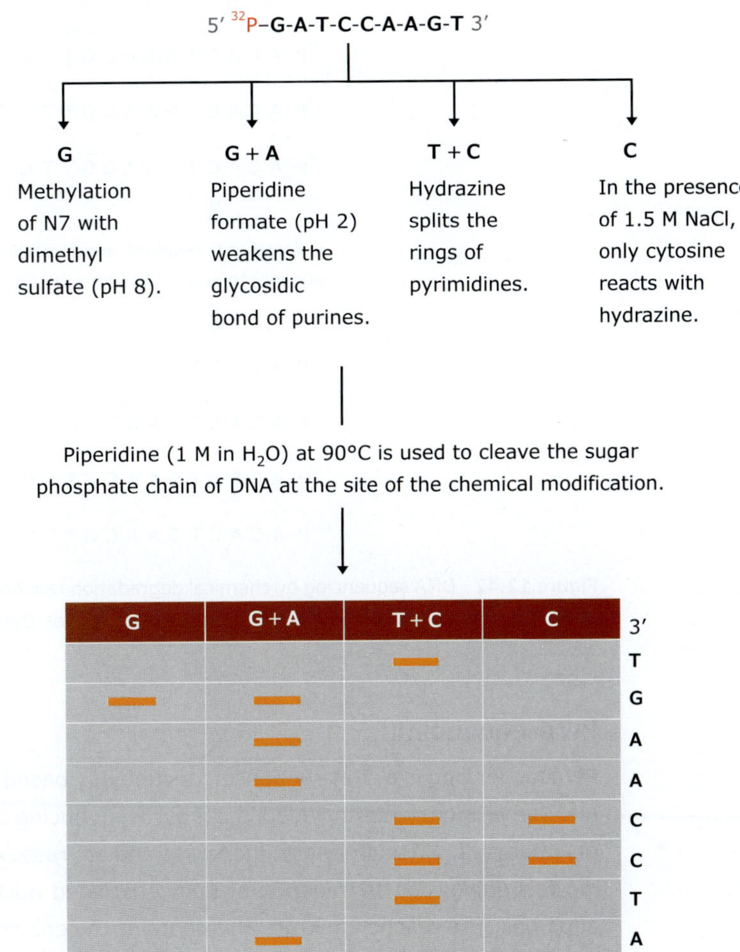

Figure 12.41 Sequencing by the Maxam-Gilbert method. The oligonucleotide in this example (5′ G-A-T-C-C-A-A-G-T 3′) is subjected to four chemical cleavage reactions (G, G+A, T+C and C). The resulting products of each of the reactions are separated by electrophoresis through a denaturing acrylamide gel and visualized by autoradiography.

To illustrate the procedure, we will follow the 'G' cleavage reaction. First, the molecules are treated with dimethyl sulfate, which attaches a methyl group to the purine ring of G *nucleotides*. Only a limited amount of dimethyl sulfate is added, the objective being to modify, on average, just one G per polynucleotide. At this stage the DNA strands are still intact, cleavage not occurring until a second chemical – piperidine – is added. Piperidine removes the modified purine after cleaving the phosphodiester bonds. The result is a set of cleaved DNA molecules, some of which are labeled one and some are not. The labeled molecules all have one end in common and one end determined by the cut sites, the latter indicating the positions of the G nucleotides in the DNA molecules that were cleaved. The cleaved molecules are electrophoresed in a polyacrylamide gel and the sequence read in a similar way to that described for chain termination sequencing.

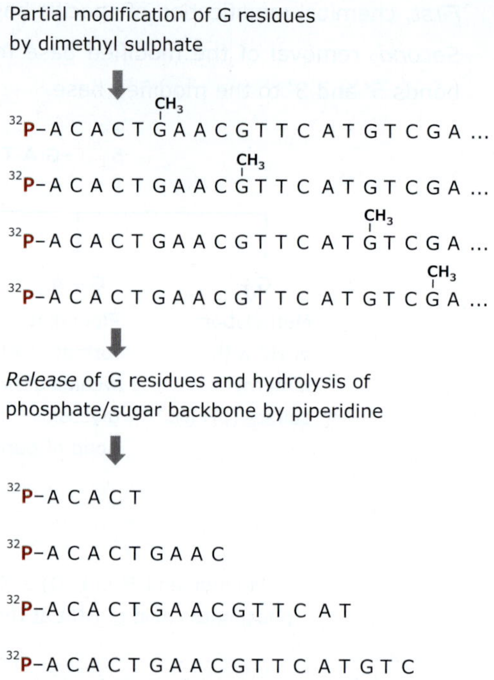

Partial modification of G residues by dimethyl sulphate

↓

^{32}P-A C A C T G(CH$_3$) A A C G T T C A T G T C G A ...

^{32}P-A C A C T G A A C G(CH$_3$) T T C A T G T C G A ...

^{32}P-A C A C T G A A C G T T C A T G(CH$_3$) T C G A ...

^{32}P-A C A C T G A A C G T T C A T G T C G(CH$_3$) A ...

↓

Release of G residues and hydrolysis of phosphate/sugar backbone by piperidine

↓

^{32}P-A C A C T

^{32}P-A C A C T G A A C

^{32}P-A C A C T G A A C G T T C A T

^{32}P-A C A C T G A A C G T T C A T G T C

Figure 12.42 DNA sequencing by chemical degradation reaction-specific for guanidine residues. The guanine base is first modified with dimethyl sulfate, which makes the chain susceptible to cleavage by piperidine.

Pyrosequencing

Pyrosequencing is a DNA sequencing technology based on the sequencing-by-synthesis principle. The four enzymes included in the pyrosequencing system are the Klenow fragment of DNA polymerase I, ATP sulfurylase, luciferase and apyrase. Apyrase is included in the pyrosequencing technology for degradation of unincorporated nucleotides and excess ATP between base additions. The reaction mixture also contains the enzyme substrates *adenosine phosphosulfate* (APS), luciferin and the sequencing template with an annealed primer to be used as starting material for the DNA polymerase. The first reaction, DNA polymerization, occurs if the added nucleotide forms a base pair with the sequencing template and, thereby, is incorporated into the growing DNA strand. The inorganic pyrophosphate, released by the DNA polymerase serves as substrate for ATP sulfurylase, which produces ATP. Finally, the ATP is used by luciferase to generate light and the light signal is detected. Hence, only if the correct nucleotide is added to the reaction mixture, light is produced. Apyrase removes unincorporated nucleotides and ATP between the additions of different bases.

$$\text{(Oligonucleotide)}_n + \text{dNTP} \xrightarrow{\textit{Polymerase}} \text{(Oligonucleotide)}_{n+1} + \text{PP}_i$$

$$\text{PP}_i + \text{APS} \xrightarrow{\textit{Sulfurylase}} \text{ATP} + \text{Sulphate}$$

$$\text{ATP} + \text{Luciferin} + \text{O}_2 \xrightarrow{\textit{Luciferase}} \text{AMP} + \text{PP}_i + \text{Oxyluciferin} + \text{Light}$$

$$\text{ATP} \xrightarrow{\textit{Apyrase}} \text{AMP} + \text{PP}_i$$

$$\text{dNTP} \xrightarrow{\textit{Apyrase}} \text{NMP} + \text{PP}_i$$

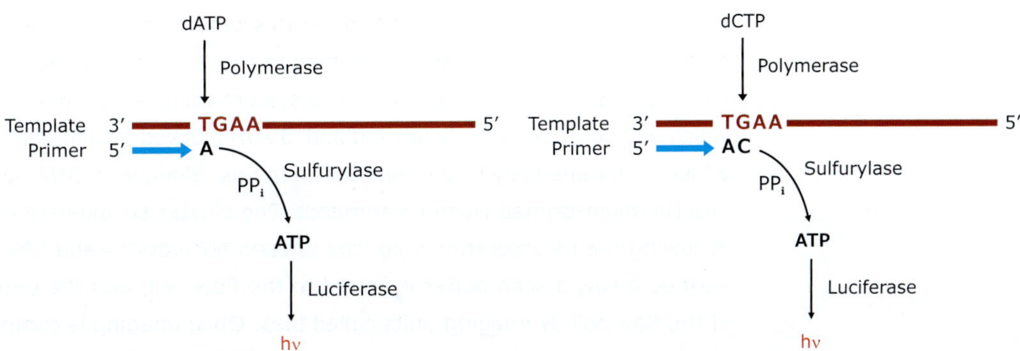

Figure 12.43 Schematic representation of the pyrosequencing enzyme system. If the added dNTP forms a base pair with the template, DNA polymerase incorporates it into the growing DNA strand and pyrophosphate is released. ATP sulfurylase converts the PP_i into ATP which serves as substrate for the light producing enzyme luciferase. The produced light is detected as evidence of that nucleotide incorporation has taken place.

Next-generation sequencing technologies

First generation sequencing technologies include sequencing by synthesis developed by Sanger and sequencing by cleavage pioneered by Maxam and Gilbert. *Next-generation sequencing* refers to non-Sanger-based high-throughput DNA sequencing technologies. Millions or billions of DNA strands can be sequenced in parallel, yielding substantially more throughput and minimizing the need for the fragment-cloning methods that are often used in Sanger sequencing of genomes. The four main advantages of next-generation sequencing over classical Sanger sequencing are:

- speed
- cost
- sample size
- accuracy

Nucleic acid sequencing is a method for determining the exact order of nucleotides present in a given DNA or RNA molecule. The chain-termination method (also commonly referred to as Sanger or dideoxy sequencing method), published in 1977, has remained the most commonly used DNA sequencing technique to date. The Human Genome Project, led by the International Human Genome Sequencing Consortium and Celera Genomics, was accomplished with *first-generation* Sanger sequencing. Since completion of the first human genome sequence, demand for cheaper and faster sequencing methods has increased greatly. This demand has driven the development of *second-generation* sequencing methods (or next-generation sequencing). Next-generation sequencing platforms perform massively parallel sequencing, during which millions of fragments of DNA from a single sample are sequenced in unison. Massively parallel sequencing technology facilitates high-throughput sequencing, which allows an entire genome to be sequenced in less than one day. In the past decade, several next-generation sequencing platforms have been developed.

To illustrate how next-generation sequencing process works, consider a single genomic DNA (gDNA) sample. The gDNA is first fragmented into a library of small segments that can be uniformly and accurately sequenced in millions of parallel reactions. The newly identified strings of bases, called *reads*, are then reassembled using a known reference genome as a scaffold (resequencing), or in the absence of a reference genome (*de novo* sequencing). The full set of aligned reads reveals the entire sequence of each chromosome in the gDNA sample.

Illumina (Solexa) sequencing

This system was initially developed in 2007 by Solexa and was subsequently acquired by Illumina, Inc. Illumina next-generation sequencing utilizes a fundamentally different approach from the classic Sanger chain-termination method. It is based on sequencing by synthesis (SBS) technology – tracking the addition of labeled nucleotides as the DNA chain is copied – in a massively parallel fashion. Illumina sequencing systems can deliver data output ranging from 300 kilobases up to 1 terabase in a single run, depending on instrument type and configuration.

In this sequencing technique, random fragments of the genome to be sequenced are immobilized in a flow cell, and then amplified *in situ*, resulting in localized clusters of around 1000 identical copies of each fragment. This system utilizes a sequencing by-synthesis approach in which all four fluorescently labeled and 3′-OH blocked nucleotides (reversible terminators) are added simultaneously to the flow cell channels, along with DNA polymerase, for incorporation into the oligo-primed cluster fragments. The cluster strands are extended by one nucleotide. Following the incorporation step, the unused nucleotides and DNA polymerase molecules are washed away, a scan buffer is added to the flow cell, and the optics system scans each lane of the flow cell by imaging units called tiles. Once imaging is completed, chemicals that affect cleavage of the fluorescent labels and the 3′-OH blocking groups are added to the flow cell, which prepares the cluster strands for another round of fluorescent nucleotide incorporation.

Ion Torrent sequencing

The Ion Torrent Sequencing is a sequencing-by-synthesis method that detects chemical changes during synthesis reactions using an electrochemical approach. Unlike optical signal-based methods (which rely on optically reading dye-labeled nucleotides), Ion Torrent detects electrical signals on a semiconductor chip. This sequencing method exploits the fact that when a dNTP is incorporated into a growing DNA strand, it forms a covalent bond and releases pyrophosphate and a positively charged hydrogen ion. The release of hydrogen ions causes changes in the pH of the solution, which are then detected by a sensor. Ion Torrent sequencing, also referred to as *ion semiconductor sequencing*, *pH-mediated sequencing*, *silicon sequencing*, or *semiconductor sequencing*, is a versatile and cost-effective technology. It is available as a **personal genomic machine** (PGM), suitable for both research and clinical laboratories.

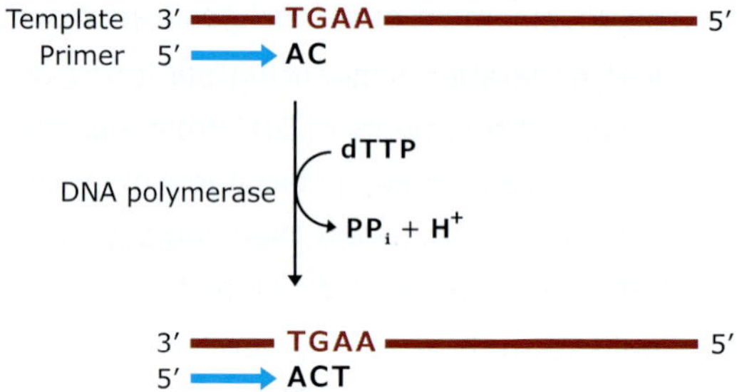

Figure 12.44 Ion Torrent technology operates on the principle that a hydrogen ion (proton) is released when a nucleotide is incorporated by DNA polymerase. The release of hydrogen ions causes changes in the pH of the solution, which are then detected by a sensor.

In Ion Torrent sequencing, the sequencing chemistry is relatively simple. The first step in this approach includes library construction, which involves DNA fragmentation and adapter ligation. These fragments are clonally amplified on small beads by emulsion PCR. Enriched beads are primed for sequencing by annealing a sequencing primer and are then deposited into the wells of an Ion Chip, a specialized silicon chip designed to detect pH changes within individual wells of the sequencer as the reaction progresses stepwise. Each microwell contains many copies of a single-stranded template DNA molecule to be sequenced, and DNA polymerase is sequentially flooded with a single species of unmodified dNTP. If the introduced dNTP is complementary to the leading template nucleotide, it is incorporated into the growing complementary strand. The hydrogen ion that is released during the reaction changes the pH of the solution, which is detected by a **pH Field Effect Transistor** (pHFET) sensor. The unattached dNTP molecules are washed out before the next cycle when a different dNTP species is introduced. If the introduced dNTP is not complementary, there is no incorporation and no biochemical reaction. If homopolymer repeats are present in the template sequence, multiple dNTP molecules will be incorporated in a single cycle. This leads to a corresponding number of released hydrogen ions and a proportionally higher electronic signal.

12.14 Biosensors

A *biosensor* is a measuring device which is used to detect chemical compounds by converting a biological response into an electrical signal. The electrical signal it produces carries the necessary information about the process under investigation. The most widely accepted definition of a biosensors is: *'a self-contained analytical device that incorporates a biologically active material in contact with an appropriate transducer for the purpose of detecting the concentration or activity of chemical species in any type of sample.'*

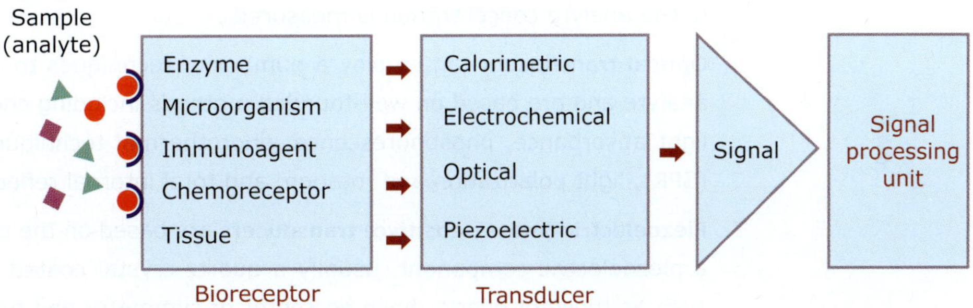

Figure 12.45 Schematic diagram showing the main components of a biosensor. The bioreceptor – interacts with the specific analyte of interest to produce signal; the transducer – transforms the signal resulting from the interaction of the analyte with the bioreceptor into another signal (i.e. transduces) that can be more easily measured and quantified; and signal processing unit – converts the signal into a workable form (amplified, processed and displayed).

The biosensor comprises three primary components:

The **biological detection system,** also known as the *bioreceptor*, serves as the initial component. This bioreceptor can take various forms, such as an enzyme, an antibody, or a similar binding molecule. It may also involve living cells or organelles.

The **transducer**, functioning as the intermediary, converts the signal generated from the interaction between the analyte and the bioreceptor into a quantifiable signal.

The **signal processing system** plays a crucial role by transforming the quantified signal into a usable format for further analysis or application.

Types of biosensors

Biosensors are usually classified into various groups either by type of transducer employed or by the kind of bioreceptor utilized. The *bioreceptors* include enzymes, antibodies, nucleic acids, microorganisms, biological tissues and organelles. Its role is to interact specifically with the target analyte and the result of this interaction is consequently transformed through transducer to measurable signal. Enzymes are the most commonly used bioreceptors in biosensors.

There are two classes of bio-recognition processes – *Bioaffinity recognition* and *Biocatalytic recognition.* Both processes involve the selective binding of an analyte with receptor. In *bioaffinity recognition*, the binding is very strong and the transducer detects the presence of the bound receptor-analyte pair. In *biocatalytic recognition*, the analytes are chemically altered to form the product molecules. Another key part of a biosensor is the *transducer*. It transforms the physical and chemical change accompanying the biorecognition event into measurable electrical signals. The transducers can be electrochemical, optical, piezoelectric and calorimetric.

Electrochemical transducers: The basic principle for this class of biosensors is that chemical reactions between immobilized bioreceptors and target analyte produce or consume ions or electrons, which affects measurable electrical properties of the solution. *Amperometric* and *potentiometric* transducers are the most commonly used electrochemical transducers. In amperometric transducers, the potential between the two electrodes is set and the current produced by the oxidation or reduction of electroactive species is measured and correlated to the concentration of the analyte of interest. *Potentiometric* transducers measure electrical potential due to changes in the distribution of charges.

Calorimetric (thermometric) transducers measure the heat of a biochemical reaction. Once the analyte comes in contact with the bioreceptor, the heat of reaction which is proportional to the analyte concentration is measured.

Optical transducers can employ a number of techniques to detect the presence of a target analyte and are based on well-founded methods including chemiluminescence, fluorescence, light absorbance, phosphorescence, photothermal techniques, surface plasmon resonance (SPR), light polarization and rotation, and total internal reflectance.

Piezoelectric (mass-sensitive) transducers are based on the coupling of the bioreceptors with a piezoelectric component, usually a quartz-crystal coated with gold electrodes. Crystals, such as those of quartz, have no center of symmetry and produce an electrical signal when stressed mechanically (i.e. by applying some pressure on them). A crystal oscillates at a certain frequency, which can be modulated by its environment. When the crystal is coated with some material, the actual frequency depends on the mass of the crystal and the coating. The resonant frequency can be measured with great accuracy hence making it possible to calculate the mass of analyte adsorbed onto the crystal surface.

Bioreceptor immobilization

Very important part of a biosensor fabrication is the immobilization of the bioreceptor in the vicinity of the transducer. Two important considerations have to be taken into account during immobilization – operational stability and long-term use. The immobilization is done either by **physical** or **chemical methods**. Chemical attachment often involves covalent bonding to transducer surface by suitable reagents. The physical adsorption utilizes a combination of van der Waals and hydrophobic forces, hydrogen bonds and ionic interactions to attach the bioreceptors to the surface of the transducer.

Physical methods

Physical methods of immobilization include *entrapment* and *adsorption*. Biorecognition molecules such as enzymes can be entrapped in polyacrylamide, calcium alginate, agarose or chitosan polymer network or gel. Chief disadvantages of this technique are irregular pore size of the gel, lack of mechanical strength and diffusional limitations encountered by substrates and products. Direct physical adsorption of enzymes on a surface is an alternative method. However, immobilization using adsorption (utilizes a combination of van der Waals forces, hydrophobic forces, H-bonds and ionic interactions to attach the biorecognition molecules on transducer or support matrices) alone generally leads to poor long-term stability.

Chemical methods

Chemical methods of enzyme immobilization include covalent binding and crosslinking using multifunctional reagents. **Covalent binding** involves the formation of strong chemical bonds between the enzyme and the support material. This method typically requires functional groups present on both the enzyme and the support surface, allowing for the formation of covalent bonds. **Crosslinking**, on the other hand, involves the formation of bridges or linkages between enzyme molecules or between the enzyme and the support material. *Glutaraldehyde*, a bifunctional aldehyde, is widely used for crosslinking due to its ability to react with amino groups, forming stable Schiff base linkages. Similarly, *cyanuric chloride*, a triazine derivative, can crosslink enzymes by reacting with amino groups to form stable triazine linkages.

Biosensor characteristics

Biosensors exhibit eight key characteristics, each essential for their effective performance:

Sensitivity: This refers to the sensor's responsiveness to changes in analyte concentration per unit.

Selectivity: The sensor's capability to exclusively identify the target analyte, despite potential interference from other chemicals.

Range: The concentration span within which the sensor maintains reliable sensitivity.

Response time: The duration required by the sensor to generate a measurable response.

Reproducibility: The precision with which the sensor consistently delivers its output.

Detection limit: The minimum analyte concentration capable of eliciting a detectable response.

Lifetime: The duration throughout which the sensor can operate without significant degradation in performance.

Stability: Indicates the sensor's ability to maintain its baseline or sensitivity over a defined period.

Enzyme-based glucose biosensor

In 1962, Leland C. Clark (father of biosensor) first developed enzyme-based electrochemical glucose biosensor. The basic operation of glucose biosensor is based on the fact that the enzyme *glucose oxidase* catalyses the oxidation of glucose to δ-gluconolactone, which then hydrolyzes to gluconic acid and hydrogen peroxide.

$$\text{Glucose} + H_2O + O_2 \xrightarrow{\text{Glucose oxidase}} \text{Gluconic acid} + H_2O_2$$

Glucose oxidase is a dimeric protein composed of two identical subunits. Each subunit folds into two domains: one domain binds to substrate β-D-glucose, while the other domain binds to FAD. The catalytic reaction involves reduction of the FAD present in the enzyme into $FADH_2$. $FADH_2$ finally reoxidizes by molecular oxygen to regenerate the FAD.

$$\text{Glucose} + \text{Glucose oxidase (FAD)} \longrightarrow \text{Glucose oxidase } (FADH_2) + \text{Gluconolactone}$$

$$\text{Glucose oxidase } (FADH_2) + O_2 \longrightarrow \text{Glucose oxidase (FAD)} + H_2O_2$$

The enzyme glucose oxidase acts as a bioreceptor molecule and once it binds with the glucose molecule, it oxidizes glucose to gluconic acid and hydrogen peroxide. To measure the glucose in aqueous solutions, three different transducers can be used:

1. An oxygen sensor that measures oxygen concentration.
2. A pH sensor that measures the acid (gluconic acid), a reaction product.
3. A peroxidase sensor that measures H_2O_2 concentration, a reaction product.

Note that an oxygen sensor is a transducer that converts oxygen concentration into electrical current. A pH sensor is a transducer that converts pH change into voltage change. Similarly, a peroxidase sensor is a transducer that converts peroxidase concentration into an electrical current. In the first biosensor (invented by L.C. Clark), biosensor was made from a thin layer of glucose oxidase entrapped on a Clark oxygen electrode using a dialysis membrane. Using this amperometric electrochemical glucose biosensor, the amount of glucose was estimated by the reduction in the dissolved oxygen concentration. The decrease in measured oxygen concentration was proportional to glucose concentration.

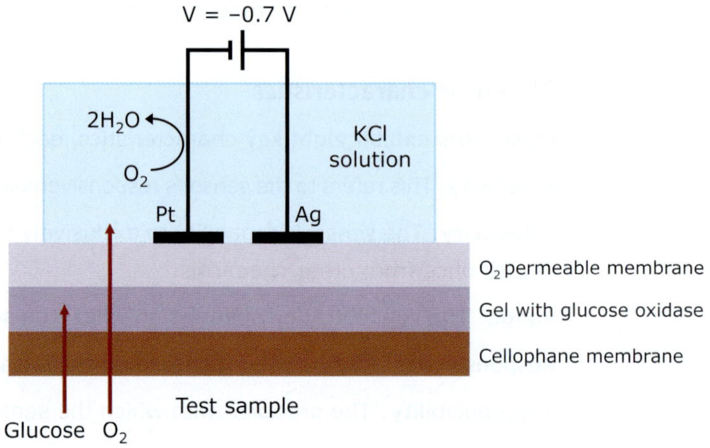

Ag anode: $4\ Ag + 4\ Cl^- \longrightarrow 4\ AgCl + 4\ e^-$

Pt cathode: $O_2 + 4\ H^+ + 4e^- \longrightarrow 2\ H_2O$

Figure 12.46 Clark glucose biosensor: O_2 in test sample diffuses through membrane into KCl electrolyte. O_2 is reduced at the Pt cathode. Four electrons used by each reduced O_2 results in current flow proportional to $[O_2]$. A voltage of −0.7 V is applied between the platinum cathode and the silver anode and this voltage is sufficient to reduce the oxygen. The cell current is proportional to the oxygen concentration and the current is measured (amperometric method of detection has been employed). The concentration of glucose is then proportional to the decrease in current (oxygen concentration).

It is also possible to use glucose dehydrogenase (GDH) instead of glucose oxidase for amperometric biosensing of glucose. However, the construction of glucose biosensors based on GDH requires a source of NAD^+ and a redox mediator to lower the overvoltage for oxidation of the NADH product.

Evolution from first to third generation biosensors

Biosensors are classified into three generations based on their design and functionality:

First-generation biosensors operate by allowing the normal product of the reaction to diffuse to the transducer, thereby causing the electrical response.

Second-generation biosensors utilize specific *'mediators'* between the reaction and the transducer to generate an improved response.

Third-generation biosensors, on the other hand, rely on the reaction itself to cause the response, without direct involvement of product or mediator diffusion.

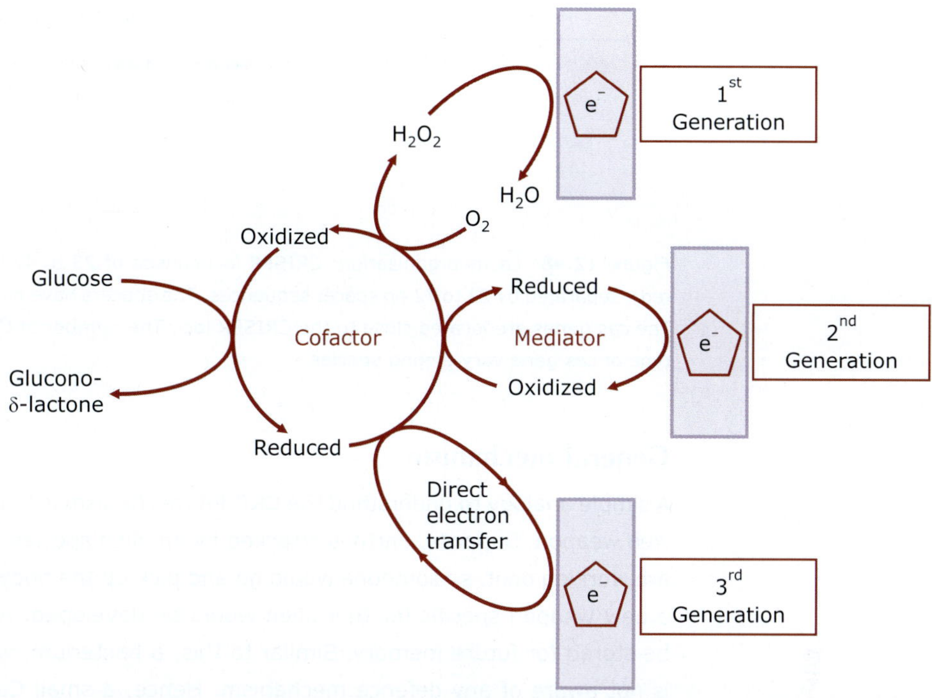

Figure 12.47 The evolution from 1st to 3rd generation electrochemical biosensors. The figure highlights modifications in the biosensor layout with each generation using glucose sensors as an example.

12.15 CRISPR/Cas systems

CRISPR/Cas systems are highly diverse adaptive microbial immune systems used by most archaea (~90%) and many eubacteria (~40%) to protect themselves from invading viruses and plasmids. These systems allow the cell to recognize and distinguish incoming 'foreign' DNA from 'self' DNA. **CRISPR/Cas** systems consist of two general parts: **CRISPRs** (clustered regularly interspaced short palindromic repeats) and **Cas** (CRISPR-associated) proteins.

CRISPRs consist of highly conserved short repeated sequences separated by similarly sized short **spacers sequences**. The size of CRISPR repeats and spacers varies between 23 to 47 bp and 21 to 72 bp, respectively. The bacterial genome may contain more than one CRISPR locus. The CRISPR loci have highly diverse and hypervariable spacer sequences, even between closely related strains. Spacers are unique sequences originating from viral or plasmid DNA. These sequences are used as recognition elements for invading viral or plasmid DNA. By adding new spacers in their own genome, bacteria able to recognize new matching viral or plasmid genomes. Another feature associated with CRISPR loci is the presence of a conserved sequence, called **leader**, located upstream of the CRISPR with respect to the direction of transcription. CRISPR activity requires the presence of a set of CRISPR-associated (**cas**) genes, usually found adjacent to the CRISPR, that code for **Cas proteins** essential to the immune response. The Cas proteins with perform a variety of functions such as nucleases, helicases and polymerases. The CRISPR immune system works through the cooperation of many diverse Cas proteins, resulting in CRISPR-Cas systems currently being grouped into two classes, six types, and over 30 subtypes.

Emmanuelle Charpentier and Jennifer Doudna are awarded the Nobel Prize in Chemistry 2020 for discovering the CRISPR/Cas9 system, a method for genome editing.

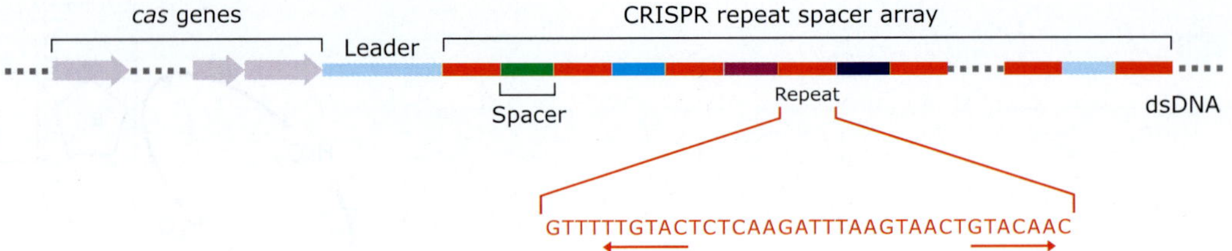

Figure 12.48 Locus organization: CRISPR loci consist of 23 to 47 bp palindromic repeat sequences (in red), separated by 21 to 72 bp spacer sequences. The spacers have no common features in their sequences. The *cas* genes are located close to the CRISPR loci. The number of CRISPR loci, their sequences, and the type of *cas* gene vary among species.

General mechanism

A simple analogy to understand the CRISPR mechanism is to imagine the design of a personalized weapon. Suppose, earth is attacked by an alien species and none of the earthly weapons are working on it, so someone would go and pick up the body parts of alien and based on that a new weapon specific for that alien would be developed. Also, some of the body parts will be stored for future memory. Similar to this, a bacterium, when attacked by bacteriophage, is not aware of any defence mechanism. Hence, a small Cas protein picks the part of DNA of phage and bring it close to the CRISPR loci, where it is integrated into the CRISPR site. A **crRNA** (CRISPR RNA) is synthesized from that loci which along with other proteins then recognize the complementary sequence present in the invading DNA of phage and degrade it. The CRISPR/Cas system works in a similar manner.

The general mechanism of CRISPR works in three distinct phases: **adaptation** (or *acquisition*), **expression and maturation** (or *biogenesis*) and **interference** (or *targeting*).

Adaptation phase

Adaptation is the first step in the process of developing a weapon against foreign invading DNA (viral or plasmid DNA). During adaptation, Cas proteins capture short segments of foreign nucleic acids (termed **protospacers**) and integrate them as 'spacers' into the CRISPR locus in order to record a molecular memory of past invaders. The invading DNAs are directionally integrated, as new CRISPR spacers, into a CRISPR array that is separated by repeat sequences. Some CRISPR/Cas systems employ an alternative mechanism of adaptation — namely, spacer acquisition from RNA, via reverse transcription by a reverse transcriptase encoded at the CRISPR/Cas locus.

Expression and maturation phase

Once the desired sequence (i.e. protospacer) from the invading DNA is integrated into the loci of CRISPR, the CRISPR locus is transcribed into a pre-CRISPR RNA (**pre-crRNA**). CRISPR loci are transcribed from an upstream promoter located in the AT-rich leader sequence. The pre-crRNA is then processed into mature **crRNAs**, each containing a transcribed spacer sequence joined to the partial repeat sequence. The pre-crRNA processing differs in different CRISPR/Cas variants. It may be mediated by a complex composed of multiple Cas proteins or a single multidomain Cas protein or by non-Cas host RNases.

Interference phase

The crRNA forms a complex with Cas proteins. The crRNA of the crRNA-Cas complex makes base pairing with the protospacer of the invading DNA. Finally, crRNA-directed cleavage of invading DNA occurs by Cas proteins at **protospacer**, a site complementary to the crRNA spacer sequence.

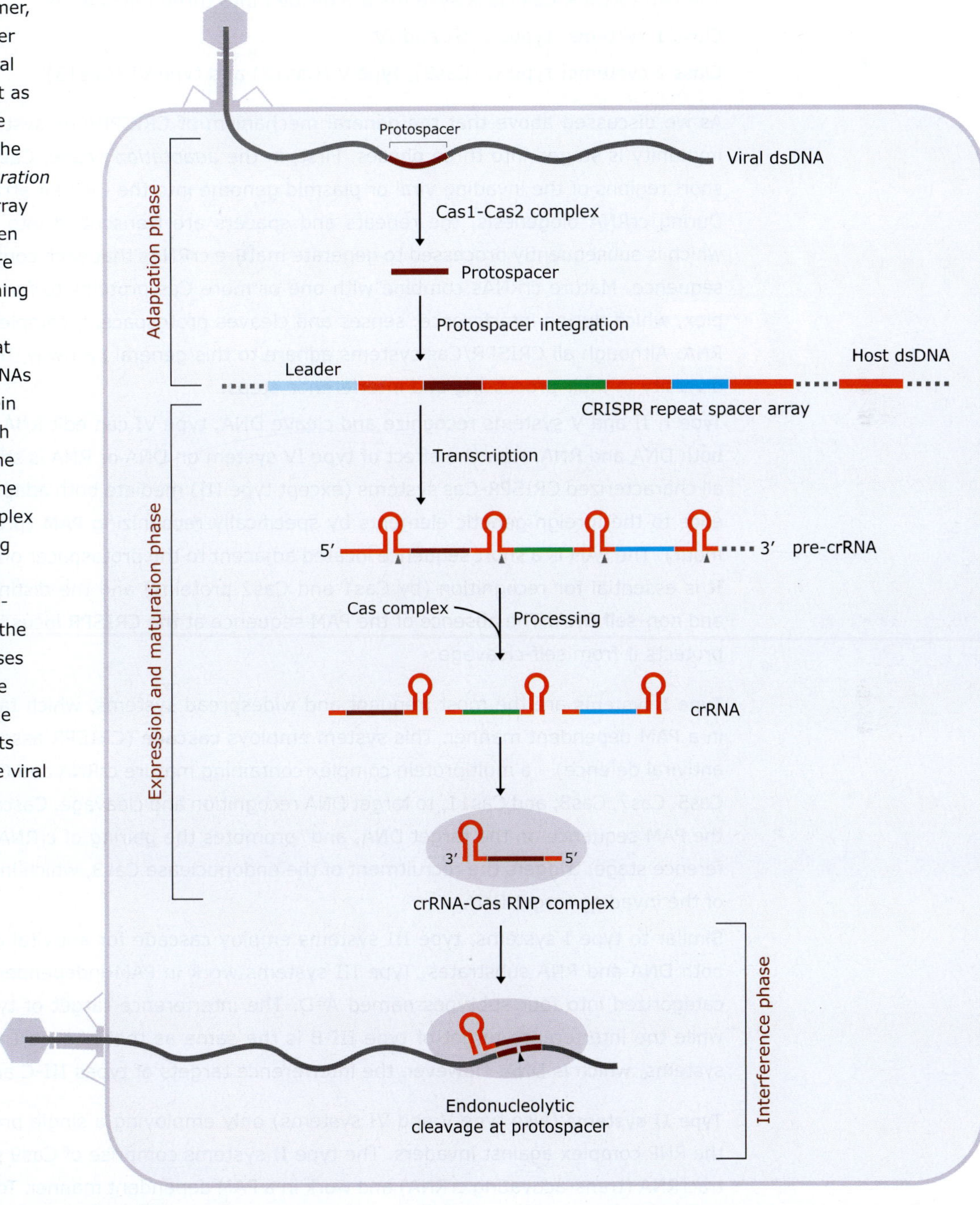

Figure 12.49 The CRISPR/Cas system has three phases. 1. In the *adaptation stage*, the Cas1–Cas2 complex, which comprises two Cas1 dimers and a single Cas2 dimer, acquires a protospacer from the invading viral DNA and integrates it as a new spacer into the CRISPR array. 2. In the *expression and maturation stage*, the CRISPR array is transcribed and then processed into mature crRNAs, each containing a transcribed spacer and part of the repeat sequence. These crRNAs form ribonucleoprotein (RNP) complexes with Cas proteins. 3. In the *interference stage*, the crRNA–Cas RNP complex identifies the invading target DNA through complementary base-pairing. Cleavage of the target sequence causes the destruction of the invading viral genome and therefore protects the bacteria from the viral infection.

Types of CRISPR/Cas systems

The diversity of Cas proteins and the presence of multiple CRISPR loci make classification a complex task. According to the current classification scheme, all CRISPR/Cas systems are divided into two distinct classes, based on the differences between the interference complexes – *class 1 systems* and *class 2 systems*.

The **class 1 systems** involve multi-subunit effector complexes (composed of multiple Cas proteins— Cas3, Cas5–Cas8, Cas10 and Cas11). By contrast, **class 2 systems** involve single subunit effectors (composed of Cas9, Cas12 or Cas13).

The two CRISPR-Cas class systems are divided into three types each,

Class 1 systems: types I, III and IV

Class 2 systems: type II (Cas9), type V (Cas12) and type VI (Cas13)

As we discussed above that the general mechanism of CRISPR/Cas system based adaptive immunity is staged into three phases. First, in the *adaptation phase*, Cas proteins integrate short regions of the invading viral or plasmid genome into the CRISPR array as new spacers. During crRNA biogenesis, the repeats and spacers are transcribed into a long pre-crRNA, which is subsequently processed to generate mature crRNAs that each contain a single spacer sequence. Mature crRNAs combine with one or more Cas proteins to form an effector complex, which during *interference*, senses and cleaves protospacers, complementary to the crRNA. Although all CRISPR/Cas systems adhere to this general pathway, they exhibit striking diversity in their processing and interference steps.

Type I, II and V systems recognize and cleave DNA, type VI can edit RNA, and type III edits both DNA and RNA. How the effect of type IV system on DNA or RNA is still unknown. Nearly all characterized CRISPR-Cas systems (except type III) mediate both adaptation and interference to the foreign genetic elements by specifically recognizing **PAM** (protospacer-adjacent motif). The PAM is a short sequence located adjacent to the protospacer on the invading DNA. It is essential for recognition (by Cas1 and Cas2 proteins) and the distinction between self and non-self DNA. The absence of the PAM sequence at the CRISPR locus in the host genome protects it from self-cleavage.

Type I systems are the most frequent and widespread systems, which target invading DNA in a PAM dependent manner. This system employs **cascade** (CRISPR-associated complex for antiviral defence) – a multiprotein complex containing mature crRNA and Cas proteins– Cas6, Cas5, Cas7, Cas8, and Cas11, to target DNA recognition and cleavage. Cascade first recognizes the PAM sequence on the target DNA, and promotes the pairing of crRNA. During the interference stage, triggers the recruitment of the endonuclease Cas3, which initiates degradation of the invading target DNA.

Similar to type I systems, type III systems employ cascade for antiviral defense and target both DNA and RNA substrates. Type III systems work in PAM-independent manner and are categorized into four subtypes named A–D. The interference target of type III-A is mRNA, while the interference target of type III-B is the same as that of type I and II CRISPR-Cas systems, which is DNA. However, the interference targets of types III-C and D are unclear.

Type II systems (also type V and VI systems) only employing a single protein and crRNA in the RNP complex against invaders. The type II systems comprise of Cas9 protein, crRNA and tracrRNA (trans-activating crRNA) and work in a PAM dependent manner. TracrRNA is a hairpin RNA transcribed from a repeat region. Cas9 binds and stabilizes the tracrRNA-crRNA duplex. tracrRNA is responsible for activating RNase III to promote the pre-crRNA processing. Mature crRNA combines with tracrRNA and Cas9 activates cleavage. Type II systems include three subtypes, II-A, II-B, and II-C.

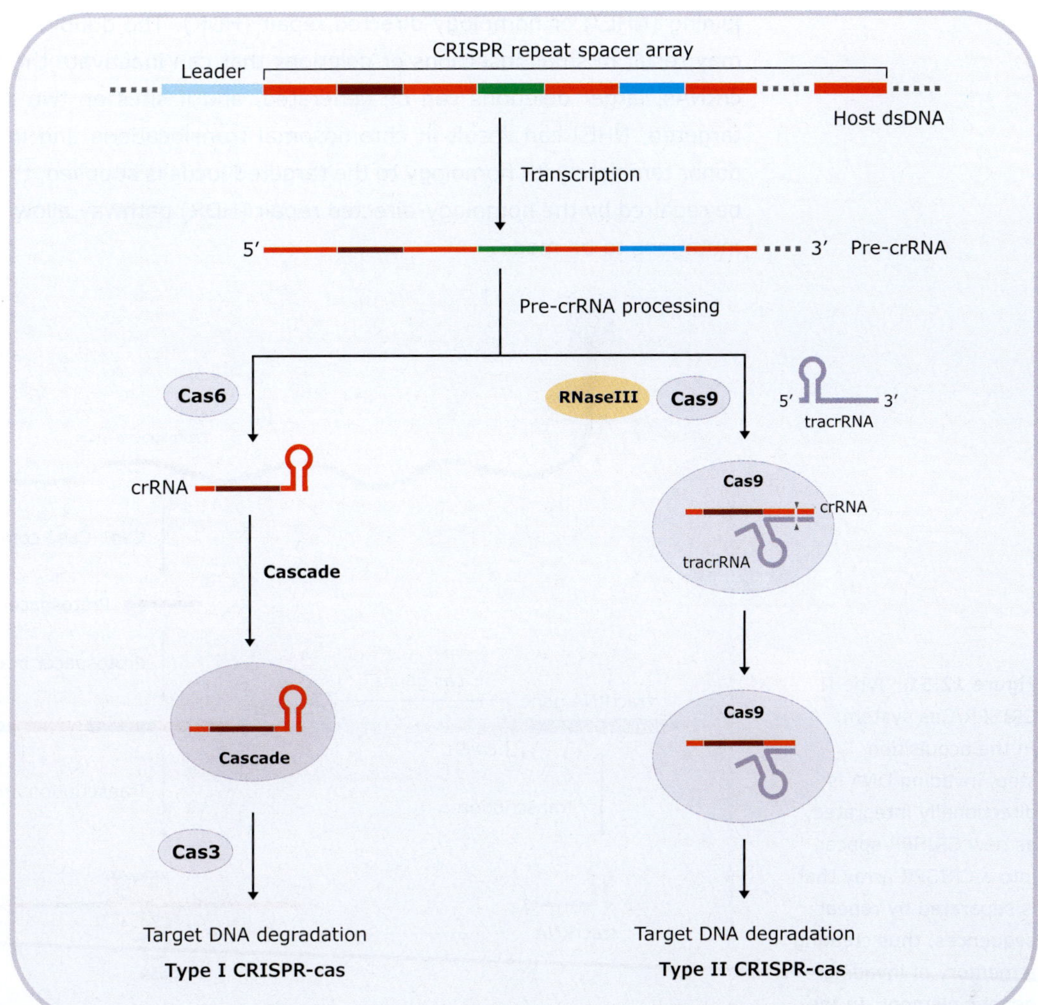

Figure 12.50 Comparison of mode of action of types I and II CRISPR-Cas systems. These two systems differ in their processing and interference steps. During the expression stage, CRISPR loci are transcribed and the resulting pre-crRNA is processed into short crRNAs by cleavage in the repeat sequences. In type I system, pre-crRNA cleavage is carried out by Cas6 endoribonucleases. In type II systems, this process involves the expression of a tracrRNA, which base-pairs with the repeats in the pre-crRNA transcript. The resulting duplexes are cleaved in the repeat sequences by RNase III in a Cas9-dependent reaction. Thus, in both CRISPR–Cas systems, cleavage of the pre-crRNA occurs in the repeat sequences, hence, mature crRNA consists of a spacer flanked by partial repeats. In type II systems, the tracrRNA remains bound to the crRNA and the mature crRNA–tracrRNA duplexes are complexed with Cas9. In type I systems, mature crRNA is bound by a Cas protein complex (i.e. Cascade). During the interference stage crRNAs function as guides for the Cas proteins, as they recognize and bind to complementary sequences in invading genome.

Double-stranded breaks (DSBs) induce endogenous DNA repair mechanisms. Within a cell, double-stranded breaks in DNA can be repaired by non-homologous end joining (**NHEJ**) or homology-directed repair (**HDR**). NHEJ is a process by which the cell can repair strand breaks without the need for a homologous template. NHEJ often leads to insertions or deletions of nucleotides (termed indels) at the site of repair. In contrast, HDR performs DNA repair using homologous DNA as a template.

CRISPR/Cas9 and targeted genome editing

The CRISPR/Cas bacterial immune system is used as a simple, RNA-guided platform for highly efficient and targeted genome editing. CRISPR genome editing technology takes advantage of DNA repair. crRNAs guide Cas9 proteins to target sequences, which are recognized through their complementarity to the crRNA. The Cas9 nuclease then introduces a double-strand break in the target DNA. This break can be repaired through two pathways— non-homologous end-joining (NHEJ) or homology-directed repair (HDR). The double-strand break repair by NHEJ may result in small insertions or deletions that can inactivate the target gene. By using two crRNAs, larger deletions can be generated, and if sites on two different chromosomes are targeted, NHEJ can result in chromosomal translocations and inversion. Alternatively, if a donor template with homology to the targeted locus is supplied, the double strand break may be repaired by the homology-directed repair (HDR) pathway allowing for precise replacement mutations to be made.

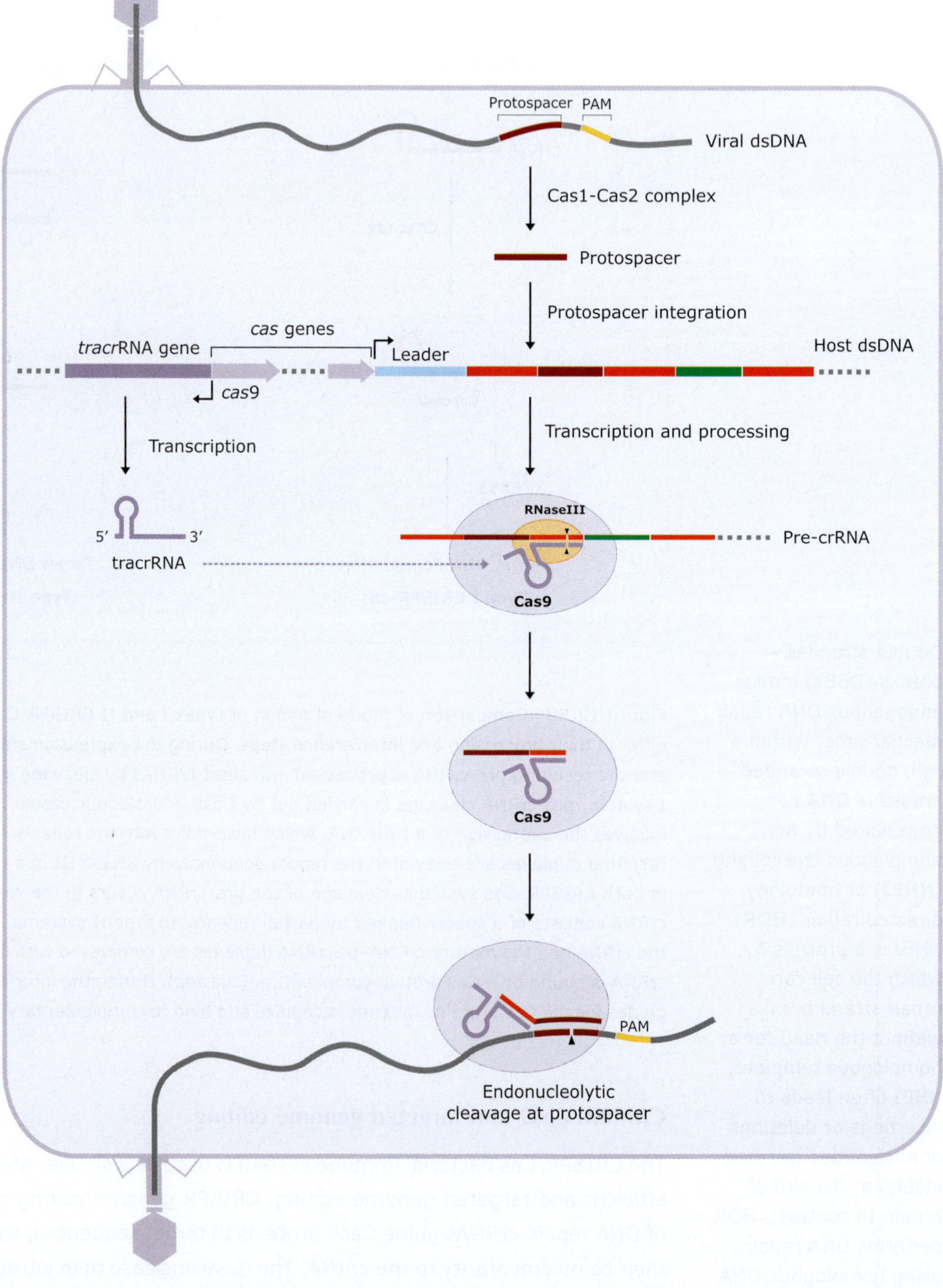

Figure 12.51 Type II CRISPR/Cas system. In the acquisition step, invading DNA is directionally integrated, as new CRISPR spacer, into a CRISPR array that is separated by repeat sequences, thus creating a memory of invading genetic element. In the expression and maturation step, the CRISPR locus is transcribed into a pre-crRNA, which is then processed into mature crRNAs that contain spacer sequences joined to partial repeats. The CRISPR locus also encodes a tracrRNA that has complementarity to the repeat regions of crRNA transcripts. In the interference stage, a crRNA–tracrRNA hybrid is formed through binding of the complementary repeat region sequences, and this RNA hybrid guides the Cas nuclease towards complementary DNA sequences, which leads to the targeting and cleaving of invading genetic elements.

Tools for programmable genome editing

Scientists can probe the function of a gene or open reading frame by mutating or deleting a locus of interest and observing the resulting phenotype. In the past decade, researchers hypothesized that by exploiting endogenous, cellular DNA repair pathways, one could create mutations or precise edits at a desired location in the genome, termed *genome editing*. Double-strand breaks are lethal to cells, thus organisms evolved mechanisms to repair these lesions. Scientists proposed that by generating a targeted, double-strand break at a site of interest, then during the repair process errors may occur, resulting in a mutation at a desired site. Additionally, endogenous double-strand break repair pathways could also stimulate the incorporation of exogenous DNA, creating very specific researcher-designed edits. Thus, nuclease-mediated approaches have been extensively explored for site-specific gene editing.

Meganucleases (or *homing nucleases*), are among the first classes of nucleases that were engineered to target specific genomic sites for gene editing purposes. Meganucleases are a group of nucleases that recognize long nucleotide sequences and induce a double-stranded break at their targeted site. The long recognition sequence of meganucleases may occur only once within a genome, thereby facilitating its use for site-specific genome editing. Meganucleases can be reengineered to target novel sequences through strategies such as protein engineering, structure-based design, and molecular evolution, although the procedure is usually labor-intensive.

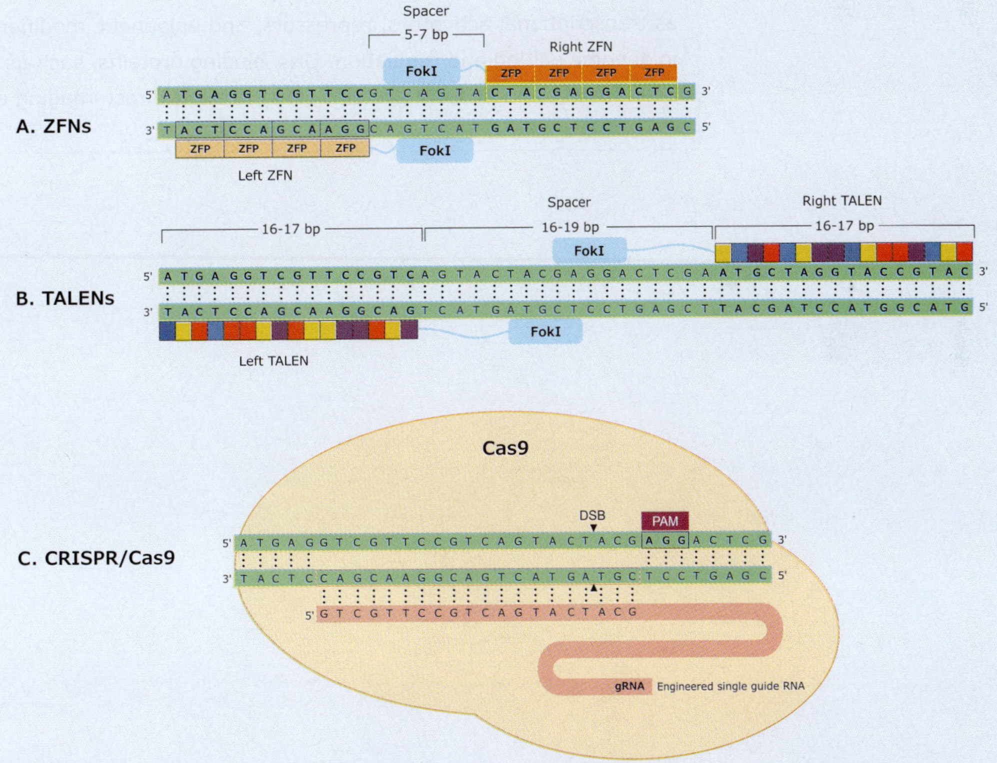

Figure 12.52 Schematic representation of components of ZFNs, TALENs and CRISPR/Cas9 systems. **A**. ZFN enzyme comprise a DNA-binding domain formed by zinc finger modules (ZF), each recognizing a unique 3-base pair sequence on the target DNA, and a DNA-cleaving domain composed of FokI. Two ZFN are designed to recognize DNA sequences that flank the desired cleavage site. In the example, each ZFN comprises four ZF which recognize 12 bp on opposite strands. Upon interaction of ZF with the target site, a FokI dimer catalyzes a targeted double-strand break. **B**. Targeted DNA cleavage is achieved by a pair of TALEN molecules. Each TALEN is comprised by a TALE and a FokI. **C**. CRISPR/Cas9 systems.

Other examples of programmable genome editing machines include *zinc-finger nucleases* (**ZFNs**) and *transcription activator-like effector nucleases* (**TALENs**), in which the DNA-binding domains of transcription factors have been fused with the nuclease domain of the restriction enzyme **FokI**, an obligate dimer. When targeted to paired adjacent sequences, the FokI domains of these programmable, site-specific nucleases form a dimer that activates the nuclease activity, thus creating a double-stranded break near their binding sites. Researchers can exploit the cell's endogenous DNA repair pathways to create mutations at the desired double-stranded break sites. However, because these tools function through protein–DNA interactions, targeting to a new site requires engineering and cloning a new protein, which precludes ZFNs and TALENs from being used for high-throughput applications.

In contrast to most known DNA-binding proteins, **Cas9** is an RNA-guided nuclease whose sequence specificity largely arises from Watson–Crick base pairing between its guide RNA and the target DNA site, in addition to a direct interaction between Cas9 and a short protospacer-adjacent motif (**PAM**) of DNA. Thus, Cas9 can be programmed to target new sites simply by changing its guide RNA sequence, making it an ideal platform for high-throughput sequence-specific gene editing, as well as other applications. Its natural endonuclease activity has been co-opted for sequence-specific editing of the genome in a wide range of organisms, including bacteria, fungi, plants and animals. To enable sequence-specific genomic regulation, nuclease-deactivated Cas9 (dCas9) has been engineered and can be fused to a variety of effectors, such as transcriptional activators, repressors, and epigenetic modifiers. In addition to applications in genome editing and regulation, DNA-binding proteins, such as ZFs, TALEs, and dCas9, have been fused to fluorescent proteins (FPs) to allow direct imaging of genomic loci in living cells.

References

Alberts B, et al (2008), *Molecular Biology of the Cell*, 5th ed. Garland Science.

Bakker E (2004), *Electrochemical sensors*, Analytical Chemistry 2004, 76 (12), 3285–3298.

Berk AJ (1989), Characterization of RNA molecules by S1 nuclease analysis. *Methods in Enzymology*, 180, 334–347

Brown TA (2010), *Gene cloning and DNA analysis*, 6th ed. Wiley-Blackwell.

Carrigan PE, Ballar P, Tuzmen S (2011), Site-directed mutagenesis, *Methods Mol Biol*. 700:107-24.

Divan A and Royds J (2013), *Tools and techniques in biomolecular science*, Oxford University Press.

Doudna J and Charpentier E (2014), The new frontier of genome engineering with CRISPR-Cas9. Science, Vol 346 Issue 6213.

Eggins B (2002) *Chemical sensors and biosensors*, Analytical Techniques in the Sciences. John Wiley & Sons, West Sussex.

Ekins R and Chu FW (1999), Microarrays: their origins and applications. *Trends Biotechnol.* 17(6):217–18

Erlich HA. Gelfand D and Sninsky JJ (1991), Recent advances in the polymerase chain reaction. *Sciences,* 252: 1643-1651.

Fields S and Sternglanz R (1994), The two hybrid system: an assay for protein-protein interactions, *Trends in Genetics*, 10, 286–292.

Fields, S and Song OK (1989), A novel genetic system to detect protein-protein interactions. *Nature,* 340: 245-246.

Freeman WM, Walker SJ and Vrana KE (1999), Quantitative RT-PCR: Pitfalls and potential. *BioTechniques,* 26: 112-125.

Galas DJ and Schmitz A (1978), DNase footprinting: a simple method for the detection of protein-DNA binding specificity. *Nucleic Acids research*, 5, 3157-3170.

Gilmartin PM and Bowler DChris (2002), *Molecular Plant Biology: A Practical Approach*, Volume 2, Oxford university press.

Horvath P and Barrangou R (2010), CRISPR/Cas, the Immune System of Bacteria and Archaea. Science, 327, 167.

Jiang W and Marraffini L (2015), CRISPR-Cas: New Tools for Genetic Manipulations from Bacterial Immunity Systems. Annu. Rev. Microbiol, 69:209–28.

Khudyakov YE and Fields HA (2010), *Artificial DNA: Methods and Applications*, CRC Press.

Leanna CA and Hannink M (1996), The reverse two-hybrid system: a genetic scheme for selection against specific protein/protein interactions. *Nucl. Acids Res,* 24: 3341-3347.

Li Y and Peng N (2019), Endogenous CRISPR-Cas System-Based Genome Editing and Antimicrobials: Review and Prospects, Front. Microbiol.

Liu Z, Dong H, Cui Y, Cong L and Zhang D (2020). Application of different types of CRISPR/Cas-based systems in bacteria, Microbial cell factories volume 19, Article number: 172.

Lodish H, Berk A, Kaiser CA, et al (2008), Molecular Cell Biology, 6th ed. New York: WH Freeman.

Makarova K, Haft D, Barrangou R and Brouns S (2011), Evolution and classification of the CRISPR–Cas systems. Nature Reviews Microbiology 9, 467-477

Makarova KS et al (2019), Evolutionary classification of CRISPR–Cas systems: a burst of class 2 and derived variants. Nature Reviews, Microbiology.

Mardis ER (2008), Next-Generation DNA Sequencing Methods, Annu. Rev. Genomics Hum. Genet, 9:387-402.

Mardis ER (2013), Next-Generation Sequencing Platforms, Annu. Rev. Anal. Chem 6:287-303.

McPherson M and Moller S (2006), *PCR – The basics*, 2nd ed. Taylor and Francis group.

Metzker ML (2010), Sequencing technologies — the next generation, Nature Reviews Genetics 11, 31-46.

Miesfeld RL (1999), *Applied Molecular Genetics*, Wiley-Liss, Inc.

Nicholl ST (2002), *An Introduction to Genetic Engineering*, 3rd ed. Cambridge University Press.

Parra I and Windle B (1993), *High resolution visual mapping of stretched DNA by fluorescent hybridization*. Nature Genetics 5, 17–21.

Puschnik A, Majzoub K, Ooi Y and Carette J (2017), A CRISPR toolbox to study virus–host interactions. Nature Reviews Microbiology 15, 351–364.

Ried T, Schröck E, Ning Y, et al. (1998), Chromosome painting: a useful art. *Hum Mol Genet* 7:1619–1626.

Ronaghi M, Ehleen M and Nyrn P (1998), A sequencing method based on real-time pyrophosphate. *Science*, 281, 363–365.

Schena M, Shalon D, Davis RW and Brown PO (1995), Quantitative monitoring of gene expression patterns with a complementary DNA microarray. *Science,* 270 (5235): 467-70.

Southern EM (2001), DNA microarrays: History and overview. *Methods Mol Biol* 170:1–15

Speicher MR and Carter NP (2005), The new cytogenetics: Blurring the boundaries with molecular biology. Nature Reviews Genetics 6, 782–792.

Stoughton RB (2005), Applications of DNA microarrays in biology. Annu Rev Biochem, 74:53–82.

Strachan T and Read AP (2004), *Human Molecular Genetics*, 3rd ed. Garland Publishing.

VanGuilder HD, Vrana KE and Freeman WM (2008), Twenty-five years of quantitative PCR for gene expression analysis. *Biotechniques*, 44, 619–624.

Wang H, Russa ML and Qi LS (2016), *CRISPR/Cas9 in Genome Editing and Beyond*, Annual Review of Biochemistry, Vol. 85:227-264.

Weissensteiner T, Nolan T, Bustin SA, Griffin HG and Griffin A (2004), *PCR Technology: Current Innovations*, 2nd ed. CRC Press.

Basic laboratory calculations

Mole concept

'**Mole**' is derived from the Latin word, which means '*a mass*.' It is the SI unit for the *amount* of a substance, abbreviated as 'mol'. 1 mole is equal to 6.022×10^{23} particles. Particles may be atoms or molecules, or ions. This number is called **Avogadro's number**. Hence, mole is a number. Just like a dozen means 12 objects, a mole means 6.022×10^{23} objects. For example, a mole of oxygen is equal to 6.022×10^{23} oxygen atoms. Similarly, a mole of glucose is equal to 6.022×10^{23} molecules of glucose. Likewise, a mole of sodium carbonate (Na_2CO_3) is equal to 6.022×10^{23} Na_2CO_3 units. Since Na_2CO_3 contains two sodium ions and one carbonate ion, therefore, a mole of Na_2CO_3 also contains $2 \times 6.022 \times 10^{23}$ sodium ions and $1 \times 6.022 \times 10^{23}$ carbonate ions.

The mass of one mole of atoms of a pure element in grams is numerically equal to the 'atomic weight' of that element in atomic mass units (amu). Atomic weight describes the relative mass of an atom. The term 'atomic weight' is widely used, although it is basically a *mass* rather than a *weight*. This is also called the **molar mass** of the element. For example, the atomic weight of iron is 55.85 amu. This tells us that the molar mass of iron is 55.85 g/mol or that one mole of iron atoms (6.022×10^{23} iron atoms) is 55.85 g of iron.

Example

How many moles of atoms does 136.9 g of iron contain?

Plan

The atomic weight of iron is 55.85 amu. It means the molar mass of iron (mass of one mole of iron atoms) is 55.85 g/mol. Therefore, moles of atoms in 136.9 g of iron = 136.9/55.85 = 2.451 mol

Example

How many moles of O_2 molecules and the number of O_2 molecules are present in 40.0 g of O_2 molecules at 25°C?

Plan

One mole of O_2 molecule contains 6.02×10^{23} O_2 molecules, and its mass is 32.0 g.

Since, 32 g contains 1 mole of O_2 molecule

Number of moles of O_2 present in 40 g of O_2 molecules = $\frac{1}{32}$ × 40 g = 1.25 mole O_2

Since, 1 mole of O_2 molecule contains 6.02 × 10^{23} O_2 molecules

Number of O_2 molecules present in 40 g of O_2 molecules = $\frac{40}{32} \times 6.02 \times 10^{23} = 7.52 \times 10^{23}$

Formula weight

The *formula weight* (actually, it is the *formula mass*) of a molecule is the sum of the atomic weights of each atom in its chemical formula. For example, water (H_2O) has a formula weight of:

2 × (1 amu) + 1 × (16 amu) = 18 amu

Suppose a substance exists as a discrete molecule in which atoms are covalently bonded. In this case, the chemical formula is the molecular formula. The **formula weight** is the molecular weight (actually, it is the *molecular mass*). For example, carbon, hydrogen, and oxygen covalently bind to form a glucose molecule with the chemical and molecular formula of $C_6H_{12}O_6$. The formula weight (FW) and the molecular weight (MW) of glucose are thus:

6 × (12 amu) + 12 × (1 amu) + 6 × (16 amu) = 180 amu

In ionic substances, ions are not covalently bonded and hence, do not exist as discrete molecules. However, they do associate with discrete ratios of ions. Thus, we can describe their formula weights but not their molecular weights.

Example

What is the formula weight of acetic acid (CH_3COOH)?

Plan

We add the atomic weights of the elements in the formula, each multiplied by the number of times the element occurs.

Number and types of atoms	× Mass of one atom	Formula weight	
2 × C	× 12 amu	24 amu of C	
4 × H	× 1 amu	4 amu of H	60.0 amu
2 × O	× 16 amu	32 amu of O	

Expressing concentrations of solutions

A **solution** is a homogeneous mixture of two or more substances. If the solution contains two components only, it is called a **binary solution**. The component in larger quantity is called the **solvent** and the other component is called the **solute**. The solvent may be organic or inorganic in nature. When water acts as a solvent, it is known as an **aqueous solution**.

The general term *concentrations* of solutions are expressed in terms of *either* the amount of solute present in a given mass or volume of *solution* or the amount of solute dissolved in a given mass or volume of *solvent*.

Percent solutions

Many reagents are prepared as a percent of solute dissolved in solution. Percent solutions are always based on 100 mL. Depending on the solute's initial physical state, its concentration can be expressed as a weight to volume percent (% w/v) *or* a volume to volume percent (% v/v).

% w/v: Refers to the weight of solute (in grams) in a total of 100 mL of solution.

% v/v: Refers to the amount of liquid solute (in mL) in a final volume of 100 mL of solution.

Example

Preparing 100 mL of 40% (w/v) polyethylene glycol solution.

Plan

40% of 100 is equal to 40. Dissolve 40 g of polyethylene glycol in distilled water and make the final volume of the solution with the polyethylene glycol is 100 mL.

Example

Preparing 47 mL of a 7% (w/v) solution of sodium chloride, NaCl.

Plan

7% of 47 is equal to 3.29. Dissolve 3.29 g of NaCl in distilled water and make the final volume of the solution 47 mL.

Molarity

Concentrations are commonly measured in *molarity*, with units of mol L^{-1} (= M). The molarity of a solution is defined as *the number of moles of solute dissolved per liter of solution*. When you make a solution of a particular molarity, you need to know three things: the desired molarity, the desired volume and the formula weight (FW) of the compound to be dissolved.

$$\text{Molarity} = \frac{\text{Number of moles of solute}}{\text{Volume of solution (in liter)}} = \frac{\text{Amount of solute / FW of solute}}{\text{Volume of solution (in liter)}}$$

For example, a 1 M solution of NaCl (FW 58.5) contains 58.5 g of NaCl dissolved in a final volume of 1000 mL (1 L) water. A 2 M solution of NaCl has twice that amount (i.e. 117 g) of NaCl dissolved in 1000 mL water.

Example

Preparing 200 mL of 0.3 M NaCl solution.

Plan

The formula weight of NaCl is 58.5 g. Since to prepare 1 M of 1 L of NaCl solution, 58.5 g NaCl is required. Hence, to prepare 0.3 M of 1 L NaCl, 58.5 × 0.3 = 17.55 g NaCl is required. If 200 mL of a 0.3 M NaCl solution is being prepared.

$$\frac{17.55\text{ g}}{1000\text{ mL}} \times 200\text{ mL} = 3.51\text{ g}$$

Therefore, to prepare 200 mL of 0.3 M NaCl solution, 3.51 g of NaCl should be dissolved in distilled water to a final volume of 200 mL.

Example

Preparing 50 mL of 20 millimolar (mM) sodium hydroxide (NaOH) solution.

Plan

The formula weight of NaOH is 40 g.

20 mM is equal to 0.02 M (20×10^{-3})

1 M of NaOH = 40 g.

0.02 M NaOH = 40 × 0.02 = 0.8 g NaOH

Therefore, to prepare 1 L (1000 mL) of 0.02 M NaOH, 0.8 g of NaOH is required.

$$50 \text{ mL of } 0.02 \text{ M} = \frac{0.8 \text{ g}}{1000 \text{ mL}} \times 50 \text{ mL} = 0.04 \text{ g}$$

To prepare 50 mL of 20 mM NaOH, 0.04 g of NaOH should be dissolved in distilled water to a final volume of 50 mL.

Example

You have 2.0 M of HCl solution. How would you calculate the moles in 6.0 mL of solution?

Plan

$$\text{Molarity} = \frac{\text{Number of moles of solute (x)}}{\text{Volume of solution (in liter)}}$$

$$2 \text{ mol/L} = \frac{x}{0.006 \text{ L}}$$

$$x = 0.006 \text{ L} \times 2 = 0.012 \text{ mol}$$

Example

Calculate the molarity of a solution that contains 3.65 grams of HCl in 2.00 liters of solution.

Plan

$$\text{Number of moles of HCl} = \frac{\text{Amount of solute}}{\text{FW of solute}} = \frac{3.65}{36.5} = 0.1$$

$$\text{Molarity of solution} = \frac{\text{Number of moles of solute}}{\text{Volume of solution (in liter)}} = \frac{0.1}{2} = 0.05 \text{ M}$$

Example

What is the molarity of a 0.85% NaCl solution?

Plan

First, define the percentage solution: 0.85% w/v solution = 0.85 g NaCl/100 mL solution.

Now convert g/mL into g/L

$$\frac{0.85 \text{ g}}{100 \text{ mL}} = \frac{X \text{ g}}{1000 \text{ mL}}$$

$$X = \frac{0.85 \times 1000}{100} = 8.50 \text{ g/L}$$

Once the number of grams of solute per liter is known, the molarity can be calculated:

Formula weight of NaCl = 23 + 35.5 = 58.5 g

$$\text{Molarity of solution} = \frac{\text{Amount of solute / FW of solute}}{\text{Volume of solution (in liter)}} = \frac{8.5 \text{ g}/58.5}{1 \text{ L}} = 0.14 \text{ M}$$

Example

Express 2.5 M NaCl as a percent solution.

Plan

The formula weight of NaCl is 58.5 g. The first step in solving this problem is to determine how many grams of NaCl are in a 2.5 M NaCl solution. This can be find out by using the following relationship

$$\frac{58.5 \text{ g}}{1 \text{ M}} = \frac{x \text{ g}}{2.5 \text{ M}}$$

$$x = \frac{58.5 \times 2.5}{1} = 146.25 \text{ g}$$

Therefore, to prepare a 2.5 M solution of NaCl, 146.25 g of NaCl will be dissolved in a total volume of 1 L.

Percent is an expression of concentration in parts per 100. To determine the relationship between the number of grams of NaCl present in a 2.5 M NaCl solution and the equivalent percent concentration, ratios can be set up that state 146.25 g is to 1000 mL as x g is to 100 mL.

$$\frac{146.25}{1000 \text{ mL}} = \frac{x \text{ g}}{100 \text{ mL}}$$

$$x = \frac{146.25 \times 100}{1000} = 14.62 \text{ g}$$

Therefore a 2.5 M NaCl solution contains 14.62 g of NaCl in 100 mL, which is equivalent to 14.62% NaCl solution.

Dilution

It is a process of reducing the concentration of a solute in a solution, usually simply by adding more solvent. Generally, a higher percent solution (more concentrated) is used in laboratories to prepare a lower percent (less concentrated) solution. Hence, a dilution of the concentrated solution is being performed.

When we dilute a solution by mixing it with more solvent, the amount of solute present does not change. But the volume and the concentration of the solution do change. Because the same number of moles of solute is divided by a larger number of liters of solution, the molarity decreases. There are several methods that can be used to calculate the concentration of a diluted solution. Since the total amount of solute is the same before and after dilution, one approach is to use the equation $C_1V_1 = C_2V_2$, where C_1 is the initial concentration of the concentrated solution, V_1 is the initial volume of concentrated solution taken to perform the dilution, C_2 is the concentration of the diluted solution, and V_2 is the final, total volume of the diluted solution.

Example

From a 1 M Tris solution, how we can prepare 400 mL of 0.2 M Tris solution?

Plan

The following equation can be used to solve for V_1

$C_1V_1 = C_2V_2$

$1 \text{ M} \times V_1 = 0.2 \text{ M} \times 400 \text{ mL}$

$V_1 = 80 \text{ mL}$

Example

How is 4 mL of 50 mM NaCl solution prepared from a 2 M NaCl stock solution?

Plan

In this example, a conversion factor must be included in the equation so that molarity (M) can be converted to millimolarity (2 M = 2000 mM).

$C_1V_1 = C_2V_2$

$V_1 \times 2000$ mM = 50 mM × 4 mL

$V_1 = 0.1$ mL

Hence, the addition of 0.1 mL of 2 M NaCl stock solution to 3.9 mL of distilled water is required to produce 4 mL final volume of 50 mM NaCl.

Example

A laboratory procedure calls for 250 mL of an approximately 0.10 M solution of NH_3. Describe how you would prepare this solution using a stock solution of concentrated NH_3 (14.8 M).

Plan

$C_1V_1 = C_2V_2$

$V_1 \times 14.8$ M = 0.10 M × 250 mL

$V_1 = 1.7$ mL

Example

How many milliliters of 18.0 M H_2SO_4 are required to prepare 1 L of a 0.90 M solution of H_2SO_4?

Plan

The volume (1 L) and molarity (0.90 M) of the final solution, as well as the molarity (18 M) of the original solution, are given. Therefore, the relationship $V_1C_1 = V_2C_2$ can be used.

$V_1 \times 18$ M = 1 L × 0.90 M

$V_1 = 50$ mL

Example

If water is added to 175 mL of a 0.45 M KOH solution until the volume is 250 mL, what will the molarity of the diluted solution be?

Plan

$C_1V_1 = C_2V_2$

0.45 M × 175 mL = $C_2 \times 250$ mL

$C_2 = 0.32$ M

Example

How much 0.075 M NaCl solution can be made by diluting 450 mL of 9.0 M NaCl?

Plan

$C_1V_1 = C_2V_2$

9.0 M × 450 mL = 0.075 M × V_2

$V_2 = 54$ L

Units, Conversion factors and Physical constants

Base units

Length	meter (m)
Mass	kilogram (kg)
Time	second (s)
Electric current	ampere (A)
Temperature	kelvin (K)
Amount of substance	mole (mol)
Luminous intensity	candela (cd)

Conversion factors

Multiplication factor	*Prefix*
10^{12}	tera (T)
10^{9}	giga (G)
10^{6}	mega (M)
10^{3}	kilo (k)
10^{2}	hecto (h)
10	deka (da)
10^{-1}	deci (d)
10^{-2}	centi (c)
10^{-3}	milli (m)
10^{-6}	micro (μ)
10^{-9}	nano (n)
10^{-12}	pico (p)
10^{-15}	femto (f)

Physical constants

Ideal gas constant (R) = 1.987 cal/mole K = 8.314 $JK^{-1}.mol^{-1}$

Planck's constant (h) = 6.62×10^{-27} erg-sec = 6.62×10^{-34} J.sec

Avogadro's number = 6.022×10^{23} mol^{-1}

Density of water = 1 g/cm^3

Electron rest mass = 9.11×10^{-31} kg

Proton rest mass = 1.67×10^{-27} kg

Speed of light (c) = 3.00×10^{8} m/sec

Gravitational constant (G) = 6.67×10^{-11} Nm^2/kg^2

Acceleration due to gravity (g) = 9.8 m/s^2

Self Test

Question 1

A biochemist is attempting to separate a DNA binding proteins (X) in a solution. Only three other proteins (A, B and C) are present. The proteins have the following properties:

	Isoelectric point	Size	Bind to DNA
Protein A	7.4	82,000	Yes
Protein B	3.8	21,500	Yes
Protein C	7.9	23,000	No
Protein X	7.8	22,000	yes

What type of protein separation techniques might she use to separate?

P. Protein X can be separated from protein A by size-exclusion chromatography.

Q. Protein X can be separated from protein B by ion-exchange chromatography.

R. Protein X can be separated from protein A by isoelectric focusing.

S. Protein X can be separated from protein C by affinity chromatography.

a. P and Q
b. Q and S
c. P, Q and S
d. R and S

Question 2

A mixture of amino acids can be analyzed by first separating the mixture into its components through ion exchange chromatography. Amino acids placed on a cation exchange resin containing sulfonate groups flow down the column at different rates because of two factors that influence their movement:

1. Ionic attraction between the sulfonate residues on the column and positively charged functional groups on the amino acids.
2. Hydrophobic interactions between amino acid side chains and the strongly hydrophobic backbone of the polystyrene resin. For each pair of amino acids listed, determine which will be eluted first from the cation-exchange column by a pH 7 buffer.

P. Asp and Lys

Q. Glu and Val

R. Gly and Leu

S. Ser and Ala

a. P–Asp, Q–Glu, R–Gly, S–Ser
b. P–Asp, Q–Val, R–Leu, S–Ala
c. P–Lys, Q–Glu, R–Leu, S–Ser
d. P–Lys, Q–Val, R–Gly, S–Ala

Question 3

A protein has a molecular mass of 400 kDa when measured by gel filtration chromatography. When subjected to gel electrophoresis in the presence of SDS, the protein gives three bands with molecular masses of 180, 160 and 60 kDa. When electrophoresis is carried out in the presence of SDS and β-mercaptoethanol, three bands are again formed, this time with molecular masses of 160, 90 and 60 kDa. What is the subunit composition of the protein?

P. The protein has four subunits – 160, 90, 90 and 60 kDa.

Q. Two 90 kDa subunits are linked by disulfide bonds.

R. The protein has three subunits – 180, 160 and 60 kDa.

S. Two 60 kDa subunits are linked by disulfide bonds.

a. P and Q

b. P and S

c. Q and R

d. R and S

Question 4

A mixture of glutamic acid and lysine is separated by paper electrophoresis in two experiments, one at pH 1 and other at pH 12. After the separation, the papers are treated with ninhydrin to reveal the location of the amino acids (colored spots). The results are shown below. Which of the following represent the *correct* result?

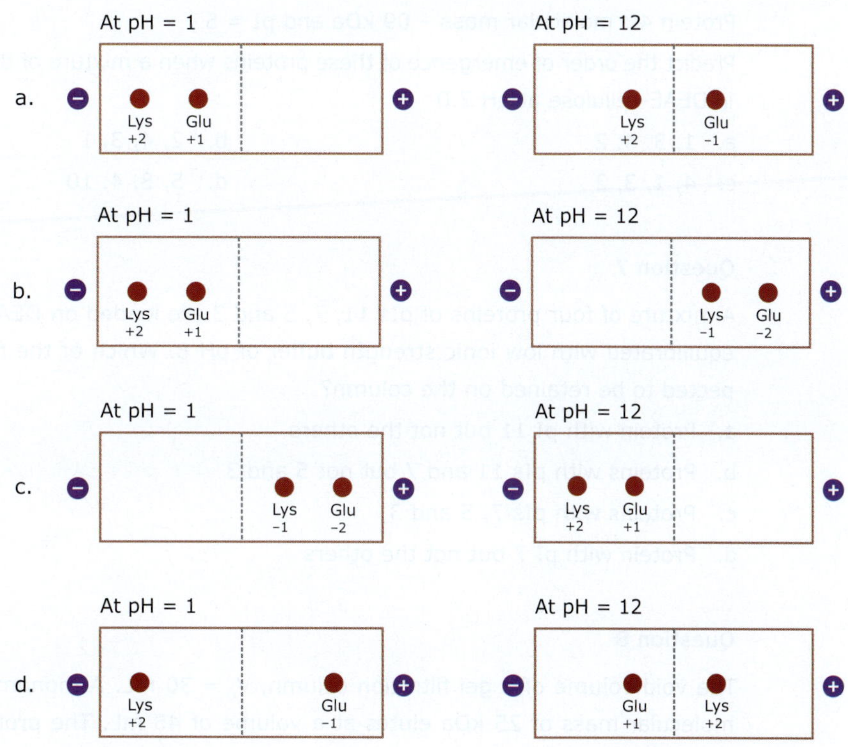

Question 5

From the data given below, identify the protein pair that would (P) give the least mobility band on a SDS polyacrylamide gel electrophoresis experiment and (Q) elute last on an anion exchange e.g. DEAE column respectively:

Protein	pI	Subunit MW	Native MW
A	9	10,000	20,000
B	8	35,000	35,000
C	6	15,000	90,000
D	5	20,000	80,000
E	3	30,000	30,000

a. P-protein C ; Q-protein A

b. P-protein A ; Q-protein E

c. P-protein B ; Q-protein E

d. P-protein B ; Q-protein A

Question 6

You have a mixture of proteins with the following properties:

Protein 1 : molecular mass - 12 kDa and pI = 10

Protein 2 : molecular mass - 62 kDa and pI = 4

Protein 3 : molecular mass - 28 kDa and pI = 8

Protein 4 : molecular mass - 09 kDa and pI = 5

Predict the order of emergence of these proteins when a mixture of the four is chromatographed in DEAE-cellulose at pH 7.0

a. 1, 3, 4, 2

b. 2, 4, 3, 1

c. 4, 1, 3, 2

d. 5, 8, 4, 10

Question 7

A mixture of four proteins of pIs 11, 7, 5 and 3 are loaded on DEAE anion-exchange column equilibrated with low ionic strength buffer of pH 8. Which of the four proteins would be expected to be retained on the column?

a. Protein with pI 11 but not the others

b. Proteins with pIs 11 and 7 but not 5 and 3

c. Proteins with pIs 7, 5 and 3

d. Protein with pI 7 but not the others

Question 8

The void volume of a gel filtration column, V_0 = 30 mL. A monomeric protein with a known molecular mass of 25 kDa elutes at a volume of 45 mL. The protein that you are trying to purify elutes at a volume of 35 mL. Which of the following is a valid conclusion?

a. Your protein is repelled by the gel filtration material

b. Your protein molecular mass = 25 kDa

c. Your protein molecular mass > 25 kDa

d. Your protein molecular mass < 25 kDa

Question 9

A mixture containing proteins-1, -2, -3, -4 and -5 with molecular weights 5,000, 10,000, 25,000, 65,000 and 100,000, respectively, were separated as a Sephadex G-50 column. The order of elution of these proteins from the column will be

a. protein-1, protein-2, protein-3, protein-4 and protein-5

b. protein-5, protein-4, protein-3, protein-2 and protein-1

c. protein-1,-2, and -3 elute first, followed by protein-5 and -4

d. protein-4 and -5 elute first, followed by protein-3, -2 and -1

Question 10

Some students used paper chromatography to separate the pigments in purple ink. They set up a chromatogram and after 15 minutes the colors had separated as shown in the diagram.

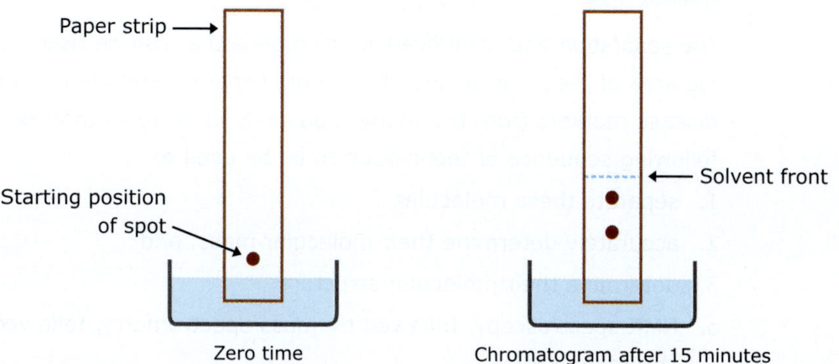

Which one of the following diagrams is most likely to indicate the appearance of the chromatogram after a further 30 minutes?

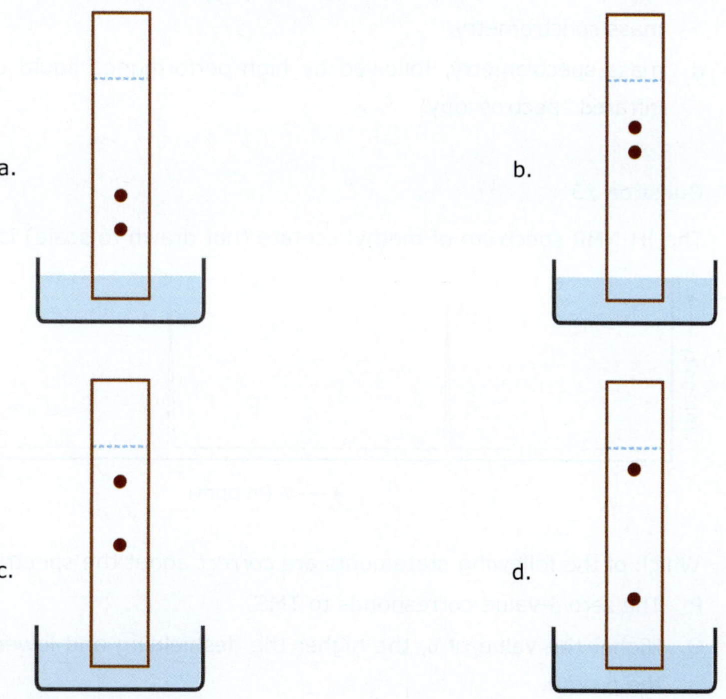

Question 11

You are given a mixture that contains glutamic acid (pI = 3.2), arginine (pI = 10.8) and valine (pI = 6.0) and you subject the mixture to electrophoresis.

P. Arginine migrates toward the cathode when the electrophoresis is carried out at a pH of 7.1.

Q. Valine and glutamic acid migrate toward the anode when the electrophoresis is carried out at a pH of 7.1.

R. Glutamic acid migrates farthest towards the anode at a pH of 7.1.

S. Since amino acids themselves are colorless, ninhydrin is used to locate the amino acids which forms a purple derivative in the presence of an amino acid.

a. Q and S

b. Q, R and S

c. P, Q and R

d. P, Q, R and S

Question 12

The separation and identification of proteins that can be used as disease markers is an exciting area of research. Researchers must separate and identify proteins that could be used as disease markers from the many thousands of proteins that exist in our bodies. Which of the following sequence of techniques could be used to

1. separate these molecules
2. accurately determine their molecular mass, and
3. determine their molecular structure.

a. NMR spectroscopy, followed by mass spectrometry, followed by high-performance liquid chromatography.

b. high-performance liquid chromatography, followed by mass spectrometry, followed by NMR spectroscopy.

c. high-performance liquid chromatography, followed by infrared spectroscopy, followed by mass spectrometry.

d. mass spectrometry, followed by high-performance liquid chromatography, followed by infrared spectroscopy.

Question 13

The ^{1}H-NMR spectrum of methyl acetate (not drawn to scale) is given below.

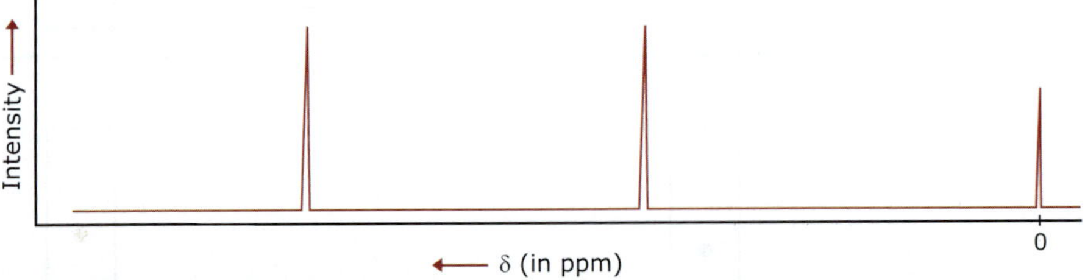

Which of the following statements are *correct* about the spectrum?

P. The zero δ-value corresponds to TMS.

Q. Higher the value of δ, the higher the deshielding and lower the electron density around the proton.

R. Number of peaks indicating the number of chemically different protons present in the molecule.

S. The two signals are of equal intensity because the number of protons responsible for them is equal.

a. P and Q
b. Q and S
c. Q, R and S
d. P, Q, R and S

Question 14

The absorption spectrum of a chromophore is primarily determined by the chemical structure of the molecule. However, the polarity of the solvent also affects the electronic spectra. Which of the following statements are *correct* about the polarity effect?

P. The π–π^* transition shows *red* shift in more polar solvents.

Q. The n–π^* transition shows a *red* shift in more polar solvents.

R. The n–π^* transition occurs at a shorter wavelength in polar solvents than in nonpolar solvents.

S. Solvent effect for n–π^* transitions is opposite to that found for π–π^* transitions. Polar solvents cause a shift to lower wavelengths relative to non-polar solvents.

a. Q and S
b. P and R
c. P, R and S
d. P, Q, R and S

Question 15

Solutions containing tryptophan and tyrosine have characteristic absorption spectra. The molar absorption coefficients at 240 nm and 280 nm are the following:

Wavelength (nm)	ε_{Tyr} (M^{-1} cm^{-1})	ε_{Trp} (M^{-1} cm^{-1})
240	11,300	1960
280	1,500	5380

A 10 mg sample of a protein is hydrolyzed to its constituent amino acids and diluted to 100 ml. The absorbance of this solution, using a 1 cm light path, is 0.717 at 240 nm and 0.239 at 280 nm. Estimate the content of tryptophan and tyrosine using units of µmol/g protein.

a. 322 and 586 µmol/g
b. 281 and 586 µmol/g
c. 586 and 686 µmol/g
d. 181 and 506 µmol/g

Question 16

A solution containing two substances, A and B, has an absorbance in a 1 cm cuvette of 0.36 at 350 nm and 0.225 at 400 nm. The molar absorption coefficients ($M^{-1} \times cm^{-1}$) of A and B at the two wavelengths are given below. The concentration of B is 3×10^{-5} M. Calculate the concentrations of A in the solution.

Compound	ε at 350 nm	ε at 400 nm
A	15,000	300
B	7,000	6500

a. 3×10^{-5} M
b. 3×10^{-3} M
c. 1×10^{-5} M
d. 1×10^{-3} M

Question 17

A protein contains 2 Trp and 4 Tyr residues. The molecular mass of the protein is 17000 Da and that of Trp and Tyr are 204 and 180 Da, respectively. Values of $E^{1\%}_{1cm}$ the absorption co-efficient of 1% (g/v) solution of Trp and Tyr in 1 cm cell at 280 nm, are 269.60 and 83.33, respectively. The absorption of 1 mg/ml protein solution in 1 cm cell at 280 nm will be

a. 0.1
b. 1.0
c. 0.7
d. 1.7

Question 18

The product of an enzyme reaction is estimated in a colorimeter. For a 1 cm pathlength cuvette, 20% of the incident light intensity is absorbed. If the path length is increased to 2 cm, the percentage of light transmitted is

a. 36
b. 40
c. 64
d. 90

Question 19

IR spectroscopy is one type of vibrational spectroscopy. It measures the vibrational frequency of bonds. Which of the following statements concerning vibrational frequencies are *correct*?

1. It is unique for each type of bond.
2. Stretching frequencies are lower than corresponding bending frequencies.
3. Bonds to lighter atoms usually vibrate at higher frequencies than heavier atoms.
4. The frequency of the stretching vibration of a bond depends on the masses of the atoms and the stiffness of the bond.
5. The C—O bond has a lower frequency than the C—N bond because an O-atom has more mass than N-atom.

a. 1, 4 and 5
b. 2, 3 and 4
c. 1, 3, 4 and 5
d. 2, 3, 4 and 5

Question 20

Separation of small peptides, electrophoresis on a paper support is effected on the basis of the charge on a peptide at different pH values. Predict the direction of migration for the peptide (Lys-Gly-Ala-Glu) at the given pH values. 'C' is mentioned for migration toward the cathode, A for migration toward the anode and 'O' if the peptide remains stationary.

	pH			
	2.0	4.0	6.0	11.0
a. Lys-Gly-Ala-Glu	C	C	O	A
b. Lys-Gly-Ala-Glu	A	C	O	A
c. Lys-Gly-Ala-Glu	C	A	A	A
d. Lys-Gly-Ala-Glu	C	C	A	A

Question 21

The most important property of compound light microscope is its power of resolution, which is numerically equivalent to D, the minimum distance between two distinguishable objects. D depends on three parameters namely, the angular aperture, α, the refractive index, N, and wavelength, λ, of the incident light. Below are given few possible options to increase the resolution of the microscope.

P. Decrease the value of λ or increase either N or α to improve resolution.

Q. Moving the objective lens closer to the specimen will decrease $\sin\alpha$ and improve the resolution.

R. Using a medium with high refractive index between the specimen and the objective lens to improve the resolution.

S. Increase the wavelength of the incident light to improve the resolution.

Which of the following combination of above statements is *correct*?

a. P and R
b. Q and R
c. P and S
d. R and S

Question 22

Using FRAP (fluorescence recovery after photo-bleaching) techniques, diffusion coefficient of three integral membrane proteins M_1, M_2 and M_3 in a kidney cell is calculated as 1 μm/s, 0.05 μm/s and 0.005 μm/s, respectively. Considering fluid-mosaic nature of biological membrane and relationship of structural organization of integral membrane protein with diffusion coefficient, which protein(s) will have highest number of integral membrane domain?

a. M_2 and M_3
b. M_2 only
c. M_3 only
d. M_1 and M_3

Question 23

The DNA mixture containing ^{14}N and ^{15}N labelled DNA from phage T4 was denatured and allowed to re-anneal. How many bands one would observe upon $CsCl_2$ density gradient centrifugation of the above mixture?

a. 0
b. 1
c. 2
d. 3

Question 24

A DNA sequencing reaction was performed with the fragment 5′-XXXXGCGATCGYYYY-3′ as the template, dideoxy GTP, all the four dNTPs, and the required primers and enzyme. XXXX and YYYY in the given DNA fragment represent primer binding sites. The set of fragments obtained during the reaction will be (the primers are not shown in the amplified fragments)

a. 5′-CGATCGC-3′ only
b. 5′-CG-3′, 5′-CGCTAG-3′, 5′-CGCTAGC-3′
c. 5′-CG-3′, 5′-CGATCG-3′, 5′-CGATCGC-3′
d. 5′-G-3′, 5′-GCG-3′, 5′-GCGATCG-3′

Question 25

The following gel pattern was obtained in an attempt to sequence a DNA molecule using the Sanger method. What is the sequence of the original DNA molecule?

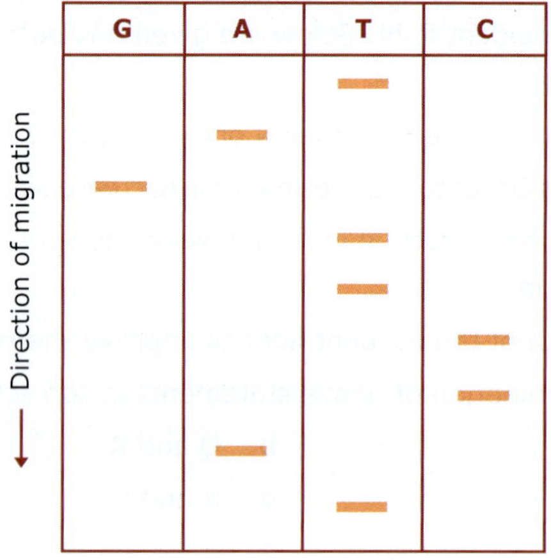

a. 5′ ATCAAGGTA 3′

b. 3′ ATCAAGGTA 5′

c. 5′ ATGGAACTA 3′

d. 5′ TACCTTGAT 3′

Question 26

A restriction fragment, obtained with a type II endonuclease, was subjected to Maxam-Gilbert sequencing with results as shown in the autoradiogram below. Find out the sequence of DNA fragment.

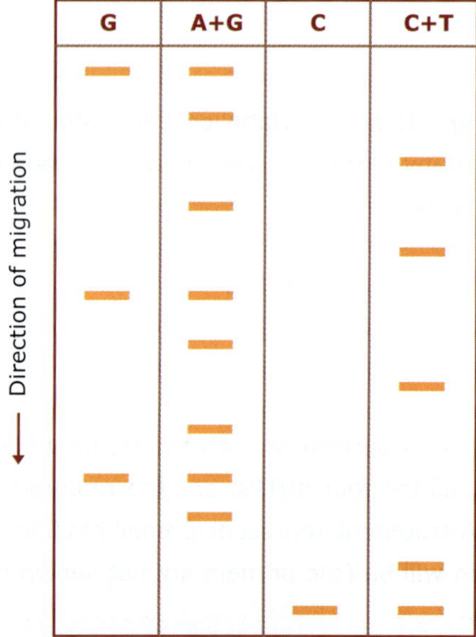

a. 5′ GATATGATAGATC 3′

b. 5′ CTAGATAGTATAG 3′

c. 5′ GGTGTGGTGGGTC 3′

d. 5′ CTATACTATCTAG 3′

Question 27

Given below are the experimental protocols to find out the exact location of repetitive DNA sequence in mitotic chromosome by FISH (fluorescence in situ hybridization). Which one of the protocols will give the correct result?

a. Mitotic chromosomes were fixed on glass slide → incubated with biotinylated telomeric DNA → denatured → incubated with fluorescently labeled.
b. Mitotic chromosomes were fixed on glass slide → denatured → incubated with FITC labeled unrelated non-repetitive DNA sequence → counterstained with propidium iodide → localization observed under fluorescence microscope.
c. Mitotic chromosomes were fixed on glass slide → denatured → incubated with biotinylated satellite DNA → localization observed under fluorescence microscope.
d. Mitotic chromosomes were fixed on glass slide → denatured → incubated with repetitive DNA sequence binding protein → FITC labeled antibody against the protein → localization observed under fluorescence microscope.

Question 28

An investigator discovers a new receptor for a known ligand and wanted to identify the binding partner of the receptor i.e. its co-receptor. The antireceptor antibody is not available but anti GFP-antibody is available. Which one of the following strategies is most likely to identify the co-receptor?

a. The GFP-receptor fusion protein is expressed in a cell line and analyzed by LC-MS/MS.
b. The GFP-receptor fusion protein is expressed in a cell line and the cells positive for GFP were sorted out, lysed and run on a polyacrylamide gel.
c. The GFP-receptor protein is coated on ELISA plate, followed by ELISA with anti-GFP antibody.
d. The receptor is cloned as a fusion protein of GFP and expressed in stimulated cells. The immunoprecipitated complex obtained by anti-GFP antibody was analyzed by LC-MS/MS.

Question 29

Five *E. coli* strains have been identified, each of which has a different mutation that disrupts the normal regulation of a particular operon. For each mutant strain, the mutation has been mapped to the promoter or the operator region; however, the exact sequence changes are not known for these mutations. It is known that the normal promoter/operator consists of a single binding site for a positively acting transcription factor located just upstream of the promoter itself. Short DNA fragments containing the promoter and the operator were subcloned from each of the five mutant strains and from the wild type, purified, and radiolabeled. These fragments were then incubated under conditions of DNA excess with either purified regulatory factor or RNA polymerase or with both polymerase and regulatory factor.

The resulting protein-DNA complexes were separated by electrophoresis, and the radioactive DNA fragments were detected by exposure to X-ray film, giving the results shown below. Electrophoresis is from top to bottom; the largest complexes run slowest.

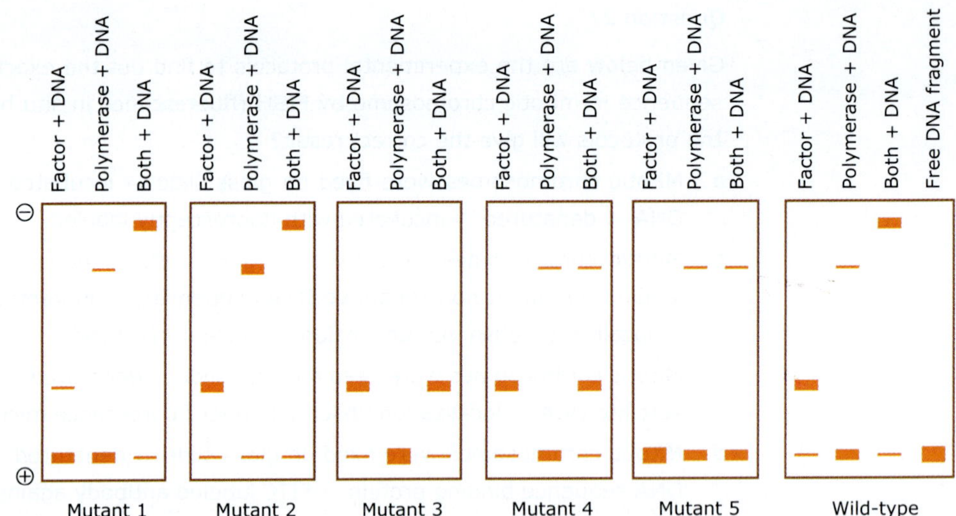

A. One of the mutations increases the affinity of the polymerase for the promoter. Transcription of the operon is not stimulated by the regulatory factor in this mutant. Which mutant is most likely to show this effect?

a. Mutant 1
b. Mutant 2
c. Mutant 3
d. Mutant 4

B. One of the mutations maps to the operator. Transcription of the operon is not stimulated by the regulatory factor in this mutant. Which mutant is most likely to show this effect?

a. Mutant 1
b. Mutant 2
c. Mutant 3
d. Mutant 5

Question 30

Four different receptors viz, A, B, C and D bind to the same ligand X. In order to determine which receptor has the highest ligand binding affinity following experiment was carried out. Cells were transfected with green fluorescence protein (GFP)-tagged receptors (A, B, C and D) individually, then incubated with red fluorescence protein (RFP) tagged –'X' and subjected to FACS analysis. Following are the results

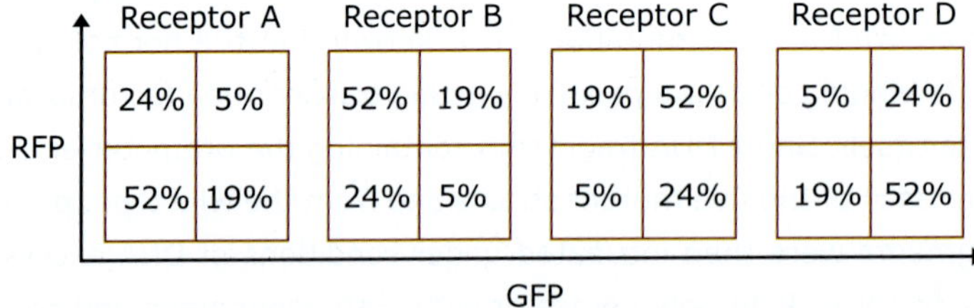

Which receptor has the highest ligand binding affinity?

a. A
b. B
c. C
d. D

Question 31

A small fraction of clear cellular lysate was run on an isoelectric focusing gel (IEF) to purify a particular protein, which showed a number of sharp bands corresponding to different pI values. The protein of interest has a pI of 5.2. Therefore, the band corresponding to pI 5.2 was cut, eluted with appropriate buffer and subjected to SDS-PAGE, which showed 3 distinct bands. Which one of the following inferences cannot be drawn from the above observations?

a. Several different proteins having same pI may be present at the single band on IEF gel.

b. SDS-PAGE showed 3 distinct bands which may represent molecular mass of different proteins.

c. The protein of interest may be composed of 3 subunits.

d. SDS-PAGE showed 3 distinct bands due to differences in net charge of different proteins.

Question 32

Optical density of a 400 base pair long 1 ml DNA solution was found to be 0·052. How many DNA molecules are present in the solution? [1 base pair = 650 dalton, optical density of 1·0 D corresponds to 50 μg DNA/ml]

a. $6{\cdot}023 \times 10^{12}$

b. $6{\cdot}023 \times 10^{18}$

c. $4{\cdot}633 \times 10^{18}$

d. $5{\cdot}2 \times 10^{13}$

Question 33

Poly-L-lysine exists in pure α-helix, β-sheet and random coiled conformations depending upon the solvent conditions. The values of mean residue ellipticity at 220 nm ($[\theta]_{220}$) are −35,700, −13,800 and +3,900 deg cm^2dmol^{-1} for α-helix, β-sheet and random coil conformations of this polypeptide, respectively. The polypeptide exists in α-helix conformation at pH 10.8 and 25°C. Addition of urea leads to a two state transition between α-helix and random coil conformation. It has been observed that $[\theta]_{220}$ of the polypeptide is −14800 deg cm^2dmol^{-1} in the presence of 6M urea. The percentage of the polypeptide in α-helix conformation is:

a. 37

b. 41

c. 47

d. 50

Question 34

Below is a theoretical size exclusion chromatography profile showing the elution of 4 monomeric, globular proteins with molecular masses 38, 65, 27 and 49 kDa.

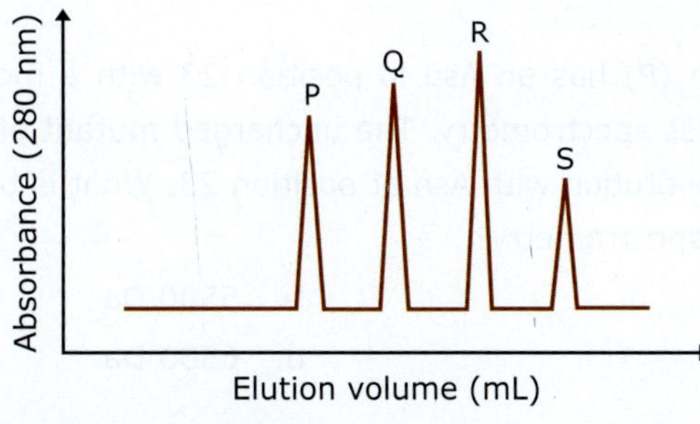

Identify the mass of the protein species responsible for each peak.

a. P-65, Q-49, R-38, S-27

b. P-65, Q-49, R-27, S-38

c. P-49, Q-38, R-65 S-27

d. P-27, Q-49, R-65, S-38

Question 35

A researcher is studying the subcellular localization of a particular protein 'X' in an animal cell. The researcher performs successive centrifugation at increasing rotor speed. The researcher starts spinning the cellular homogenate at 600g for 10 min, collects the pellet, spins the supernatant at 10,000g for 20 min, collects the pellet, spins the supernatant at 100,000g for 1 hour, collects both the pellet and the final supernatant. On subjecting various pellets and the final supernatant to Western blotting with anti-protein-X antibody, the protein X is observed to be maximally expressed in pellet after centrifugation at 10,000g. Based on the above observation, what will be the most likely localization of protein X.

a. Nucleus
b. Ribosomes
c. Mitochondria
d. Microsomes

Question 36

A mixture of two proteins was subjected to following three chromatographic columns:

1. Cation exchange,
2. Size exclusion (Sephadex 100) and
3. Reverse phase.

Following elution profiles were obtained.

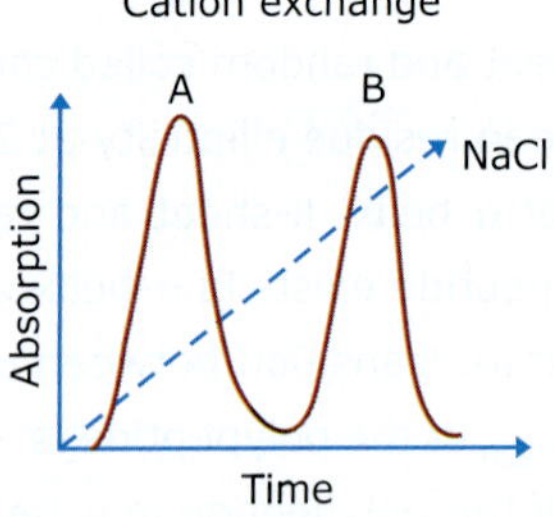

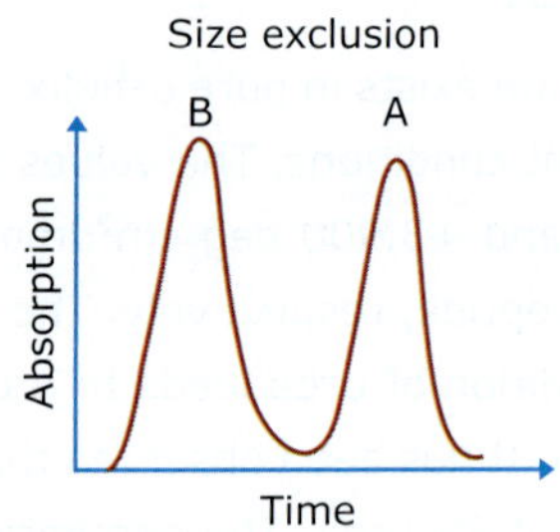

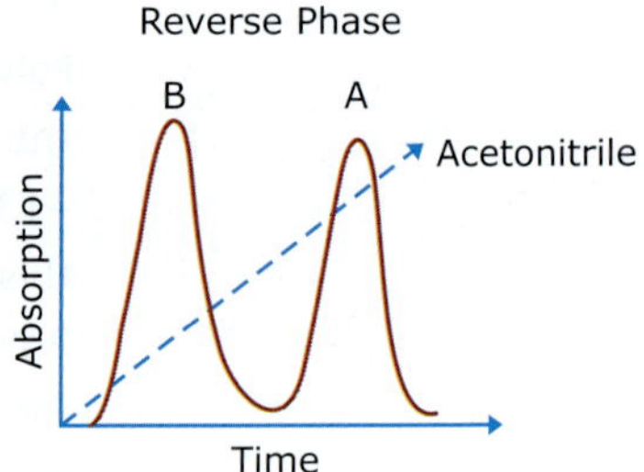

Which of the following statements is *correct*?

a. A is larger and more hydrophobic than B.
b. B is more anionic and more hydrophobic than A.
c. A is more hydrophobic and smaller than B.
d. A is more cationic and smaller than B.

Question 37

An uncharged protein (P) has an Asp in position 23 with a molecular weight of 6501 Da, as determined by mass spectrometry. The uncharged mutant of this protein (P') contains a single amino acid substitution with Asn at position 23. What is the molecular weight of P' as determined by mass spectrometry?

a. 5000 Da
b. 5500 Da
c. 6000 Da
d. 6500 Da

Question 38

The sedimentation coefficient of a DNA molecule is 22 S. A dilute solution of the DNA is spun in a centrifuge at 40,000 rpm starting from a distance of 6 cm from the axis of rotation. How far will the DNA move in 20 min?

a. 1.75 cm
b. 0.28 cm
c. 0.38 cm
d. 2.80 cm

Question 39

A protein is studied in the analytical ultracentrifuge. At high speed (60,000 rpm), the sedimentation velocity v is measured as 1.58×10^{-4} cm sec^{-1} at a radius of 10.0 cm. Estimate the sedimentation coefficient of the protein, expressing the results in Svedberg Units (S) where 1 S = 10^{-13} sec. If the sample cell in the ultracentrifuge is 1.5 cm deep, estimate approximately how long it takes to sediment from top to bottom.

a. 2S, 7000 sec
b. 3S, 5000 sec
c. 1S, 10000 sec
d. 4S, 9500 sec

Question 40

The molecular mass of a purified protein is 60 kDa, according to gel filtration chromatography. When performing the chromatography in the presence of 6 M urea, the resulting molecular mass is 30 kDa. When 6 M urea and 10 mM mercaptoethanol are present, a single product with a molecular mass of 15 kDa can be detected. What is the structure of the protein like?

a. A dimer of 30 kDa subunits linked by disulfide bonds.
b. A tetramer of 15 kDa subunits with all subunits linked to each other by disulfide bonds.
c. A tetramer of 30 kDa subunits linked by disulfide bonds.
d. A tetramer of 15 kDa subunits with two subunits forming 30 kDa dimers through disulfide bonds.

Answers

1. c
2. a
3. a
4. b

   At pH 1, the average charges on both amino acids is positive, all migrate towards the cathode, although Lys migrates more rapidly than Glu. At pH 12, the average charge on two amino acids is negative, and thus all migrate toward the anode. In this case, however, Glu migrates more rapidly than Lys.
5. c
6. a
7. c
8. c
9. d
10. c
11. d
12. b
13. d

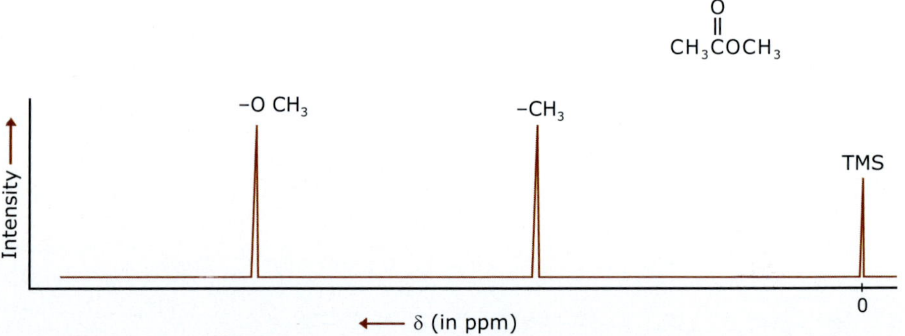

Besides showing the number of different kinds of protons, the ^1H–NMR spectrum also tells how many protons of each kind are present in a molecule. This is shown by the intensity of the signal. The intensity of a signal is measured in terms of the area under that peak. The area under an NMR signal is directly proportional to the number of protons giving rise to that signal.

14. c

The wavelengths of absorption of chromophores in electronic spectra are often influenced by the solvent. When a polar solvent is used, the dipole-dipole interaction with the solvent molecules lowers the energy of the excited state more than that of the ground state. This is due to the fact that excited states are more polar than ground states. The energy difference between the excited and ground states is reduced. This leads to a small red shift of the absorption maximum in polar solvents. Thus, the π–π^* transition shows a red shift of the order of 10–20 nm when the solvent is changed from hexane (nonpolar) to ethanol (polar).

Solvent effect for n–π^* transitions is opposite to that found for π–π^* transitions. Polar solvents cause a shift to lower wavelengths (blue shift) relative to non-polar solvents. The lone-pair electrons in the non-bonding orbital hydrogen bond or otherwise interact strongly with the polar solvent, leading to a lowering in energy of the non-bonding orbital whereas π^* orbital is affected much less. The result is an increase in the transition energy on going from a less polar to a more polar solvent. For example, in hexane solution, acetone shows absorption maximum at 279 nm whereas in aqueous solution, the absorption maximum is at 264.5 nm.

15. b

According to Beer-Lambert law, $A = \varepsilon.c.l$

$\varepsilon_{240}^{Tyr} = 11{,}300\ M^{-1}\ cm^{-1}$	$\varepsilon_{240}^{Trp} = 1960\ M^{-1}\ cm^{-1}$
$\varepsilon_{280}^{Tyr} = 1500\ M^{-1}\ cm^{-1}$	$\varepsilon_{280}^{Trp} = 5380\ M^{-1}\ cm^{-1}$

The absorbance of a solution of more than one independent species is additive.

$$[Tyr] = \frac{(5380 \times 0.717) - (1960 \times 0.239)}{(11{,}300 \times 5380) - (1500 \times 1960)} = \frac{3.39 \times 10^3}{57.9 \times 10^6} = 5.85 \times 10^{-5}\ M$$

$$[Trp] = \frac{(11{,}300 \times 0.239) - (1500 \times 0.717)}{57.9 \times 10^6} = 2.81 \times 10^{-5}\ M$$

The solution is at a concentration of 0.1 g/*l*. Thus, the content of tryptophan and tyrosine is 281 and 586 μmol/g protein, respectively.

The absorbance of a solution of more than one independent species is additive. For two components, M and N,

$$A_\lambda = A_\lambda^M + A_\lambda^N = \varepsilon_\lambda^M \times l \times [M] + \varepsilon_\lambda^N \times l \times [N]$$
$$= (\varepsilon_\lambda^M \times [M] + \varepsilon_\lambda^N \times [N]) \times l$$

If measurements are made at two (or more) wavelengths where the ratios of extinction coefficients differ, the resulting equations,

$$A_1 = (\varepsilon_1^M[M] + \varepsilon_1^N[N]) \times l$$
$$A_2 = (\varepsilon_2^M[M] + \varepsilon_2^N[N]) \times l$$

can be solved for the concentrations of the absorbing solutes,

$$[M] = \frac{1}{l}\,\frac{\varepsilon_2^N A_1 - \varepsilon_1^N A_2}{\varepsilon_1^M \varepsilon_2^N - \varepsilon_2^M \varepsilon_1^N}$$

$$[N] = \frac{1}{l}\,\frac{\varepsilon_1^M A_2 - \varepsilon_2^M A_1}{\varepsilon_1^M \varepsilon_2^N - \varepsilon_2^M \varepsilon_1^N}$$

A wavelength at which two or more components have the same extinction coefficient is known as an *isosbestic wavelength.* Also called an isosbestic point, it therefore has the property

$$\varepsilon_\lambda^M = \varepsilon_\lambda^N = \varepsilon_{iso}$$

At an isosbestic wavelength, the absorbance can be used to determine the total concentration of the two components

$$A_{(\lambda=iso)} = \varepsilon_{iso} \times l \times [M] + \varepsilon_{iso} \times l \times [N] = \varepsilon_{iso} \times l \times ([M] + [N])$$

Measurements at an isosbestic wavelength plus one other wavelength where the extinction coefficients differ for the two components provide a particularly simple solution to the Beer–Lambert Law equations:

$$[M] + [N] = \frac{A_{iso}}{\varepsilon_{iso} l}$$

$$\frac{[M]}{[N]} = \frac{\varepsilon_1^N A_{iso} - \varepsilon_{iso} A_1}{\varepsilon_{iso} A_1 - \varepsilon_1^M A_{iso}}$$

16. c

Because both compounds show absorption at both wavelengths, we can set up two simultaneous equations.

$A_{350} = \varepsilon_{A350} \times [A] + \varepsilon_{B350} \times [B]$

$0.36 = 15000 \times [A] + 7000 \times [B]$

$0.36 - 7000 \times [B] = 15000 \times [A]$

$15000 \times [A] = 0.36 - 0.21$

$$[A] = \frac{0.15}{15000} = \frac{15}{1500000} = 1 \times 10^{-5}\ \text{M}$$

17. b

The relationship between *molar extinction coefficient* (ε_{molar}) and *percent extinction coefficient* ($\varepsilon_{percent}$) is as follows:

$(\varepsilon_{molar}) \times 10 = (\varepsilon_{percent}) \times$ (molecular weight of protein)

ε_{molar} of Trp will be as follow:

$$\varepsilon_{molar} = \frac{\varepsilon_{percent} \times \text{MW of amino acid}}{10} = \frac{269.60 \times 204}{10} = \frac{54998.4}{10} \cong 5500\ \text{M}^{-1}\ \text{cm}^{-1}$$

ε_{molar} of Tyr will be as follow:

$$\varepsilon_{molar} = \frac{\varepsilon_{percent} \times \text{MW of amino acid}}{10} = \frac{83.33 \times 180}{10} = \frac{14999.4}{10} \cong 1500\ \text{M}^{-1}\ \text{cm}^{-1}$$

If the absorptivity is not known, but the amino acid composition is known, the molar extinction coefficient at 280 nm can be predicted from this equation:

Molar extinction coefficient (ε_{280}) = (number of Trp × 5500) + (number of Tyr × 1500)

= 2 × 5500 + 4 × 1500 = 11000 + 6000 = 17000.

$$\text{Absorbance (A)} = \frac{\varepsilon.c.l}{MW} = \frac{17000 \times 1 \times 1}{17000} = 1$$

18. c

The Lambert–Beer law is not a linear relationship. Consider a beam of light passing through a 1 cm cuvette containing 1 mg/liter of a light-absorbing compound. Suppose 80% of the incident light is transmitted (20% of the incident light is absorbed). Now let us place a second identical cuvette in the light path directly behind the first cuvette.

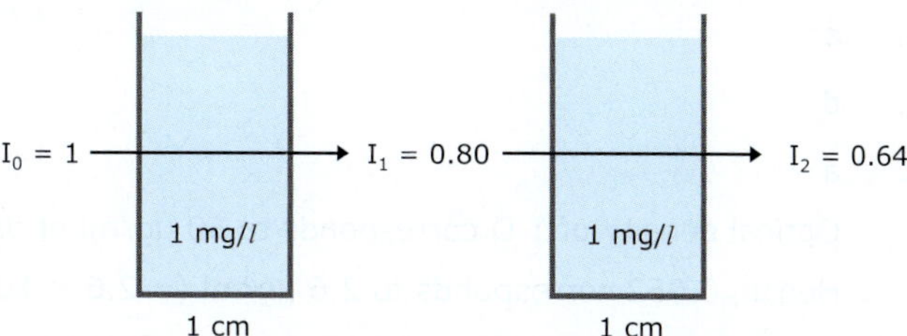

What is the intensity of the light transmitted through both cuvettes? The Lambert–Beer law is not a linear relationship. Thus, I_2 is not 0.64. Each centimeter of path length does not absorb a constant amount of light. Instead, each centimeter absorbs 20% of the incident light. However, absorbs 20% of 0.8, which is 0.16. It transmits 80%. It transmits 80% of 0.8, which is 64% of the original incident light. We could obtain exactly the same result if we use single cuvette with 2 cm thickness.

19. c

20. a

The peptide carries a net charge of +1.5 at pH 2.0 (lys side chain, +1; α-amino group, +1; Glu side chain, 0; and terminal carboxyl, -0.5, since the pH coincides with its pK value). At pH 4.0, the net charge is +0.5; the Glu side chain is half ionized (-0.5), but the terminal carboxyl is almost completely ionized (-1). At pH 6.0, the net charge is 0 due to a +2 charge contributed by the Lys residue and a -2 charge contributed by the Glu residue. At pH 11.0, the α-amino group is deprotonated (charge of 0) and the Lys side chain is half-protonated (charge of +0.5); thus, the net charge is -1.5.

21. a

22. c

23. d

24. c

25. a

Moving up the gel corresponds to reading the sequence of the extended primer from 5′ to 3′. The sequence is 5′ TACCTTGAT 3′.

The template sequence is the reverse complement of this sequence.

5′ ATCAAGGTA 3′.

26. b

5′ CTAGATAGTATAG 3′

Bands appearing in the G and C lanes can be read directly. Bands in the A+G lane that are not duplicated in the G lane are read as A. Bands in the C+T lane that are not duplicated in the C lane are read as T. The sequence is read from the bottom of the gel to top.

27. c

28. d

29. A. b B. d

30. c

31. d

32. a

Optical density of 1 D corresponds to 50 μg/ml of dsDNA.

Hence, 0.052 corresponds to 2.6 μg/ml (= 2.6×10^{-6} g/ml) of dsDNA.

Average molecular mass of 1 base pair is 650 daltons.

Hence, molecular mass of 400 bp = 650 × 400 = 260000 daltons.

$$\text{Number of moles of DNA} = \frac{2.6 \times 10^{-6}\ \text{gram}}{260000} = 1 \times 10^{-11}$$

Since, 1 mole contains 6.023×10^{23} molecules.

Thus, 1×10^{-11} moles will contain $6.023 \times 10^{23} \times 1 \times 10^{-11} = 6.023 \times 10^{12}$ molecules.

33. c

$$\alpha\text{-helix}\ (\%) = \frac{[(-\text{MRE}_{\text{observed}}) + \text{MRE}_{\text{random coil}}]}{[\text{MRE}_{\text{Pure }\alpha\text{-helix}} + \text{MRE}_{\text{random coil}}]} \times 100 = \frac{14800 + 3900}{35700 + 3900} \times 100 = \frac{18700}{39600} \times 100 = 47.22\%$$

34. a

35. c

36. c

37. d

Protein P has Asp at 23rd position with a molecular mass of 6501 Da.

Protein P′ has Asn at 23rd position.

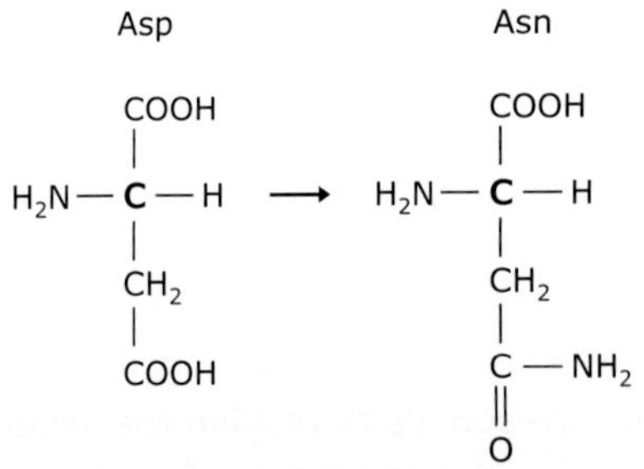

```
   Asp                      Asn

   COOH                     COOH
    |                        |
H2N—C—H      →         H2N—C—H
    |                        |
   CH2                      CH2
    |                        |
   COOH                      C—NH2
                             ||
                             O
```

After mutation, the functional group –COOH is changed into $-CONH_2$. The functional group —OH is changed into $—NH_2$.

Weight of OH = 16 + 1 = 17 and weight of NH_2 = 14 + 2 × 1 = 16

Difference in weight = 17 – 16 = 1

It means that molecular mass of P′ will be changed from 6501 – 1 = 6500 Da.

38. b

Sedimentation coefficient, $s = \frac{v}{\omega^2 r}$

Sedimentation velocity, $v = \omega^2 r \times s$

Angular velocity, $\omega = \frac{2\pi \times rpm}{60}$ rad/s $= \frac{2 \times 3.14 \times 40000}{60} \cong 4190$ rad/s

Sedimentation velocity, $v = \omega^2 r \times s$

$= (4190)^2 \times 6 \times 22 \times 10^{-13}$ (where, 1 S = 10^{-13} sec)

$= 1.75 \times 10^7 \times 132 \times 10^{-13}$

$= 0.000232$ cm/s

Distance travelled in 20 minutes $= v \times t$

$= 0.000232 \times 20 \times 60$; (1 minute = 60 sec)

$= 0.2784 \text{ cm} \cong 0.28$ cm

39. d

Sedimentation coefficient, $s = \frac{v}{\omega^2 r} = \frac{1.58 \times 10^{-4} \text{ cm s}^{-1}}{\left[\frac{2\pi \times 60,000}{60}\right]^2 \text{s}^{-2} \times 10 \text{ cm}} = 4 \times 10^{-13}$ sec

Hence, s = 4 S (where, 1 S = 10^{-13} sec)

Time taken to sediment is 1.5 cm. At first, we assume a constant velocity, then time

$t = \frac{1.5 \text{ cm}}{1.58 \times 10^{-4} \text{ cm s}^{-1}} = 9500$ sec

40. d

Urea denatures proteins, leading to their dissociation into their subunits and mercaptoethanol reduces disulfide bonds.

The original protein is a dimer of 2×15 kDa dimers with disulfide bonds.

Index

D

E

F

G

H

I

J

K

L

M

N

O

P

Q

R

S

T

U

V

W

X

Y

Z